W9-AQC-293

10.25

*Consulting Editor*

**Russell C. Brinker**
*New Mexico State University*

# Engineering Mechanics of
## Deformable Bodies

# Engineering
# Mechanics

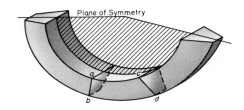

Plane of Symmetry

**INTERNATIONAL**

*Scranton,*

# of
# Deformable Bodies

## second edition

**EDWARD F. BYARS** *and* **ROBERT D. SNYDER**

*Department of Theoretical and Applied Mechanics*
*West Virginia University*

**TEXTBOOK COMPANY**

*Pennsylvania*

*Second Printing, March 1970*

**Copyright ©, 1969 and 1963, by International Textbook Company**

*All rights reserved. No part of the material protected by this copyright notice may be reproduced or utilized in any form or by any means, electronic or mechanical, including photocopying, recording, or by any informational storage and retrieval system, without written permission from the copyright owner. Printed in the United States of America.*

*Library of Congress Catalog Card Number:* 70-76403

*Standard Book Number* 7002 2242 1

# Preface to the Second Edition

The reception of our first edition was so gratifying that we were reluctant to make many drastic changes for the second edition. However, we have taken into consideration the various comments and criticisms of the first edition by its users in an effort to improve the text. Noticeably, we have added material on singularity functions and energy methods, rewritten the always difficult chapter on strain, and updated the discussion of experimental techniques and equipment. Also, approximately 90 percent of the problems have been changed, modified, or replaced.

We would like to take this opportunity to thank the users of the first edition for their comments, criticisms, and suggestions for improvements.

EDWARD F. BYARS
ROBERT D. SNYDER

Morgantown, West Virginia
February, 1969

# Preface to the First Edition

We hope that the contents of this book will justify our audacity in adding to the already swollen ranks of texts on mechanics of materials. This text is intended for use in the undergraduate first course in solid mechanics, usually called Strength of Materials or Mechanics of Materials. The title "Engineering Mechanics of Deformable Bodies" was thought to be more descriptive of what the course includes. The text presupposes a workable knowledge of statics and integral calculus.

Some of the topics included in this book and their order and method of presentation constitute a departure from the traditional approach to strength of materials. However, we feel that this departure is necessary in order to treat the subject with the generality and breadth which today's complex engineering problems demand. Basic concepts and analytical procedures are emphasized throughout the book in developing the theories of the behavior of deformable bodies and in applying these theories to engineering problems.

Part I of the book is devoted to the basic concepts of stress and strain and a discussion of the mechanical properties of engineering materials. This section includes such topics as the generalized Hooke's law, transformation of stress and strain at a point, and experimental stress-strain curves and properties. Where applicable, the discussion of each topic includes plastic as well as elastic considerations.

We feel that the information in Part I must be presented to the student *before* he can begin to develop rational design procedures. The fact that this text does not begin with a full treatment of loads and a review of statics does not imply that we mean to de-emphasize the free-body analysis or the importance of statics. Chapter 6 on Static Analysis of Load-Resisting Members is of great importance and should be emphasized accordingly.

Part II is an intermediate part which consists of a brief chapter on types of loads and a chapter on statics, including a thorough treatment of shear and bending-moment diagrams and other types of internal-reaction diagrams.

Part III is devoted primarily to the analysis of the usual simple structural elements, such as members resisting axial, torsion, bending, and buckling loads. This section begins with a brief discussion of modes of failure and design criteria in which the words "failure" and "safety" are clearly defined. In the succeeding chapters, which deal with the analysis of individual types of load-resisting members, primary emphasis is placed on the analytical procedures rather than on any of the particular expressions that are derived. Great care is taken to insure that the student gains a clear understanding of how the stress distribution influences the load-carrying capacity of a structural element. The basic analyses are not restricted to ideally elastic behavior or idealized inelastic behavior, and in all cases the limitations of derived expressions are clearly stated and emphasized. Some of the topics associated with design, such as riveted connections, are omitted because they are better covered in later design courses rather than in a basic engineering-science course.

Part IV does not necessarily follow the first three parts in the sequence of a course. If time permits, individual topics in Part IV may be covered earlier with associated material. For example, items in Chapter 14 may be covered along with Chapter 4 on Mechanical Properties.

This book can be used for a one-semester course or a two-semester course. The text contains enough subject material to permit the available time to dictate the scope and depth of coverage. The problems are designed to illustrate important points in the theory and to compel the student to think. We have included enough problems to enable a school to use the book at least several years.

Throughout the text we have tried to keep in mind the student's viewpoint. For example, we have assumed that very little is directly *obvious* to the student, and we have attempted to avoid the use of such presumptuous phrases as "it can easily be seen that" or "it is obvious that." Our aim is to make the presentation easy for the student at first in order to stimulate his interest in the subject. By "easy" we mean ease of understanding, not lack of rigor; e.g., while the analyses and derivations are quite rigorous, care was taken to prevent the mathematics from overshadowing the fundamental theories of mechanics of deformable bodies and their practical application to engineering problems.

The authors wish to thank each other for enduring the many trials, tribulations, and each other in the process of this development. We would like to express our indebtedness to West Virginia University for supplying the facilities and creative atmosphere which made the writing of this book a more pleasant undertaking. Finally we would like to say that more gratitude than words can express is due our secretary, Mrs. Janet Cunningham, who did more than any other one person (including the authors) in getting this book into print.

EDWARD F. BYARS
ROBERT D. SNYDER

Morgantown, West Virginia
June, 1963

# To the Student

This course will introduce you to the analysis of non-rigid or deformable bodies. What are "deformable" bodies? These are simply engineering members (or pieces, structures, parts, components, or whatever you want to call them) which change geometrically and mechanically when loaded. We will start in this course with relatively simple members. The more complex members and the more complex loadings will come in later design courses.

It has been only during the twentieth century that the study of deformable bodies has gained its rightful recognition as a basic engineering science. Before this time much design was done on a "trial and error" basis where "rules of thumb" were used extensively and the reliability of the results depended on whose thumb was used. In today's competitive engineering world, where economy and reliability are paramount, and with the present use of high-speed machines operating under great ranges of temperature and other conditions, the obsolete rules of thumb no longer suffice and the engineers must have a basic knowledge of deformable bodies upon which to build sound design theories.

In this course we will present and discuss some of the fundamental principles concerning the behavior of deformable bodies. These are some of the "tools of the trade" with which you can undertake the big job of determining the load-carrying ability of various types of engineering components.

As we analyze these various simple components, we will continually ask ourselves three basic questions:

1. How is the body loaded?
2. What is its shape (geometry)?
3. What is it made of (material)?

We will discuss these questions many times before the semester is over.

Even though we are considering this course as a basic engineering science, we must be realistic and practical. Real engineering conditions and responses often do not fit our idealizations. Some examples of idealizations of conditions are: considering pin joints to be smooth, considering a load to be applied at a point, neglecting the weight of a member, neglecting small temperature gradients in the bodies, and neglecting the dimension changes of loaded bodies when writing equilibrium equations.

Some examples of idealizations of responses are: considering the materials to be homogeneous and isotropic, assuming that the behavior of a body's surface is indicative of its internal behavior, and assuming in many cases that the material behaves similarly under dissimilar loading conditions.

Some of these idealizations do not appreciably affect the validity of the solution, but in other cases idealizations can drastically limit the usefulness of a particular solution. Watch out for these as we go along, and be aware of their significance.

E. F. B.
R. D. S.

# Contents

**Part I**
**Stress, Strain, and Mechanical Properties**

1. Stress . . . 3

   Introduction. Normal Stress and Shear Stress. Stress on an Element. Analysis of Plane Stress. Mohr's Circle for Plane Stress. Absolute Maximum Shear Stress. Variation of Stress Throughout a Body. Problems.

2. Strain . . . 26

   Deformations. Displacement. Definition of Strain. Analysis of Plane Strain. Mohr's Circle for Plane Strain. Strain Measurement and Rosette Analysis. Problems.

3. Stress-Strain Relationships . . . 47

   Introduction. The Stress-Strain Curve. Mathematical Modeling of Idealized Stress-Strain Curves. Poisson's Ratio. Generalized Hooke's Law. Relationship Between Elastic Constants, $E$, $G$, and $\mu$. Closure. Problems.

4. Experimental Mechanical Properties of Engineering Materials . . . 65

   Elastic and Plastic Deformations. Ultimate Strength and Rupture Stress. Yield Point and Yield Strength. Resistance to Energy Loads. True Stress. The Nature of Solids. The Nature of Elastic and Plastic Action. Effects of Other Variables on the Stress-Strain Relationship. Problems.

**Part II**
**Load Analysis**

5. Types of Loads . . . 87

   Introduction. Representation of Loads. Types of Loads. Other Load Classifications. Problems.

6. Static Analysis of Load-Resisting Members . . . 95

   Introduction. Method of Sections. Variation of Internal Reactions. Conventional Shear-Force and Bending-Moment Diagrams for Beams. Singularity Functions. Problems.

### Part III
### Relationships Between Loads and Factors Important in Design

7. Design Procedure . . . 127

> Introduction. Failure and Safety. Modes of Failure and Design Criteria. Steps in Design Procedure. Problems.

8. Axially Loaded Members . . . 134

> Introduction. Basic Load-Stress Relationship. Centroidal Loading. Thin-Walled Cylindrical Pressure Vessels. Deformation of Members in Pure Tension or Compression. Statically Indeterminate Axially Loaded Members. Simplified Shear and Bearing Stresses. Problems.

9. Torsion . . . 173

> Introduction. Kinematic Response of Cylindrical Members. Basic Load-Stress Relationship for Cylindrical Torsion Member. Shear-Strain Distribution In Cylindrical Torsion Member. Angular Deformation or Angle of Twist. Summary of Cylindrical Torsion Members. Noncircular Torsion Members. Hollow Thin-Walled Torsion Members. Problems.

10. Bending . . . 205

> Introduction. Pure Bending of Beams with Symmetrical Sections. Basic Load-Stress Relationship for Bending. Average Longitudinal and Transverse Shear Stresses. Elastic Bending of Unsymmetrical Sections. Summary of Beam Stresses. Basic Geometric Relationships for Deflection. Deflection by Direct Integration. Moment-Area Method for Deflection. Deflection by Superposition. Statically Indeterminate Beams. Problems. Supplementary Problems.

11. Buckling . . . 273

> Nature of Buckling. Uniqueness of Failure. Elastic Buckling. Discussion of Elastic Buckling. End Conditions. Inelastic Buckling. Discussion of Engesser Equation. Empirical Column Formulas. Problems.

### Part IV
### Other Design Considerations

12. Additional Beam Topics . . . 309

> Introduction. Shear Center. Bending of Curved Beams. Nonhomogeneous Beams. Problems.

13. Strain Energy and Theories of Failure . . . 329

> Strain Energy and Failure. Total Strain Energy. Elastic Strain Energy. Volumetric and Distortion Components of Elastic Energy. Theories of Failure. Use of the Theories of Failure. Problems.

14. Work and Energy Methods . . . 346

   Introduction.  Work Elastic Strain Energy of Structural Members.
   Castigliano's Theorem.  Problems.

15. Nonstatic Loads, Strain Concentrations, and Time Dependent
    Properties . . . 361

   Introduction. Stress Concentrations. Fatigue. Creep and Relaxation.
   Impact Loads.  Problems.

16. Experimental Mechanics . . . 386

   Introduction.  Electrical Resistance Strain Gages.  Strain-Gage Circuitry.  Measuring Instruments.  Strain-Gage Transducers.  Photoelasticity.  Birefringent Coatings.  Brittle Lacquer Coatings.  Problems.
   Moire Analysis.

   Appendix . . . 409

   Answers to Selected Problems . . . 423

   Index . . . 441

# Engineering Mechanics of
## Deformable Bodies

### part I

# Stress, Strain, and Mechanical Properties

## chapter 1

# *Stress*

## 1-1. Introduction

The primary reason for the existence of an engineering material, member, or structure is to resist loads. Occasionally, other factors such as appearance are important. As far as this study of deformable bodies is concerned, however, the main function of an engineering member is to resist loads. Engineering mechanics of deformable bodies deals with the relationships between the loads applied to the bodies and the resulting internal effects and dimension changes of the bodies. In statics you studied forces, force systems, and equilibrium of force systems acting on bodies. But these bodies were considered to be *rigid*, and no consideration was given to dimension changes or to the intensities of internal forces in the bodies. In this course, internal-force intensities and dimension changes of *deformable* bodies will be studied in detail. Free-body analysis will remain of the utmost importance here, as in statics, because equilibrium equations must be used to relate the external forces to the internal forces. Generally, the internal-force intensities are known as stresses and the dimension changes are called deformations.

Before we derive the important relationships between loads and the resulting stresses and deformations, and before we show how they are applied in design, we want you to gain a thorough understanding of stress and deformation and to see how they are related to one another. Once these basic concepts are well understood, then you will be able to follow the derivations in Part III of this book in which stresses and deformations are related to loads.

## 1-2. Normal Stress and Shear Stress

When a solid body is subjected to external forces, or loads, cohesive or adhesive forces will be produced inside the body. These internal forces are necessary to keep the body together. If the internal forces were not present, the body would be unable to keep its shape or support any external loads.

Consider the body $A$ in Fig. 1-1($a$). This body is being acted upon by a balanced system of external forces; that is, the body is in equilibrium under the action of the external forces $F_1$, $F_2$, $F_3$, $F_4$, and $F_5$. These

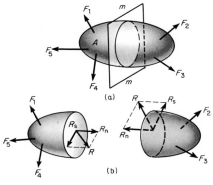

FIG. 1-1

forces are tending to pull the body $A$ apart. So internal forces are necessary to keep the body together. In order to indicate and analyze what occurs inside the body, let us imagine that the body is cut into two parts along a plane, as $mm$ in Fig. 1-1($a$), and that the two portions are separated, as in ($b$). The internal forces acting on each portion at the plane are then shown. The force $R$ represents the resultant internal force which is necessary to keep the two portions of the body together at section $mm$. This force $R$ can be resolved into two components, namely, $R_n$ normal to the plane and $R_s$ parallel to the plane.

It must be realized that the force $R$ in Fig. 1-1($b$) is not a single force, but is actually the resultant of the many, many minute cohesive forces which are exerted by the individual material particles across the cut section comprising the body. In Fig. 1-2 is illustrated the idea of forces on individual particles. Here each particle having a cross-sectional area $\Delta a$ resists two component loads, namely $\Delta R_n$ normal to $\Delta a$ and $\Delta R_s$ parallel to $\Delta a$. The magnitude and direction of the resultant force $\Delta R$ may vary from particle to particle.

The average force intensity on an individual particle is equal to $\Delta R$ divided by $\Delta a$. Although these particles can be quite small in size, they are nevertheless finite. Furthermore, in most materials on a microscopic level there are voids or cracks between the particles. However, it is con-

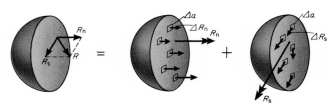

FIG. 1-2

venient to think of the material as being "smeared out" in a fashion such that the cut section is a smooth continuous surface of voidless material. Thus, while in reality the resultant $R$ is a finite sum of finite forces $\Delta R$,

$$R = \sum_{i=1}^{N} \Delta R_i$$

it is mathematically convenient to think in terms of infinitesimal areas $da$ and infinitesimal forces $dR$ such that

$$dR_n = \sigma\, da$$
$$dR_s = \tau\, da$$

so that

$$R_n = \int_{\text{area}} \sigma\, da$$
$$R_s = \int_{\text{area}} \tau\, da$$

The intensities $\sigma$ and $\tau$ are called the normal and shear stress respectively, and are usually thought of as

$$\sigma = \lim_{\Delta a \to 0} \frac{\Delta R_n}{\Delta a} = \frac{dR_n}{da} \qquad (1\text{-}1)$$

$$\tau = \lim_{\Delta a \to 0} \frac{\Delta R_s}{\Delta a} = \frac{dR_s}{da} \qquad (1\text{-}2)$$

since $\Delta a$ can be made as small as we like on our "smeared out" or voidless material surface.

There are two general types of normal stresses, namely, tensile and compressive. Tensile normal stress is produced by a force acting away from the plane area; a pulling force. Compressive normal stress is produced by a force acting toward the plane area; a pushing force. Shear stresses, however, are produced by forces parallel to the plane area, and there is no need for further classification.

The fundamental dimensions of stress are force divided by length squared. The most commonly used units in engineering in the United States are pounds per square inch (psi).

It should be noted that stress, either normal or shear, must be associated with a plane area and is not simply a force vector.[1] Normal stress is derived from, and should always be associated with, a force perpendicular to a plane area. Shear stress is derived from, and should be associated with, a force parallel to a plane area. Care must be taken to avoid the use of "stress" as a synonym for "force."

## 1-3. Stress on an Element

**Normal and Shear Stresses.** Our definition of stress given above is based on an infinitesimal force $dR$ acting on an infinitesimal area $da$.

---

[1] In a more sophisticated mathematical treatment, stress is a quantity of higher order called a tensor.

Thus,

$$dR_n = \sigma\, da \qquad dR_S = \tau\, da$$

Let us now expand our idea of infinitesimal forces and areas into the concept of a system of infinitesimal forces acting on the *sides* of an infinitesimal three-dimensional element. Each side of this element has an infinitesimal plane area *da*.

Figure 1-3 shows an infinitesimal element with dimensions *dx*, *dy*, and *dz* in the interior of body *A*. The element has been oriented so that its sides are parallel to the arbitrarily chosen *x*, *y*, and *z* axes. One side of this element has been exposed by the cutting plane *mm*, and the interior cohesive force *dR* acting on the area *da* (with dimensions *dy* and *dz*) is shown. The force *dR* can be resolved into components $dR_x$, $dR_y$, and $dR_z$ parallel to the *x*, *y*, and *z* axes, respectively.

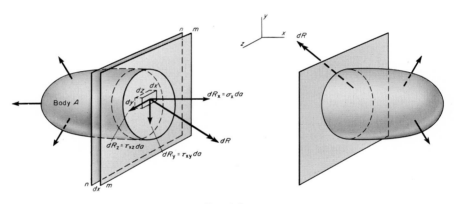

FIG. 1-3

The force $dR_x$ produces a normal stress on the area perpendicular to the *x* axis; this stress is denoted by $\sigma_x$. Then,

$$dR_x = \sigma_x\, da$$

The force $dR_y$ produces a shear stress parallel to the *y* axis on the area perpendicular to the *x* axis. This stress is denoted by $\tau_{xy}$, where the first subscript denotes the axis perpendicular to the area and the second subscript denotes the direction of the shear force. Therefore,

$$dR_y = \tau_{xy}\, da$$

The force $dR_z$ produces a shear stress parallel to the *z* axis on the area perpendicular to the $x$ axis. If this stress is denoted by $\tau_{xz}$,

$$dR_z = \tau_{xz}\, da$$

If we now pass similar cutting planes through body $A$ so as to expose all six sides of the element, the element can be "taken out." In isolating the element from the body, a very important assumption will be made. *It will be assumed that the infinitesimal forces on any infinitesimal plane area undergo no change in magnitude if the cutting plane is shifted an infinitesimal distance perpendicular to the original cutting plane.* For instance, it is assumed that in Fig. 1-3 the force $dR$ would be the same on an infinitesimal area of plane $nn$ as it is on such an area of plane $mm$. This assumption, which is discussed more thoroughly in Sec. 1-7, may be made whenever we consider the stress at any one particular "point" in or on a body.[2] The assumption is illustrated in Fig. 1-4(*a*), where it is assumed that the $x$, $y$, and $z$ forces on area *efgh* are equal in magnitude to the corresponding ones on area *abcd*, those on area *bfgc* are equal to the corresponding ones on area *aehd*, and those on area *aefb* are equal to the corresponding ones on area *dhgc*. Notice that the normal forces occur in *collinear* pairs of equal magnitude and opposite sense. The shear forces occur in *parallel* pairs of equal magnitude and opposite sense, but the forces are an infinitesimal distance apart; so the shear forces produce couples.

**Triaxial and Biaxial States of Stress.** In Fig. 1-4(*b*) the infinitesimal forces on the element are expressed in terms of the stresses which they produce. The corresponding forces on the hidden planes are omitted for clarity. The condition shown in Fig. 1-4 is a general triaxial state of stress. Further analysis of this state of stress is rather involved and will not be pursued here.

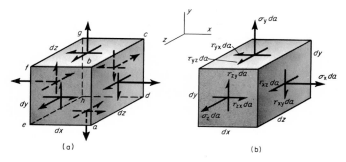

FIG. 1-4

[2] In effect we are supposing that a homogeneous state of stress exists in the *neighborhood* of the element under consideration.

Many of the problems encountered in engineering deal with the analysis of stress at a free surface of a body, that is, a surface on which no forces are acting. If area *aefb* in Fig. 1-4(*a*) were a free surface, the forces on it and on the parallel surface *dhgc* would be zero. Also note that if these forces are zero, then all forces parallel to the *z* axis would have to be zero in order to maintain rotational equilibrium about the *x* axis and the *y* axis. When all the forces on an element are coplanar, there is a biaxial state of stress and the condition is often referred to as *plane stress*. A biaxial state of stress is shown in Fig. 1-5.

Notice that all the forces acting on the element in Fig. 1-5(*a*) are parallel to the *xy* plane and form a coplanar force system. Since coplanar force systems can be illustrated completely and easily in two dimensions, two-dimensional illustrations such as Figs. 1-5(*b*) and 1-5(*c*) are used throughout this book to represent plane stress. Although this is often done for the sake of simplicity, you must not lose sight of the three-dimensional aspects of the analysis.

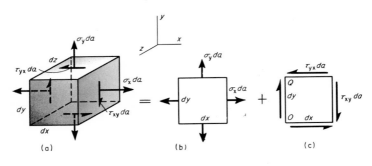

FIG. 1-5

Figure 1-5(*b*) shows the normal forces acting on the element, and Fig. 1-5(*c*) shows the shear forces. If Fig. 1-5(*b*) were superposed onto Fig. 1-5(*c*), the resulting force system would be identical with that in Fig. 1-5(*a*). This idea of resolving a compound force system into several simple force systems is utilized frequently in this text to simplify some analyses.

The forces in Fig. 1-5(*b*) constitute a balanced system, since the normal forces occur in *collinear* pairs of equal magnitude and opposite sense. In Fig. 1-5(*c*) the shear forces occur in *parallel* pairs of equal magnitude and opposite sense. However, for this system to be in equilibrium, the summation of moments must be zero. If we arbitrarily select point $O$ as a center of moments, we obtain the following relationship:

$$\sum \overset{\curvearrowright}{M_O} = 0$$

$$\tau_{xy} \, da \, dx \; - \; \tau_{yx} \, da \, dy \; = \; 0$$

$$\tau_{xy}(dy \, dz) \, dx \; - \; \tau_{yx}(dx \, dz) \, dy \; = \; 0$$

$$(\tau_{xy} \; - \; \tau_{yx}) \, dx \, dy \, dz \; = \; 0$$

Since $dx$, $dy$, and $dz$ are not zero,

$$\tau_{xy} \; = \; \tau_{yx}$$

This simple relationship shows that for plane stress the following statements are true:

*a*) Shear stresses must occur simultaneously on perpendicular planes; that is, $\tau_{xy}$ could not exist unless $\tau_{yx}$ also exists. Since these stresses are equal in magnitude, no distinction is necessary and the subscripts may be interchanged or even omitted in representing plane stress.

*b*) The shear forces on mutually perpendicular planes must produce two couples of equal magnitude and opposite sense. To meet this requirement, the shear-force vectors at a corner must either converge as at corner $Q$ in Fig. 1-5(*c*), or diverge as at corner $O$. Never can one shear vector be directed toward a corner while a perpendicular one is directed away from the same corner.

A similar analysis of the general triaxial state of stress represented in Fig. 1-4(*b*) would show that $\tau_{zy} = \tau_{yz}$ and $\tau_{zx} = \tau_{xz}$.

We can now say that any biaxial state of stress can be completely defined by three quantities, namely, two normal stresses and one shear stress. Therefore any of the following combinations would identify a biaxial state of stress:

For the $xy$ plane, $\sigma_x$, $\sigma_y$, $\tau_{xy}$

For the $xz$ plane, $\sigma_x$, $\sigma_z$, $\tau_{xz}$

For the $yz$ plane, $\sigma_y$, $\sigma_z$, $\tau_{yz}$

In each case of plane stress there would be no force perpendicular to the plane of stress.

## 1-4. Analysis of Plane Stress

In Fig. 1-6 (*a*) is shown an example of the most general case of biaxial stress for the $xy$ plane; in (*b*) the directions and magnitudes of $\sigma_x$, $\sigma_y$, and $\tau$ are arbitrary, but it is assumed that they are known. These stresses indicate the state of stress at one point $A$ in a loaded body, and they apply only to the shown orientation of the element; in this case the sides are in horizontal and vertical planes (the $xz$ and $yz$ planes). If the orientation of the element at this point in the body were changed by rotating it about the $z$ axis, as indicated in Fig. 1-6(*c*), we would be considering stresses on other planes, such as the $nz$ and $sz$ planes, and these stresses would probably have different values than the original stresses $\sigma_x$, $\sigma_y$, and $\tau$.

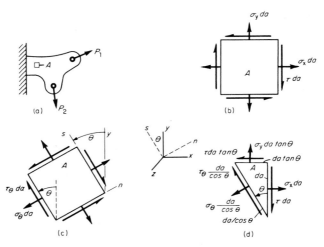

Fɪɢ. 1-6

We must realize that the material is unaware of what we have called the $x$ and $y$ axes. That is, the material has to resist the loads regardless of how we wish to name them or whether they are horizontal, vertical, or otherwise. Furthermore, the material will break when the intensity of the forces become too great regardless of how *we* look at the body. Thus a fundamental problem in engineering design is to determine the maximum normal stress or maximum shear stress at any particular point in the body. Assuming that at the point in question we can determine $\sigma_x$, $\sigma_y$, and $\tau$ in Fig. 1-6(*b*), there is no reason to believe a priori that these are the maximum normal stress and shear stress occurring at that point. The maximum stresses are probably associated with some other orientations of the infinitesimal element. It is, therefore, important to be able to obtain from the assumed known stresses $\sigma_x$, $\sigma_y$, and $\tau$, the stresses $\sigma_\theta$ and $\tau_\theta$ associated with other orientations at this point.

For example, suppose that we wish to find the stresses for some other orientation of the element at the same point, such as the orientation shown in Fig. 1-6(*c*), where $\theta$ is an arbitrary angle. All stresses on the element for this orientation are unknown. So it is necessary to draw a free-body diagram which will show the known stresses and the desired unknown stresses. Such a free-body diagram is shown in Fig. 1-6(*d*), in which the free body is obtained by cutting the original element at the arbitrary angle $\theta$ and forming a triangular wedge. This wedge has two faces in common with the original element in (*b*) and one face in common with the element considered in (*c*).

Equilibrium equations for the free-body diagram in (*d*) will enable us to obtain the stresses $\sigma_\theta$ and $\tau_\theta$ on the element in Fig. 1-6(*c*) in terms of the known stresses $\sigma_x$, $\sigma_y$, and $\tau$ and the angle $\theta$. Summation of forces in the $n$ direction for the free-body diagram in $(d)$, gives:

$$\sum \overset{+\nearrow}{F_n} = 0$$

$$-\sigma_\theta \frac{da}{\cos \theta} - \tau (da \tan \theta) \cos \theta - (\tau\, da) \sin \theta$$

$$+ (\sigma_x\, da) \cos \theta + \sigma_y (da \tan \theta) \sin \theta = 0$$

Dividing by $da$ and multiplying by $\cos \theta$, we get

$$\sigma_\theta = \sigma_x \cos^2 \theta + \sigma_y \sin^2 \theta - 2\tau \sin \theta \cos \theta$$

From trigonometry,

$$\sin^2 \theta = \frac{1 - \cos 2\theta}{2}$$

$$\cos^2 \theta = \frac{1 + \cos 2\theta}{2}$$

$$2 \sin \theta \cos \theta = \sin 2\theta$$

Therefore,

$$\sigma_\theta = \sigma_x \frac{1}{2}(1 + \cos 2\theta) + \sigma_y \frac{1}{2}(1 - \cos 2\theta) - \tau \sin 2\theta$$

$$\sigma_\theta = \frac{\sigma_x + \sigma_y}{2} + \left(\frac{\sigma_x - \sigma_y}{2}\right) \cos 2\theta - \tau \sin 2\theta \tag{1-3}$$

By summing forces in the $s$ direction in a like manner, you should now prove to yourself that

$$\tau_\theta = \left(\frac{\sigma_x - \sigma_y}{2}\right) \sin 2\theta + \tau \cos 2\theta \tag{1-4}$$

Equations 1-3 and 1-4 are called the transformation equations for plane stress. If the stresses at *a point* for a given orientation are known, then the stresses at *that point* for any other orientation can be obtained from these transformation equations. In deriving these equations, tensile forces were considered positive, compressive forces were negative, shear forces that produce clockwise couples were positive, and the angle $\theta$ was positive when measured counterclockwise from the $y$ axis. The arbitrary angle of orientation $\theta$ *must* be measured in the plane of the known stresses.

Notice that Eqs. 1-3 and 1-4 express the desired unknown stresses $\sigma_\theta$ and $\tau_\theta$ in terms of the known stresses $\sigma_x$, $\sigma_y$, and $\tau$ and the angle $\theta$. For any particular known state of stress, the quantities $\sigma_x$, $\sigma_y$, and $\tau$ are unique and single-valued, and they are therefore constants in Eqs. 1-3 and 1-4. So the dependent variables $\sigma_\theta$ and $\tau_\theta$ are actually functions of only one independent variable, $\theta$. This means that for any known state of plane stress, we can find the value of $\theta$ for which the normal stress $\sigma_\theta$ is a maximum or minimum by differentiating Eq. 1-3 with respect to $\theta$ and setting the derivative equal to zero. Thus,

$$\frac{d\sigma_\theta}{d\theta} = \left(\frac{\sigma_x - \sigma_y}{2}\right)(-\sin 2\theta)(2) \ - \ \tau(\cos 2\theta)(2) = 0 \qquad (1\text{-}3a)$$

$$\left(\frac{\sigma_x - \sigma_y}{2}\right)\sin 2\theta = - \tau \cos 2\theta$$

$$\tan 2\theta = \frac{-\tau}{\left(\dfrac{\sigma_x - \sigma_y}{2}\right)} \qquad (1\text{-}5)$$

Likewise the value of $\theta$ which will maximize $\tau_\theta$ can be obtained by setting the derivative of Eq. 1-4 equal to zero. The result is

$$\tan 2\theta = \frac{\left(\dfrac{\sigma_x - \sigma_y}{2}\right)}{\tau} \qquad (1\text{-}6)$$

Equations 1-5 and 1-6 will each yield two values of $\theta$. One value indicates the orientation for the maximum stress, and the other indicates the orientation for the minimum stress.

Closer examination of Eqs. 1-5 and 1-6 shows that tan $2\theta$ from Eq. 1-5 is the negative reciprocal of the value from Eq. 1-6. This means that the angle $2\theta$ for which $\sigma_\theta$ is a maximum or minimum will differ by 90° from the angle $2\theta$ for which $\tau_\theta$ is a maximum. Therefore the orientation for maximum normal stress will differ from that for maximum shear stress by 45°. Also note from Eq. 1-3a that setting the derivative of $\sigma_\theta$ equal to zero is equivalent to setting $\tau_\theta$ from Eq. 1-4 equal to zero. Thus, no shear stress is present for the orientation which maximizes or minimizes the normal stress.

## 1-5. Mohr's Circle for Plane Stress

**Construction of Mohr's Circle.** Although the equations in Sec. 1-4 are useful, a graphical representation of them can be used more easily and better illustrates their physical significance. If we square Eqs. 1-3 and 1-4, we get the following relationships:

$$\left[\sigma_\theta - \left(\frac{\sigma_x + \sigma_y}{2}\right)\right]^2 = \left[\left(\frac{\sigma_x - \sigma_y}{2}\right)\cos 2\theta - \tau \sin 2\theta\right]^2$$

$$\tau_\theta{}^2 = \left[\left(\frac{\sigma_x - \sigma_y}{2}\right)\sin 2\theta + \tau \cos 2\theta\right]^2$$

Now by adding the above two equations, expanding, and rearranging, we obtain

$$\left[\sigma_\theta - \left(\frac{\sigma_x + \sigma_y}{2}\right)\right]^2 + \tau_\theta^2 = \left(\frac{\sigma_x - \sigma_y}{2}\right)^2 + \tau^2 \tag{1-7}$$

By slight rearrangement, Eq. 1-7 can be put into the form of an equation of a circle. Thus,

$$\left[\sigma_\theta - \left(\frac{\sigma_x + \sigma_y}{2}\right)\right]^2 + [\tau_\theta - 0]^2 = \left[\sqrt{\left(\frac{\sigma_x - \sigma_y}{2}\right)^2 + \tau^2}\right]^2 \tag{1-7a}$$

$$[\ x -\quad (a)\quad]^2 + [y - b]^2 = [\quad\quad c\quad\quad]^2 \tag{1-8}$$

The equation of a circle as presented in most mathematics books is written directly below the combined transformation equation 1-7a above where $a$ is the $x$ coordinate of the center of the circle, $b$ is the $y$ coordinate of the center, and $c$ is the radius. Equation 1-7a represents the equation of a circle for which the $x$ coordinate of the center is $(\sigma_x + \sigma_y)/2$, the $y$ coordinate of the center is 0, and the radius is the radical term. As before, $\sigma_\theta$ and $\tau_\theta$ are the unknown variables and correspond to $x$ and $y$ respectively, in the circle equation. Now, by adopting the proper sign convention and assuming that $\sigma_x > \sigma_y > 0$, we can plot Eq. 1-7 as a circle based on rectangular axes and the variables $\sigma_\theta$ and $\tau_\theta$. Figure 1-7(a) shows Mohr's circle of stress, which is so called because it was developed by Otto Mohr (1835–1918).[3]

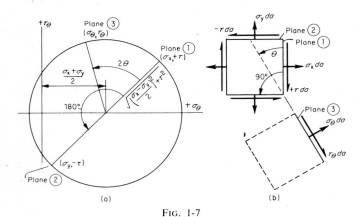

FIG. 1-7

Each of the infinite number of points on the circle in Fig. 1-7(a) represents one of the infinite number of possible orientations of plane ③ in

[3]An outstanding professor of structural mechanics in Germany.

(*b*). Each point on the circle has two coordinates, $\sigma_\theta$ and $\tau_\theta$, which represent magnitudes and signs of the stresses for the corresponding orientation of the plane. Tensile forces are assumed to be positive, compressive forces are negative, and shear forces that produce clockwise couples are positive. One Mohr's circle completely represents a state of *biaxial* stress at one particular point in a loaded body. The point on the circle labeled "plane ① " gives the stresses on the vertical plane marked "plane ① " on the element; and the point on the circle labeled "plane ② " gives the stresses on the horizontal plane marked "plane ② " on the element. These two points are 180° apart on the circle, while the corresponding planes on the element are 90° apart. In general, an angular measurement of $2\theta$ on the circle corresponds to an angular measurement of $\theta$ on the element. *Remember: $\theta$ on the element, $2\theta$ on the circle.*

Comparison of Eqs. 1-7 and 1-8 shows that *b* must equal zero. Therefore, the center of the circle must *always* be on the horizontal $\sigma_\theta$ axis, and the circle can be plotted if the coordinates of any two diametrically opposite points are known. This means that Mohr's circle for plane stress at a point in a loaded body can be plotted if the stresses are known on any two perpendicular planes at the point, that is, on two sides of an element such as the one in Fig. 1-7(*b*).

As an example, assume that the stresses at a critical point in some engineering member are shown in Fig. 1-8(*a*), and we want to sketch Mohr's circle. We first locate the points which represent the stresses on plane ① and plane ② of the element. Since these points are at opposite ends of a diameter of the circle, a line connecting the two points is a diameter, and the intersection of the horizontal axis with the diameter is the center. We now sketch the circle.

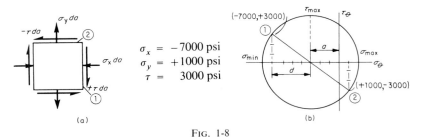

$$\sigma_x = -7000 \text{ psi}$$
$$\sigma_y = +1000 \text{ psi}$$
$$\tau = \phantom{+}3000 \text{ psi}$$

(a)                                        (b)

FIG. 1-8

A complete analysis of the circle can be made from this rough sketch by using geometry and trigonometry. An accurate graphical construction is usually unnecessary. The distance *a* in Fig. 1-8(*b*) from the origin to the center is found as follows:

$$a + d = 7000 \qquad \text{(magnitude of } \sigma_\theta \text{ for plane ①)}$$

$$d = a + 1000 \qquad \text{(from similar triangles)}$$

Therefore, $a = 3000$ and $d = 4000$. The coordinates of the center of the circle are $-3000$ and 0, and its radius is $\sqrt{4000^2 + 3000^2} = 5000$.

**Analysis of Mohr's Circle.** Once Mohr's circle is drawn for known values of $\sigma_x$, $\sigma_y$, and $\tau$ at a point, the maximum stresses at the point can be found. The coordinate of the right-hand extremity of that circle is the algebraic maximum normal stress; the coordinate of the left-hand extremity is the minimum normal stress; and the radius is the maximum shear stress. It is important to note that the points representing the maximum and minimum normal stresses are on the horizontal $\sigma_\theta$ axis, and this condition means that there is no shear stress on the planes of maximum and minimum normal stress at the point in the body. These normal stresses, which act on planes of no shear, are called *principal* stresses. They occur on planes which are perpendicular to one another (180° apart on the circle) and are called *principal* planes.

The minimum normal stress is just as important as the maximum. The algebraic minimum stress could have a magnitude greater than that of the maximum principal stress if the state of stress were such that the center of the circle is to the left of the origin, as in Fig. 1-8. The magnitude of the maximum or minimum principal stress is the horizontal coordinate of the center of the circle plus or minus the radius. In Fig. 1-8($b$), the maximum normal stress is $(-3000 + 5000) = 2000$ psi; the minimum is $(-3000 - 5000) = -8000$ psi; and $\tau_{max} = 5000$ psi.

The angle of orientation of the principal planes with respect to known planes can be found from the circle. *Remember: $\theta$ on the element corresponds to $2\theta$ on the circle.* In Fig. 1-9, the angle $2\theta$ on the circle from a

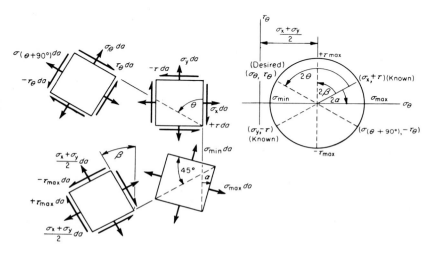

FIG. 1-9

point representing an orientation for which the stresses are known to **any** other point representing an orientation for which the stresses are desired **is** referenced *from* the known point and measured in the same direction as **the** angle $\theta$ on the element between the corresponding known plane and **the** desired plane. Satisfy yourself about this condition by going from a **known** plane to a principal plane in Fig. 1-9. Also try going from a known **plane** to any other desired plane. Then try going from a known plane to a **plane** of maximum shear stress.

You should see from the circle that a point for $\tau_{\max}$ is 90° away from a point for a principal stress. Thus a plane of maximum shear will *always* make an angle of 45° with a principal plane. You should also note that since stresses on perpendicular faces of any element are given by the co-ordinates of diametrically opposite points on the circle, the sum of the two normal stresses on any two perpendicular faces of the element is a constant. This sum is an invariant for any particular state of stress.

EXAMPLE 1-1. At some critical point in an engineering member the state of biaxial stress in Fig. 1-10($a$) exists. Here $\sigma_x = 18,000$ psi, tensile; $\sigma_y = 4,000$ psi, compressive; and $\tau = 12,000$ psi in the directions shown. $a$) Find the values of the principal stresses, and sketch the properly oriented element on which they act. Label all forces on the element, and label the angle of orientation on the sketch. $b$) Do the same for the maximum shear stress. $c$) Determine the stresses on an element rotated 50° clockwise from the given element. Sketch this element and label it completely.

*Solution:* In the computations, stresses will be expressed in kips per square inch (abbreviated ksi), a kip being equal to 1000 lbs. Thus, 18,000 psi = 18 ksi. A fairly accurate freehand sketch is begun by plotting the known stresses on "plane ①"(−4, +12) and "plane ②" (+18, −12). The line connecting points ① and ② is a diameter of the Mohr circle, and where it crosses the horizontal axis is the center. The circle is now sketched, as in Fig. 1-10($e$). The horizontal distance between the two known points is 18 + 4 = 22 (sum of the magnitudes of the $\sigma$ co-ordinates of the known points). The center is located 22/2 = 11 horizontally from either known point. The radius is $\sqrt{11^2 + 12^2} = 16.3$

$a$) The maximum principal stress is the $\sigma$ coordinate of the center plus the radius, or 7 + 16.3 = 23.3 ksi, tension. The minimum principal stress is 7 − 16.3 = 9.3 ksi, compression.

The maximum principal stress on the circle is at an angle $2\alpha$ counterclockwise from the known point ①, where $2\alpha$ is the angle whose tangent is 12/11, or 47.5° An element rotated counterclockwise from the given element through an angle $\alpha$, or 47.5/2 = 23.8°, is now sketched as in Fig. 1-10($b$). Notice that face ③ on the element is the orientation of the maximum principal stress corresponding to point ③ on the circle. The minimum principal stress is on the face ④, corresponding to point ④ on the circle, and is 90° from the face for the maximum. There are no shear forces on this element.

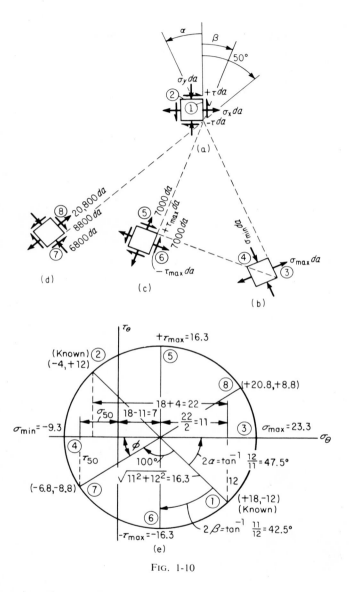

FIG. 1-10

*b*) The magnitude of the maximum shear stress is the radius, or 16.3 ksi. From the circle we see that the point ⑥ is at an angle $2\beta$ clockwise from the known point ①, and $2\beta = \tan^{-1}(11/12) = 42.5°$. Therefore, the original element is now rotated clockwise through an angle $\beta$, or $42.5/2 = 21.3°$, and the new element is sketched as in Fig. 1-10(*c*). Face ⑥ of this element has a negative (counterclockwise) shear force acting on it, since this force corresponds to the stress $-\tau_{max}$ at point ⑥ on the circle. Face ⑤ of the element has a positive (clockwise) shear force acting on it, since this force corresponds to the stress $+\tau_{max}$ at point ⑤ on the circle, and it is 90° from face ⑥. Notice that both point ⑤ and point ⑥ on the circle have a $\sigma$ coordinate of 7 ksi, tension. These tensile forces must be shown on the element in Fig. 1-10(c).

The orientation of the planes of maximum shear stress in Fig. 1-10(c) could have been found by a 45° rotation of the principal planes in (b). This rotation is indicated by the dashed line which bisects the angle between the principal planes in Fig. 1-10(b) and is parallel to plane ⑤ in (c). This condition agrees with the fact that points ⑤ and ⑥ on the circle are 2 × 45°, or 90°, from points ③ and ④.

c) The magnitudes of the stresses on an element rotated 50° clockwise from the given element are the coordinates of points ⑦ and ⑧ on the circle. Point ⑦ is located on the circle 2 × 50°, or 100°, clockwise from the known point ① Since $2\alpha$ has already been found to be 47.5°, the angle $\phi$ is $180 - (100 + 2\alpha) = 32.5°$. From Fig. 1-10(e), $\cos 32.5° = (\sigma_{50} + 7)/16.3$ and $\sin 32.5° = \tau_{50}/16.3$. So $\sigma_{50} = 6.8$ ksi, compression; and $\tau_{50} = -8.8$ ksi, the negative value indicating a counterclockwise shear force on face ⑦ of the element in Fig. 1-10(d). The $\sigma$ coordinate of point ⑧ is $7 + (7 +$ the magnitude of $\sigma_{50})$, or $+20.8$ ksi, and the $\tau$ coordinate of point ⑧ has the same magnitude as that of point ⑦ but is positive. The normal stress on the face ⑧ is 20.8 ksi, tension; and the shear force on that face is plus (clockwise).

## 1-6. Absolute Maximum Shear Stress

The discussion in Sec. 1-5 was confined to Mohr's circle for *one* plane of stress. Triaxial states of stress can be analyzed by use of Mohr's circles, but the technique is much more involved and is not presented in this book.[4] However, we will consider one particular aspect of three dimensional stress analysis.

In order to determine the *absolute maximum* shear stress on an element subjected to *plane* stress, it may sometimes be necessary to analyze a triaxial state of stress. Consider the element shown in Fig. 1-11(a), which is subjected to a biaxial state of stress.

We can rotate the element about an axis perpendicular to the xy plane of stress so that its sides are in the principal planes, as shown in Fig. 1-11(b); in (c) the element is shown in three dimensions, and the principal stresses are labeled $\sigma_1$, $\sigma_2$, and $\sigma_3$. Remember for plane stress $\sigma_3$ is zero.

Mohr's circle can be plotted for *each* biaxial state of stress, that is, for

---

[4] Mohr's circle for triaxial stress is covered in *Advanced Mechanics of Materials*, 2nd edition, by F. B. Seely and J. O. Smith, John Wiley & Sons, Inc., pp. 59–61; or in *Advanced Strength of Materials* by J. P. Den Hartog, McGraw-Hill Book Company, Inc., 1952, pp. 308–314.

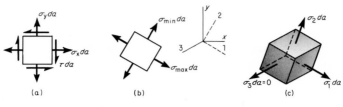

Fig. 1-11

the *1–2* plane, the *1–3* plane, and the *2–3* plane. Figure 1-12(*a*) shows Mohr's circles for the case where $\sigma_1$ is greater than $\sigma_2$ and both are tensile. Figure 1-12(*b*) is for the case where $\sigma_1$ is a tensile stress and $\sigma_2$ is a compressive stress of different magnitude. Figure 1-12(*c*) is for the case where both $\sigma_1$ and $\sigma_2$ are compressive and $\sigma_2$ has the greater magnitude. The absolute maximum shear stress is equal to one-half of the algebraic difference be-

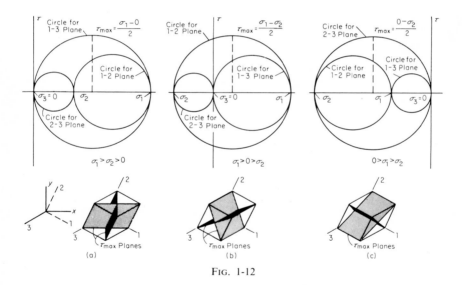

Fig. 1-12

tween the maximum and minimum principal stresses. In Fig. 1-12(*a*), the maximum principal stress is $\sigma_1$ and the minimum is $\sigma_3$ or zero. In (*b*), the maximum principal stress is $\sigma_1$ and the minimum is $\sigma_2$. In (*c*), the maximum principal stress is $\sigma_3$ or zero and the minimum is $\sigma_2$.

Although we have considered only cases where the third principal stress, $\sigma_3$, is zero, it can be proved that for *any* triaxial state of stress where $\sigma_3$ is not zero, the maximum shear stress is always one-half of the difference between the maximum and minimum principal stresses. Thus,

$$\tau_{max} = \frac{1}{2}\,(\sigma_{max} - \sigma_{min}) \tag{1-9}$$

## 1-7. Variation of Stress Throughout a Body

The analysis of plane stress presented in the preceding sections was based on the following assumption stated in Sec. 1-3: The infinitesimal forces on an infinitesimal plane area undergo no change in magnitude if the cutting plane is shifted an infinitesimal distance perpendicular to the original cutting plane. This assumption enabled us to represent plane stress at a point $P(x, y)$ by use of a diagram such as that in Fig. 1-13($a$).

In the mathematical theory of elasticity, no such simplifying assumption is made in representing the state of stress on an element and it is necessary to take into consideration the changes in magnitudes of the normal and shear forces. A change in a normal force will produce a change in the corresponding normal stress, and a change in a shear force will affect the corresponding shear stress. In Fig. 1-13($b$), for instance, the normal stress $\sigma_x$ existing at the point $P(x, y)$ undergoes a change of $\dfrac{\partial \sigma_x}{\partial x}\dfrac{dx}{2}$, where $\dfrac{\partial \sigma_x}{\partial x}$ represents the rate of change of $\sigma_x$ in the $x$ direction. Thus the change in $\sigma_x$ from the point $P$ to a vertical face of the element is the product of the rate of change $\dfrac{\partial \sigma_x}{\partial x}$ and the distance $\dfrac{dx}{2}$ over which the change occurs. In a similar manner, $\tau_{xy}$ undergoes a change of $\dfrac{\partial \tau_{xy}}{\partial x}\dfrac{dx}{2}$. The changes in both the $x$ direction and the $y$ direction are shown in Fig. 1-13($b$), where, for the sake of mathematical rigor, the positive $y$ direction is chosen downward. In the absence of any other forces, the equations of equilibrium for

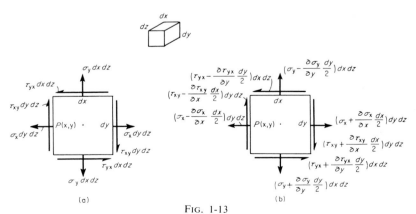

(a)                (b)

FIG. 1-13

Fig. 1-13(*b*) are as follows:

$$+ \rightarrow$$
$$\sum F_x = 0$$

$$\left(\sigma_x + \frac{\partial \sigma_x}{\partial x}\frac{dx}{2}\right)(dy\,dz) - \left(\sigma_x - \frac{\partial \sigma_x}{\partial x}\frac{dx}{2}\right)(dy\,dz) + \left(\tau_{yx} + \frac{\partial \tau_{yx}}{\partial y}\frac{dy}{2}\right)(dx\,dz)$$

$$- \left(\tau_{yx} - \frac{\partial \tau_{yx}}{\partial y}\frac{dy}{2}\right)(dx\,dz) = 0$$

$$+ \downarrow$$
$$\sum F_y = 0$$

$$\left(\sigma_y + \frac{\partial \sigma_y}{\partial y}\frac{dy}{2}\right)(dx\,dz) - \left(\sigma_y - \frac{\partial \sigma_y}{\partial y}\frac{dy}{2}\right)(dx\,dz) + \left(\tau_{xy} + \frac{\partial \tau_{xy}}{\partial x}\frac{dx}{2}\right)(dy\,dz)$$

$$- \left(\tau_{xy} - \frac{\partial \tau_{xy}}{\partial x}\frac{dx}{2}\right)(dy\,dz) = 0$$

$$\sum M_p = 0$$

$$\left(\tau_{xy} + \frac{\partial \tau_{xy}}{\partial x}\frac{dx}{2}\right)(dy\,dz)\frac{dx}{2} + \left(\tau_{xy} - \frac{\partial \tau_{xy}}{\partial x}\frac{dx}{2}\right)(dy\,dz)\frac{dx}{2}$$

$$- \left(\tau_{yx} - \frac{\partial \tau_{yx}}{\partial y}\frac{dy}{2}\right)(dx\,dz)\frac{dy}{2} - \left(\tau_{yx} + \frac{\partial \tau_{yx}}{\partial y}\frac{dy}{2}\right)(dx\,dz)\frac{dy}{2} = 0$$

Simplifying and factoring *dx dy dz* out of each equation, we obtain the following results:

$$\frac{\partial \sigma_x}{\partial x} + \frac{\partial \tau_{yx}}{\partial y} = 0 \qquad\qquad (1\text{-}10)$$

$$\frac{\partial \sigma_y}{\partial y} + \frac{\partial \tau_{xy}}{\partial x} = 0 \qquad\qquad (1\text{-}11)$$

$$\tau_{xy} - \tau_{yx} = 0 \qquad\qquad (1\text{-}12)$$

Equations 1-10 and 1-11 show that a rate of change in normal stress must be accompanied by a rate of change in shear stress, while Eq. 1-12 gives the result obtained earlier in Sec. 1-3.

Also, it can be shown that the stress transformation equations (Eqs. 1-3 and 1-4), which were derived in Sec. 1-4 are valid even when the changes in stress are taken into consideration. This follows since our stress transformation and Mohr's circle analyses were concerned solely with the stresses "at a point" rather than with equilibrium of finite elements. Hence, the results obtained in the preceding sections apply whether we choose to represent a state of plane stress as in Fig. 1-13(*a*) or as in Fig. 1-13(*b*). For simplicity in a course of this level, we will choose the former type of diagram.

## PROBLEMS

**1-1 to 1-6.** Using the "wedge" method of analysis, determine the normal and shear stresses on the indicated planes shown in the accompanying illustrations, without appealing to the transformation equations.

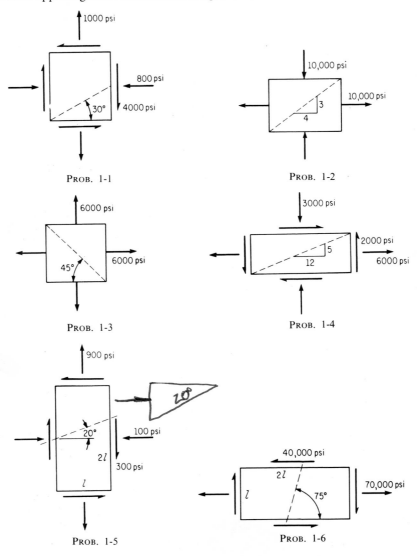

PROB. 1-1

PROB. 1-2

PROB. 1-3

PROB. 1-4

PROB. 1-5

PROB. 1-6

**1-7 to 1-18.** Sketch Mohr's circle for each of the states of plane stress shown in the accompanying illustrations.

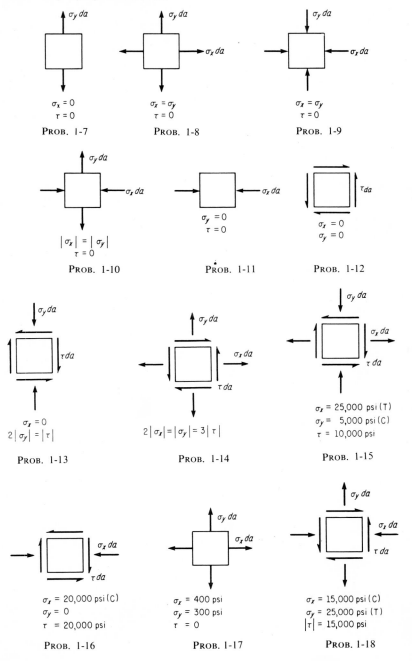

$\sigma_y \, da$

$\sigma_x = 0$
$\tau = 0$

PROB. 1-7

$\sigma_y \, da$     $\sigma_x \, da$

$\sigma_x = \sigma_y$
$\tau = 0$

PROB. 1-8

$\sigma_y \, da$     $\sigma_x \, da$

$\sigma_x = \sigma_y$
$\tau = 0$

PROB. 1-9

$\sigma_y \, da$    $\sigma_x \, da$

$\left| \sigma_x \right| = \left| \sigma_y \right|$
$\tau = 0$

PROB. 1-10

$\sigma_x \, da$

$\sigma_y = 0$
$\tau = 0$

PROB. 1-11

$\tau \, da$

$\sigma_x = 0$
$\sigma_y = 0$

PROB. 1-12

$\sigma_y \, da$    $\tau \, da$

$\sigma_x = 0$
$2 \left| \sigma_y \right| = \left| \tau \right|$

PROB. 1-13

$\sigma_y \, da$    $\sigma_x \, da$    $\tau \, da$

$2 \left| \sigma_x \right| = \left| \sigma_y \right| = 3 \left| \tau \right|$

PROB. 1-14

$\sigma_y \, da$    $\sigma_x \, da$    $\tau \, da$

$\sigma_x = 25,000 \text{ psi (T)}$
$\sigma_y = 5,000 \text{ psi (C)}$
$\tau = 10,000 \text{ psi}$

PROB. 1-15

$\sigma_x \, da$    $\tau \, da$

$\sigma_x = 20,000 \text{ psi (C)}$
$\sigma_y = 0$
$\tau = 20,000 \text{ psi}$

PROB. 1-16

$\sigma_y \, da$    $\sigma_x \, da$

$\sigma_x = 400 \text{ psi}$
$\sigma_y = 300 \text{ psi}$
$\tau = 0$

PROB. 1-17

$\sigma_y \, da$    $\sigma_x \, da$    $\tau \, da$

$\sigma_x = 15,000 \text{ psi (C)}$
$\sigma_y = 25,000 \text{ psi (T)}$
$\left| \tau \right| = 15,000 \text{ psi}$

PROB. 1-18

**1-19 to 1-22.** For the states of plane stress in Probs. 1-15, 1-16, 1-17, and 1-18, determine $\sigma_{max}$, $\sigma_{min}$, and $\tau_{max}$; and sketch the properly oriented elements on which they act.

NOTE: In each of the following problems, determine the principal stresses and the principal planes in which they act. Also, assuming that $\sigma_z = 0$ in each case, determine the maximum shear stress for each case. All stresses are in pounds per square inch.

**1-23.** $\sigma_x = +4000$, $\qquad \sigma_y = -8000$, $\qquad \tau = -2000$
**1-24.** $\sigma_x = +60,000$, $\qquad \sigma_y = +180,000$, $\qquad \tau = -30,000$
**1-25.** $\sigma_x = +12,000$, $\qquad \sigma_y = 0$, $\qquad \tau = +18,000$
**1-26.** $\sigma_x = -7000$, $\qquad \sigma_y = -7000$, $\qquad \tau = 0$
**1-27.** $\sigma_x = -14,500$, $\qquad \sigma_y = +6700$, $\qquad \tau = +4360$
**1-28.** $\sigma_x = -300$, $\qquad \sigma_y = +300$, $\qquad \tau = -200$
**1-29.** $\sigma_x = +8000$, $\qquad \sigma_y = +8000$, $\qquad \tau = +8000$
**1-30.** $\sigma_x = +4000$, $\qquad \sigma_y = -10,000$ $\qquad \tau = +5000$
**1-31.** $\sigma_x = 0$, $\qquad \sigma_y = 0$, $\qquad \tau = -6000$
**1-32.** $\sigma_x = +20,000$, $\qquad \sigma_y = +10,000$, $\qquad \tau = -10,000$
**1-33.** $\sigma_x = -8000$, $\qquad \sigma_y = -12,000$, $\qquad \tau = +4000$
**1-34.** $\sigma_x = +120$, $\qquad \sigma_y = +40$, $\qquad \tau = +480$

NOTE: In the following problems, determine the unknown quantities by use of Mohr's circle. All stresses are in pounds per square inch.

**1-35.** Given: $\sigma_x = -6000$, $\sigma_y = +2000$, $\tau_{xy} = +2000$. If $\sigma_\theta = -4000$, find $\theta$ and $\tau_\theta$.

**1-36.** Given: $\sigma_x = +8000$, $\sigma_y = -2000$, $\tau_{xy} = +2000$. If $\tau_\theta = +3000$, find $\sigma_\theta$ and $\theta$.

**1-37.** Given: $\sigma_x = +22,000$, $\tau_{xy} = -18,000$, $\sigma_y = +3000$. If $\tau_\theta = +9000$, find $\theta$ and $\sigma_\theta$.

**1-38.** Given: $\sigma_y = 0$, $\tau_{xy} = -5000$, $\sigma_x = -12,500$. If $\sigma_\theta = -7500$, find $\tau_\theta$ and $\theta$.

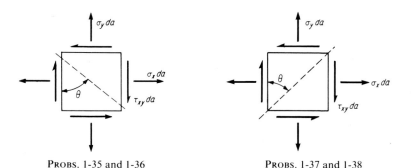

PROBS. 1-35 and 1-36 $\qquad\qquad\qquad$ PROBS. 1-37 and 1-38

**1-39.** A block made of heavy-grained wood will crack along the grain if the shear stress along the grain ever exceeds 800 psi. If $\sigma_y = 500$ psi, what range of values may $\sigma_x$ assume without the block cracking?

PROB. 1-39          PROB. 1-40

**1-40.** Two wedges of material are held together by some adhesive compound to form a cube. The joint will break if the tensile stress on the joint exceeds 1500 psi. What is the maximum value $\sigma_y$ may have, if $\sigma_x = 800$ psi and $\tau_{xy} = 500$ psi?

**1-41 to 1-46.** For Probs. 1-1 to 1-6 determine the stresses on the indicated planes by use of Mohr's circle.

**1-47.** At some point in a loaded body the stress $\sigma_x$ on the vertical face is 25,000 psi, tension while the shear stress on this face is zero. At this same point the stress $\sigma_\theta$ on the inclined face at an angle of 30° with the vertical is 15,000 psi, tension, while the shear stress on this inclined face is 17,320 psi, positive. This set of conditions can be illustrated by considering two elements, as in (a), or one

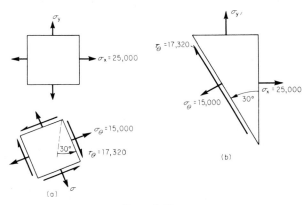

PROB. 1-47

wedge, as in (b). Determine the maximum and minimum normal stresses and the maximum shear stress. Show your answers on a properly oriented element or elements.

## chapter 2

# *Strain*

## 2-1. Deformations

As stated earlier, mechanics of deformable bodies concerns the relationships between external loads and the resulting internal force intensities and dimension changes. Chapter 1 dealt with the internal force intensities, which were defined as stresses. This chapter will be a similar treatment of dimension changes.

The primary function of an engineering member is to resist loads. When loads of any type are applied to a member, the member will always undergo dimension changes. In other words, the loads alter the size and/or the shape of the body. Such dimension changes may or may not be visible to the naked eye depending on the degree to which the loads alter the body.

A dimension change is called a deformation and will be denoted by the letter *e*. A deformation which causes an increase of a length is commonly called an elongation, or extension; and a deformation which causes a decrease of a length will be called a contraction, or compression. A deformation which results in an angular distortion is called a shear deformation.

It is possible for an element in any solid body to undergo deformations of various types and in many directions when loads are applied. Hence, the general analysis of deformation is a three-dimensional problem. Although most of the discussions in this chapter will be confined to the deformations occurring in one plane, you should not lose sight of the fact that deformations may be occurring simultaneously in directions perpendicular to the plane under consideration.

## 2-2. Displacement

Let us begin our discussion by considering a solid body of arbitrary initial size and shape, as in Fig. 2-1(*a*). At some point *P* on the surface of the body, we draw two very *short* intersecting lines *PR* and *PQ*. Let us suppose now that the body is acted upon by some force system, as in Fig. 2-1(*b*), so that the body, and therefore every point in the body, is displaced to a new position; for example, *P* to *P'* and *R* to *R'*. If the body were *rigid*, the points *P*, *R*, and *Q* would have retained their same relative positions in the body. The length *P'R'* would be the same as *PR*, *P'Q'* would be equal to *PQ*, and the angle $\theta'$ would have the same magnitude as

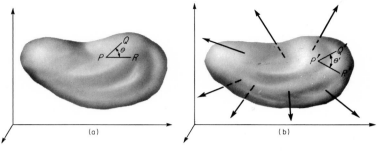

FIG. 2-1

$\theta$. Although the body as a whole may have undergone translation and rotation, there would have been *no deformation*. However, in a real body, which of course is deformable, the lengths of line elements generally would not remain unchanged, and elongations or contractions would occur. Also, the angles between line elements would change, and the body would no longer have its original shape but would become distorted. In fact, the straight lines *PR* and *PQ* would probably become curved arcs. However, if the body is not violently deformed we can consider that very short straight lines remain straight. Thus we see that when a force system acts on a deformable body, the resulting displacements may be classified into four general types:

*a*) Translation of the whole body
*b*) Rotation of the whole body
*c*) Changes of lengths; i.e., elongations or contractions
*d*) Distortions

The first two types of displacement result in movement of the body as a whole and are called rigid-body displacements, while the latter two cause the body to deform. The following presentation is concerned only with the displacements which produce deformations.

## 2-3. Definition of Strain

**Axial Strain.** Consider two orthogonal very short material lines eminating from a point *P* in the interior of a deformable body. These two lines suffice to define the plane to which we are restricting our consideration and hence we use two dimensional figures as in Fig. 2-2. Assume that the body is loaded in such a manner that no distortion occurs at point *P*; i.e., the right angle *RPQ* is preserved but the *lengths* are not, so that *PR* and *PQ* deform into *PR'* and *PQ'*, respectively, as shown in Fig. 2-2(*b*). For convenience we have chosen the *x*, *y* coordinates to be parallel to *PR* and *PQ*, respectively. *PR* has undergone an elongation of $e_x$, the notation

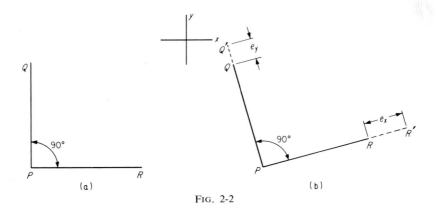

FIG. 2-2

here being for the elongation $e$ of an elemental line segment originally parallel to the $x$ axis. Similarly $PQ$ has undergone an elongation $e_y$.

By definition, the average axial strain $\epsilon_{avg}$ is the axial deformation (elongation or contraction) per unit length. Hence

$$\epsilon_{avg} = \frac{e}{L}$$

and, in particular,

$$(\epsilon_{avg})_x = \frac{e_x}{L_{PR}}$$

$$(\epsilon_{avg})_y = \frac{e_y}{L_{PQ}}$$

Now, as in Chapter 1, we assume that the material is "smeared out" to smooth over all the voids and irregularities. Then the original material lines $PR$ and $PQ$ can be made as small as we please and we define the axial strains at $P$ as

$$\epsilon_x = \lim_{L_{PR} \to 0} \frac{e_{PR}}{L_{PR}} = \frac{de_x}{dx} \qquad \epsilon_y = \lim_{L_{PQ} \to 0} \frac{e_{PQ}}{L_{PQ}} = \frac{de_y}{dy} \qquad (2\text{-}1)$$

Notice that $\epsilon_x$ is not necessarily equal to $\epsilon_y$ any more than $e_x$ was equal to $e_y$. Axial strain resulting from an elongation is commonly called a tensile strain, while that resulting from a contraction is called a compressive strain.

**Shear Strain.** Consider now two orthogonal very short material lines eminating from a point $P$ in the interior of a deformable body as before. Assume now that the body is loaded in such a manner that the lengths of $PR$ and $PQ$ remain unchanged but that the right angle has

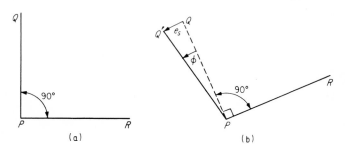

FIG. 2-3

been increased by an amount $\phi$, as shown in Fig. 2-3. Then

$$\sin \phi = \frac{e_s}{L_{PQ'}} = \frac{e_s}{L_{PQ}}$$

where $e_s$ is the shear deformation. In most engineering applications $\phi$ will be quite small so that

$$\phi \cong \sin \phi = \frac{e_s}{L_{PQ}}$$

where $\phi$ is expressed in radians and is called the average shear strain. Once again we assume the material to be voidless or smeared out so that $PQ$ and $PR$ can be as small as we wish and we define the shear strain, denoted by $\gamma$, at point $P$ as

$$\gamma = \lim_{L_{PQ} \to 0} \frac{e_s}{L_{PQ}} = \frac{de_s}{dL} \tag{2-2}$$

But from above we see that $\gamma$ at $P$ will correspond to $\phi$, the angle of distortion. Thus the shear strain is usually interpreted physically as an angle increase between two originally orthogonal very short line elements.

It is very important for you to understand the distinction between deformation and strain. Deformation is a change of linear dimension and has units of length, such as inches or feet. Strain is deformation per unit length and is therefore a dimensionless quantity. However, axial strain is usually expressed in terms of inches per inch or micro-inches (millionths of an inch) per inch. Axial strain is also commonly expressed as percent deformation. The axial strain for engineering materials in ordinary use seldom exceeds 0.002 in./in., which is equivalent to 2000 micro-inches/inch or 0.2 percent. Shear strain, since it is an angle change, is often expressed in radians.

EXAMPLE 2-1.    The axial strain in any direction due to thermal expansion is given by the relationship $\epsilon = \alpha \Delta T$, where $\alpha$ is the coefficient of thermal expansion and $\Delta T$ is the change in temperature. What are the final size and shape of the thin homogeneous plate shown in the illustration, if *a*) the entire plate undergoes a

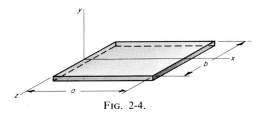

FIG. 2-4.

temperature increase of *M* degrees; *b*) the plate is subjected to a temperature gradient given by $\Delta T = Kx$? *c*) What is the shear strain $\gamma_{xz}$ in the plate for case (*a*)? *d*) What is the shear strain for case (*b*) assuming the *z* edge to be fixed?

*Solution: a*) If the entire plate undergoes a uniform temperature change of *M* degrees, for free expansion the axial strain in any direction will be uniform and of value

$$\epsilon = \alpha M$$

Hence the elongation in the *x* and *z* directions will be

$$e_x = \epsilon a \qquad e_z = \epsilon b$$

respectively, and the plate will still be rectangular with dimensions

$$L_x = a(1 + \alpha M) \qquad L_z = b(1 + \alpha M)$$

shown in Fig. 2-5(*a*).

*b*) If the plate undergoes a temperature gradient given by $\Delta T = Kx$, then at any particular point $P(x,z)$ the axial strain will be given by $\epsilon = \alpha Kx$. From Eq. 2-1

$$de_x = \epsilon dx \qquad de_z = \epsilon dz$$

Integration yields

$$e_x = \alpha K \frac{x^2}{2} \qquad e_z = \alpha Kxz$$

Since $\alpha$ is usually many orders of magnitude less than unity, the shape of the plate is shown in Fig. 2-5 (*b*), where the deformations are exaggerated.

*c*) As remarked in *a*), under a uniform free expansion the plate remains rectangular everywhere and hence no distortion occurs. Thus

$$\gamma_{xz} = 0 \text{ everywhere}$$

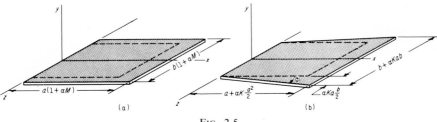

FIG. 2-5

d) For case *b*) recall

$$e_x = \alpha K \frac{x^2}{2} \qquad e_z = \alpha Kxz$$

and from Fig. 2-5 (*b*) we see clearly that distortion occurs. Also note that the maximum distortion occurs along the $z = \pm b/2$ edges of the plate and the angle of distortion is

$$\tan \phi = \frac{\alpha Ka(b/2)}{a + \dfrac{\alpha Ka^2}{2}} = \frac{\alpha K(b/2)}{1 + \dfrac{\alpha Ka}{2}}$$

Now, since $\alpha << 1$ so that the deformations and distortions are small, we can say that

$$\tan \phi \cong \phi = \gamma_{\max} = \alpha K(b/2)$$

and in general the shear strain for pairs of lines parallel to the *x-z* axes is given by

$$\gamma_{xz} = \frac{de_z}{dx} = \alpha Kz$$

## 2-4. Analysis of Plane Strain

In Sec. 2-3, axial strain and shear strain were analyzed individually. Generally, a point *P* in a body will undergo axial strain and shear strain simultaneously as indicated in Fig. 2-6. Thus, in general, a state of plane strain at a point is characterized by two axial strains and one shear strain. Since we are interested only in the relation between the initial and final configurations, we can assume that the final configuration was achieved by an axial straining process followed by a shear straining process, or, vice versa. *So long as the deformations are small, the order of the straining processes is immaterial,* so for convenience we will assume the former. Also, since we are considering the strains at a point, we will label the original lengths *dx* and *dy*, as in Fig. 2-7.

*A pair of infinitesimal line elements may be considered to define*

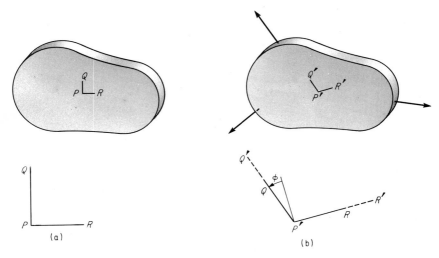

FIG. 2-6

*a point P in a body.* The line elements are made parallel to the axes of a specific two-dimensional coordinate system, such as the *x-y* system. Therefore, any two orthogonal line elements can be used to show only the strains associated with that particular set of axes. In Fig. 2-7 the shear strain $\gamma_{xy}$ and the axial strains $\epsilon_x$ and $\epsilon_y$ are associated only with line elements initially parallel to the *x* and *y* axes.

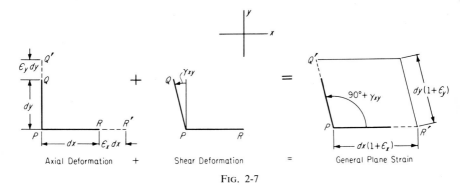

Axial Deformation     +     Shear Deformation     =     General Plane Strain

FIG. 2-7

**Transformation of Axial Strain.** Many engineering problems require the determination of the strains or deformations associated with a particular coordinate system. As a result, it is often necessary to relate the strains associated with one coordinate system to the strains associated with another coordinate system. For example, suppose we know the strains $\epsilon_x$, $\epsilon_y$, and $\gamma_{xy}$ associated with the *x-y* coordinate system, and we wish to know the strains associated with the *n-s* coordinate system, which is oriented at some angle $\theta$ with the *x-y* system, as shown in Fig. 2-8(*a*). In order to determine the strains in the *n-s* system, we must relate the deformations parallel to the *n* and *s* axes to the deformations of line elements parallel to the *x* and *y* axes. Figure 2-8(*b*) shows undeformed line elements associated with the *x-y* system and having such proportions that

$$dx = dn \cos \theta \qquad dy = dn \sin \theta$$

or

$$\cos \theta = \frac{dx}{dn} \qquad \sin \theta = \frac{dy}{dn}$$

Figure 2-8(*c*) shows these elements in a state of plane strain.

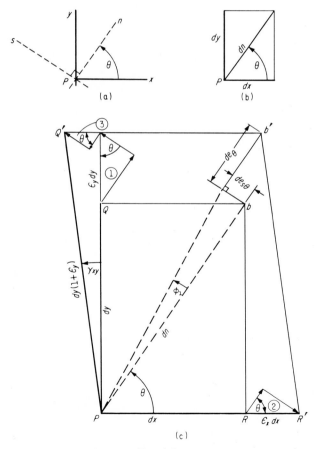

Fig. 2-8

The axial strain in the *n* direction, denoted by $\epsilon_\theta$, would be the axial deformation $de_\theta$ in that direction divided by $dn$. This deformation can be found by appropriately adding or subtracting the component deformations in the *n* direction, which are represented by the sides of triangles ①, ②, and ③ parallel to the *n* axis. That side of triangle ① parallel to $dn$ is $\epsilon_y dy \sin \theta$; that of triangle ② is $\epsilon_x dx \cos \theta$, and that of triangle ③ is $dy(1 + \epsilon_y) \gamma_{xy} \cos \theta$. Hence,

$$\epsilon_\theta = \frac{de_\theta}{dn} = \frac{\epsilon_y dy \sin \theta + \epsilon_x dx \cos \theta - dy(1 + \epsilon_y) \gamma_{xy} \cos \theta}{dn} \qquad (2\text{-}3)$$

By substituting $\cos \theta$ for $dx/dn$ and $\sin \theta$ for $dy/dn$ and neglecting the product $\gamma_{xy}\epsilon_y$, which is extremely small compared with $\gamma_{xy}$ or $\epsilon_y$, we obtain

$$\epsilon_\theta = \epsilon_x \cos^2 \theta + \epsilon_y \sin^2 \theta - \gamma_{xy} \cos \theta \sin \phi \qquad (2\text{-}4)$$

From Eq. 2-4 we can obtain the axial strain along any axis lying in the plane of analysis. This equation is called a strain transformation equation. Note that to use this equation we must employ a sign convention consistent with that assumed in the derivation. The angle $\theta$ is positive when measured counterclockwise from the *x* axis; tensile axial strains are positive, and compressive axial strains are negative; *a positive shear strain is one that causes an increase in the angle at P when the x axis is taken as the "horizontal" reference line.*

**Transformation of Shear Strain.** Figure 2-8(c) also shows that the line $dn$ parallel to the *n* axis has undergone a small change of orientation, denoted by $\phi_1$. This angle $\phi_1$, being very small, can be expressed in radians by dividing the deformation $de_{s\theta}$ by the length $dn$. The deformation $de_{s\theta}$ can be obtained by adding vectorially the component deformations *perpendicular* to the *n* axis. Thus, referring to the triangles of Fig. 2-8(*c*),

$$\phi_1 = \frac{de_{s\theta}}{dn} = \frac{\epsilon_y dy \cos \theta - \epsilon_x dx \sin \theta + dy(1 + \epsilon_y) \gamma_{xy} \sin \theta}{dn} \qquad (2\text{-}5)$$

$$\phi_1 = \epsilon_y \sin \theta \cos \theta - \epsilon_x \cos \theta \sin \theta + \gamma_{xy} \sin^2 \theta \qquad (2\text{-}6)$$

However, $\phi_1$ does *not* represent the shear strain associated with the *n-s* coordinate system. Shear strain is the *change* in the 90° angle of the element. In order to determine the shear strain associated with the *n-s* coordinate system, we must determine the angle change at *P* between the *n* and *s* axes.

Figure 2-9(*a*) shows that the line element parallel to the *n* axis will undergo a change in orientation denoted by $\phi_1$, and the line parallel to the *s* axis will undergo a change in orientation denoted by $\phi_2$. The angle at *P* before distortion was 90°. The angle at *P* after distortion will be $90° + \phi_2 - \phi_1$. Therefore, the change in the angle at *P* will be $\phi_2 - \phi_1$, and this quantity will be the shear strain $\gamma_\theta$ associated with the *n-s* system.

We must now determine $\phi_2$. Figure 2-9(*b*) shows two elements associated with the *x-y* coordinate system. The element with dimensions *dx* and *dy* has already been analyzed. The second element with dimensions *dx'* and *dy'* is proportioned so that

$$\sin \theta = \frac{dx'}{ds} \qquad \cos \theta = \frac{dy'}{ds}$$

Figure 2-9(*c*) shows this second element subjected to the same state of plane strain as the first element. The angle $\phi_2$ may be expressed in radians by dividing the deformation $de_{s\theta}$ by the length *ds*. The components of this deformation are the lengths of the sides of the triangles numbered ①, ②, and ③ *perpendicular* to the *s* axis. Thus,

$$\phi_2 = \frac{de_{s\theta}}{ds} = \frac{-\epsilon_y dy' \sin \theta + \epsilon_x dx' \cos \theta + dy'(1 + \epsilon_y)\gamma_{xy} \cos \theta}{ds} \quad (2\text{-}7)$$

$$\phi_2 = \epsilon_x \sin \theta \cos \theta - \epsilon_y \cos \theta \sin \theta + \gamma_{xy} \cos^2 \theta \quad (2\text{-}8)$$

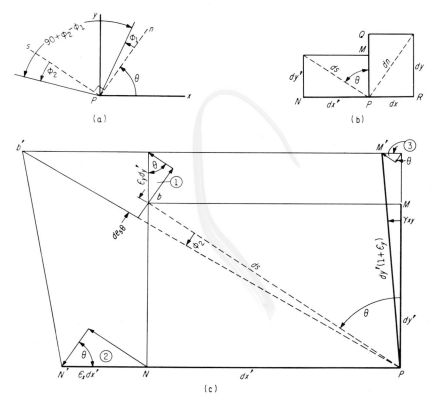

(a)

(b)

(c)

Fig. 2-9

The shear strain $\gamma_\theta$ is, therefore, from Eqs. 2-6 and 2-8

$$\gamma_\theta = \phi_2 - \phi_1 = 2(\epsilon_x - \epsilon_y)\sin\theta\cos\theta + \gamma_{xy}(\cos^2\theta - \sin^2\theta) \qquad (2\text{-}9)$$

From Eq. 2-9, which is another strain transformation equation, we can determine the shear strain associated with any set of axes lying in the plane of analysis. The sign convention must be the same as that for Eq. 2-4.

## 2-5. Mohr's Circle for Plane Strain

By use of the double-angle identities of trigonometry, the strain transformation equations (Eqs. 2-4 and 2-9) can be written as follows:

$$\epsilon_\theta = \frac{\epsilon_x + \epsilon_y}{2} + \left(\frac{\epsilon_x - \epsilon_y}{2}\right)\cos 2\theta - \frac{\gamma_{xy}}{2}\sin 2\theta \qquad (2\text{-}10)$$

$$\frac{\gamma_\theta}{2} = \left(\frac{\epsilon_x - \epsilon_y}{2}\right)\sin 2\theta + \frac{\gamma_{xy}}{2}\cos 2\theta \qquad (2\text{-}11)$$

You should note the similarity between the stress transformation equations (Eqs. 1-3 and 1-4) and the strain transformation equations (Eqs. 2-10 and 2-11).

Since we were able to express the stress transformation equations in the form of an equation representing a circle (Eq. 1-7a), we can do the same thing with the strain transformation equations. By transposing, squaring, adding, and simplifying, we obtain the following result:

$$\left[\epsilon_\theta - \left(\frac{\epsilon_x + \epsilon_y}{2}\right)\right]^2 + \left[\frac{\gamma_\theta}{2} - 0\right]^2 = \left[\sqrt{\left(\frac{\epsilon_x - \epsilon_y}{2}\right)^2 + \left(\frac{\gamma_{xy}}{2}\right)^2}\right]^2 \qquad (2\text{-}12)$$

We can plot Eq. 2-12 as a circle on a rectangular coordinate system in which the abscissa is $\epsilon_\theta$ and the ordinate is $\gamma_\theta/2$. Such a plot is called Mohr's circle for plane strain, and its properties are very similar to those of Mohr's circle for plane stress.

As an example, suppose that at some point in a loaded body we know the strains associated with a particular set of axes which we will call the $x$ and $y$ axes. For convenience, we will assume that all the strains are positive and that $\epsilon_x$ is the larger axial strain. Figure 2-10(a) shows Mohr's circle for strain plotted for the assumed conditions. Figure 2-10(b) shows the orientations of the $n$ and $s$ axes and axes labeled *1* and *2* with respect to the original $x$ and $y$ axes. The line elements associated with the various axes are shown in Fig. 2-10(b) merely to illustrate the distortion resulting from the respective shear strains.

You should study Fig. 2-10 very carefully. The following important features can be observed from this figure:

1. The center of the circle is always on the $\epsilon_\theta$ axis.
2. Each point on the circle represents two quantities for the point in

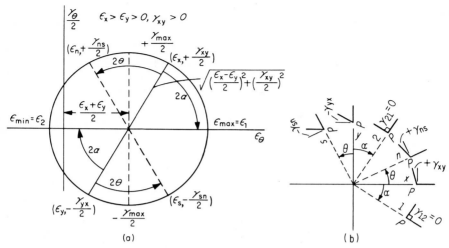

FIG. 2-10

the loaded body, namely, an axial strain along a particular axis and one-half of the value of the shear strain associated with that same axis. In every case *a positive shear strain produces an increase in the angle at P and a negative shear strain produces a decrease in that angle.* Satisfy yourself about this fact for each situation in Fig. 2-10(*b*) by rotating the book so as to make the axis in question a "horizontal" reference line. Note that the first letter of the subscript for $\gamma$ corresponds to this horizontal reference line.

3. Any angle in the *x-y* coordinate system corresponds to twice that angle on the circle; that is, $\theta$ *on the body, $2\theta$ on the circle.* Also recall that in order to use Eqs. 2-4 and 2-9 correctly, $\theta$ had to be measured from the *x* axis with the counterclockwise direction being positive. However, in using the circle, we can refer any desired axis or coordinate system to either the *x* axis or the *y* axis, as long as we correctly reference $2\theta$ on the circle.

4. Two diametrically opposite points on the circle represent the *axial strains* and the *shear strain* associated with a *set* of perpendicular axes, such as the *x* and *y* axes, the *n* and *s* axes, and the *1* and *2* axes.

5. The algebraic sum of the axial strains at a point is the same for all sets of axes; that is, $\epsilon_x + \epsilon_y = \epsilon_n + \epsilon_s = \epsilon_1 + \epsilon_2$.

6. The maximum and minimum axial strains, denoted by $\epsilon_1$ and $\epsilon_2$, occur along the axes *1* and *2*. These strains are called the principal strains, and the axes *1* and *2* are called the principal axes. No shear strain is associated with the principal axes. Line elements parallel to the principal axes will have no angular distortion and no shear strain.

7. The magnitude of the maximum shear strain is twice the radius of the circle; that is, the diameter of the circle represents the magnitude of the maximum shear strain.

EXAMPLE 2-2. Some point on the surface of a loaded body was found to have a tensile strain of 0.0001 in./in. in the $x$ direction and a compressive strain of 0.0007 in./in. in the $y$ direction, and the angular distortion $\gamma_{xy}$ was $-0.0006$ radian. *a*) Determine the values and directions of the principal strains. *b*) Determine the magnitude of the maximum shear strain. *c*) Determine the axial and shear strains associated with the $n$ axis which is 30° counterclockwise from the $x$ axis, as indicated in Fig. 2-11(*a*).

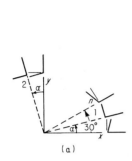
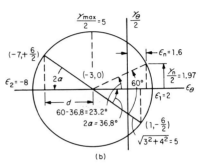

FIG. 2-11

*Solution:* First, the points representing the known strains are plotted and joined by a line, which is a diameter of the Mohr circle. The center of the circle is midway between these points, and the circle is shown in Fig. 2-11(*b*). In this circle, actual strains are multiplied by $10^4$ for use as distances. Thus,

$$\epsilon_x = \quad 0.0001 \text{ in./in.} = 1 \text{ unit}$$

$$\epsilon_y = -0.0007 \text{ in./in.} = -7 \text{ units}$$

$$\gamma_{xy} = -0.0006 \text{ radian} = -6 \text{ units}$$

$$d = \frac{1 + 7}{2} = 4$$

The coordinates of the center of the circle are $-3$ and 0, and its radius is $\sqrt{4^2 + 3^2} = 5$.

*a*) The principal strains are equal to the center coordinate plus and minus the radius. They are

$$\epsilon_1 = -3 + 5 = +2 \qquad \epsilon_2 = -3 - 5 = -8$$

So $\epsilon_1 = 0.0002$ in./in., tensile, and $\epsilon_2 = -0.0008$ in./in., compressive. The direction of $\epsilon_1$ is at an angle $\alpha$ counterclockwise from the $x$ axis, and $\epsilon_2$ is at an angle $\alpha$ counterclockwise from the $y$ axis. From the circle,

$$\tan 2\alpha = \frac{3}{4} \qquad 2\alpha = 36.8° \qquad \alpha = 18.4°$$

*b*) The magnitude of the maximum shear strain is equal to twice the radius of the circle. So

$$\gamma_{\max} = 2(\gamma/2)_{\max} = 2(5) = 10 \text{ units}$$

or $\gamma_{\max} = 0.0010$ radian.

*c*) The point representing the *n* axis on the circle is located 60° counterclockwise from the point representing the *x* axis. A radial line from this point to the center is then drawn. The angle between this radial line and the $\epsilon_\theta$ axis is

$$60° - 2\alpha = 60° - 36.8° = 23.2°$$

We now see that $\epsilon_n$ is equal to the coordinate of the center of the circle plus the base of the 23.2° triangle, and $\gamma_n/2$ is the vertical side of this triangle. Thus,

$$\epsilon_n = -3 + 5 \cos 23.2° = -3 + 4.6 = 1.6 \text{ units}$$

or $\epsilon_n = 0.00016$ in./in., tensile.

$$\gamma_n/2 = 5 \sin 23.2° = 1.97 \text{ units}$$

and $\gamma_n = 0.000394$ radian, positive.

## 2-6. Strain Measurement and Rosette Analysis

The definitions of stress and strain "at a point" are analytical abstractions. For instance, we have defined stress as a force intensity on an infinitesimal element. As such, it would be physically impossible for us to see or measure stress. However, a finite load can be measured, and a large portion of Part III of this book is concerned with relating measurable loads to the theoretical stresses existing in engineering structures.

In this chapter we have defined strain as a deformation per unit length, and we have analyzed plane strain at a point. Although it is physically possible to measure *finite* deformations on the surface of a body, a *finite* deformation can only be measured over a *finite* length. So the idea of strain at a *point* in a body is an analytical abstraction. However, if a deformation is measured over a relatively small length, an average deformation per unit length can be evaluated and interpreted as the approximate value of a strain at the "finite point" of measurement. On this basis, an axial strain can be approximated by first measuring the elongation or contraction over a small length and then dividing this deformation by the length over which the deformation occurred.

It is much more difficult to approximate a shear strain, since it would be necessary to measure the angular distortion occurring at a point on the surface of a body. Generally, the various mechanical, optical, and electrical methods of measuring *axial* deformations are far simpler and more reliable than are the methods for measuring angular distortions. In Chapter 16 some of the more common methods of strain measurement are discussed.

In most problems involving strain analysis, the engineer is interested in the values and directions of the principal strains. In cases where the directions of the principal strains are known, their values can be approximated by measuring the deformations along the principal axes. However, there are a great many problems in which neither the directions nor the values of the principal strains are known. The principal strains at a point can be determined from any three known axial strains at that point. This is indeed fortunate, since it eliminates the somewhat difficult task of measuring shear strains.

In Fig. 2-12, suppose that the axial strains along the axes *a*, *b*, and *c* are known, as well as the angles $\theta_a$, $\theta_b$, and $\theta_c$. By using these known

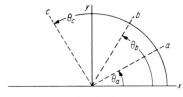

FIG. 2-12

values and applying Eq. 2-4, we obtain three simultaneous equations involving the three unknowns $\epsilon_x$, $\epsilon_y$, and $\gamma_{xy}$. These equations follow:

$$\epsilon_a = \epsilon_x \cos^2 \theta_a + \epsilon_y \sin^2 \theta_a - \gamma_{xy} \cos \theta_a \sin \theta_a$$

$$\epsilon_b = \epsilon_x \cos^2 \theta_b + \epsilon_y \sin^2 \theta_b - \gamma_{xy} \cos \theta_b \sin \theta_b$$

$$\epsilon_c = \epsilon_x \cos^2 \theta_c + \epsilon_y \sin^2 \theta_c - \gamma_{xy} \cos \theta_c \sin \theta_c$$

The solution of these equations will yield $\epsilon_x$, $\epsilon_y$, and $\gamma_{xy}$, and we will then be able to plot Mohr's circle, from which the principal strains can be determined.

A device for measuring three or more simultaneous axial strains is called a rosette strain gage. Generally the axes of strain measurement are arranged in a definite pattern, so as to make the solution of the equations numerically simple. For example, in a rectangular rosette the axial strains are measured along three axes that are 45° apart. In a delta rosette the three axes are 60° apart. There are various graphical and semigraphical methods for solving strain rosettes which eliminate the need for the solution of simultaneous equations.[2]

EXAMPLE 2-3.  A rectangular rosette strain gage was applied to the surface of a beam, as shown in Fig. 2-13(a). The strains along the axes *a*, *b*, and *c*, respectively, were $\epsilon_a$ = +0.0008 in./in., $\epsilon_b$ = −0.0006 in./in., and $\epsilon_c$ = −0.0004 in./in. Determine the directions and values of the principal strains at point *O*.

[2]A good presentation of a graphical solution of strain rosettes may be found in Chapters 6 and 7 of *The Strain Gage Primer* by Perry and Lissner, Second Edition, McGraw-Hill Book Company, 1962.

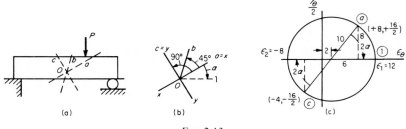

FIG. 2-13

*Solution:* Let the *a* axis of the gage be coincident with the *x* axis of an arbitrary coordinate system, as illustrated in Fig. 2-13(*b*). Then $\theta_a = 0°$, $\theta_b = 45°$, and $\theta_c = 90°$. Substituting these values and the measured strains in Eq. 2-4, we obtain the following results:

$$\epsilon_a = \epsilon_x(1) + \epsilon_y(0) - \gamma_{xy}(1)(0) = \epsilon_x$$

$$\epsilon_x = 0.0008 \text{ in./in.}$$

$$\epsilon_c = \epsilon_x(0) + \epsilon_y(1) - \gamma_{xy}(0)(1) = \epsilon_y$$

$$\epsilon_y = -0.0004 \text{ in./in.}$$

$$\epsilon_b = \epsilon_x(1/2) + \epsilon_y(1/2) - \gamma_{xy}\sqrt{1/2}\sqrt{1/2}$$

$$\gamma_{xy} = -2\epsilon_b + (\epsilon_x + \epsilon_y)$$

$$= -2(-0.0006) + (0.0008 - 0.0004) = +0.00016 \text{ radian}$$

We can now plot Mohr's circle and determine the principal strains, as shown in Fig. 2-13(*c*). In this circle, values of strain are multiplied by $10^4$ for use as distances. The computations follow:

$$\tan 2\alpha = \tfrac{8}{6} \qquad 2\alpha = 53.2° \qquad \alpha = 26.6°$$

$$\epsilon_1 = 2 + 10 = 12 \qquad \epsilon_2 = 2 - 10 = -8$$

Thus, $\epsilon_1 = 0.0012$ in./in. and $\epsilon_2 = 0.0008$ in./in. The maximum strain occurs along an axis 26.6° clockwise from axis *a*. The minimum strain occurs along an axis 26.6° clockwise from axis *c*.

You must realize that a strain measured by means of a rosette represents only an average value of the strain over the measuring length. It is therefore physically impossible to determine the *exact* value of a strain at a *point*. The results obtained in Example 2-3 represent only a good approximation of the principal strains at point *O*.

## PROBLEMS

**2-1.** A rubber band is stretched from 2 in. to 3 in. What axial strain is imposed?

**2-2.** An unloaded straight vine is 30 ft long. It stretches 8 in. when Tarzan is at the bottom of his swing. What axial strain is imposed?

**2-3.** A turnbuckle has a $\frac{1}{4}$ in. diameter screw, 20 threads per inch (each end) and is used to tighten a steel rod which is welded between two rigid walls 48 in. apart. If the rod is snug (unstrained) to start, how much axial strain will the rod be subjected to if the turnbuckle is tightened two full turns?

**2-4.** In Prob. 2-3, what would the axial strain in the rod be if, *a*) after tightening two turns, the temperature were lowered 30°F? *b*) if the temperature were raised 30°F? See appendix for coefficient of thermal expansion.

**2-5.** The bolt-end-sleeve combination shown in the illustration is adjusted until the nut is just in contact with the sleeve. The bolt has a pitch of 16 threads per inch. The nut is then tightened two full turns. If any deformations of the bolt head and nut are neglected, what is the relationship between the deformation of the bolt stem and the deformation of the sleeve?

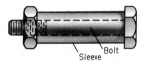

PROB. 2-5

**2-6.** A very stiff bar (assumed to be rigid) is suspended from two wires of unequal length, as shown in the illustration. Initially, point *B* is 1 in. higher than point *A*. The bar is then loaded in such a manner that the bar eventually becomes horizontal. *a*) What is the relationship between the deformation of the wire at *A* and that of the wire at *B*? *b*) What is the displacement of point *C* in terms of the deformation of the wire at *A*?

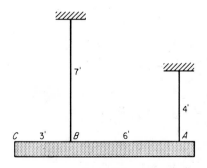

PROB. 2-6

**2-7.** The very stiff rods *AC* and *CE* are held in the positions shown in the illustration by the pins at *A* and *C* and the taut flexible wire *BD*. If point *E* is then given a horizontal displacement of 3 in. to the right, what is the change in the average strain in the wire?

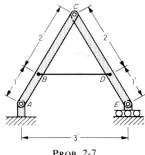

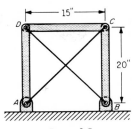

PROB. 2-7          PROB. 2-8

**2-8.** The pin-connected rods $BC$, $CD$, and $DA$, which can be assumed to be rigid, are held in the positions shown in the illustration by the taut flexible wires $AC$ and $BD$. If point $D$ is then given a horizontal displacement of 5 in. to the right, what is the change in the average strain in each of the wires?

**2-9.** A thin metal hoop has an inside diameter of 30 in. at room temperature. It is then heated until it easily slips on a mandrel 30.5 in. in diameter. If the mandrel is assumed to be rigid, what is the circumferential strain in the hoop when it cools down to room temperature?

**2-10.** A thin spherical balloon 0.01 in. thick and having an inside radius of 12 in. undergoes an internal pressure change such that the radius becomes 12.06 in. Compute the average change in surface (membrane) strain in the balloon.

**2-11.** The flexible cable shown wrapped around the cone in the illustration has an unstretched length of $2\pi R$. What is the *change* in the average strain of the cable if it is forced down to a depth of $h/2$?

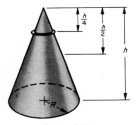

PROB. 2-11

**2-12.** What are the average strains in the wires $AP$, $BP$, and $CP$ shown in the illustration, if the point $P$ is displaced a distance $X$ to the left and a distance $Y$

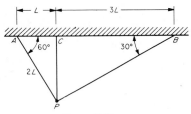

PROB. 2-12

downward? If $X$ and $Y$ are small compared with $L$, what are approximate values for these strains?

**2-13.** The rigid bar of length $L$ shown in the illustration is supported by two vertical deformable wires. The bar is rotated about the vertical axis $a$-$a$ through an angle $\theta$ measured in the horizontal plane, but the bar is also constrained so that

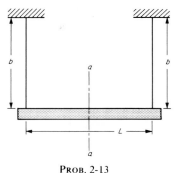

PROB. 2-13

its elvation does not change. What will be the average axial strain in each wire due to this rotation? If $\theta$ is small, what will be an approximate value for this strain?

**2-14.** Two rigid straight rods of length $L$ are separated by a plate whose dimensions are $L$ and $h$ as shown in the illustration. The upper rod $AB$ is then moved a small distance $e_0$ to the right, but remains horizontal and straight, while

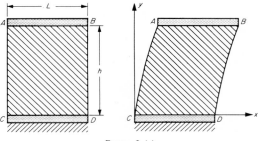

PROB. 2-14

the rod $CD$ remains stationary. What is the shear strain $\gamma_{xy}$ at the points $(L/3, h/2)$ and $(0,0)$ if $a$) the original vertical lines remain straight; $b$) the original vertical lines deform into parabolas for which the equation is $x = ky^2$.

**2-15.** A torsional spring composed of an outer metal ring, a disk of hard rubber, and an inner ring is shown in the illustration. The outer ring is rotated through a small angle $\phi$ relative to the inner ring. Assume that the original radial lines remain straight during the deformation. Find an expression for the shear strain in the rubber.

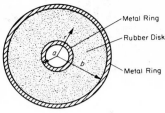

PROB. 2-15

**2-16.** The bar of length $L$ shown in the illustration is subjected to a deformation which causes a variable strain such that $\epsilon_x = e_0 \cos (\pi x/2L)$. *a*) Find the

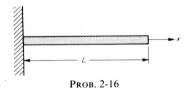

PROB. 2-16

change in the length of the bar. *b*) Sketch a curve showing the *deformation* as a function of $x$.

**2-17.** Using the strain transformation equation (Eq. 2-4 or Eq. 2-10), prove that the sum of any two perpendicular axial strains is a constant and is equal to $\epsilon_x + \epsilon_y$.

NOTE: For the given state of plane strain in each of the problems from Prob. 2-18 to Prob. 2-24, determine the values of the principal strains and their directions. The $x$ and $y$ axes are horizontal and vertical, respectively, and all strains are in (in./in.) $\times 10^6$, or in micro-inches per inch.

**2-18.** $\epsilon_x = 400$,     $\epsilon_y = 2000$,     $\gamma_{xy} = 1800$

**2-19.** $\epsilon_x = -600$,     $\epsilon_y = 600$,     $\gamma_{xy} = -800$

**2-20.** $\epsilon_x = -2800$,     $\epsilon_y = 2000$,     $\gamma_{xy} = -1400$

**2-21.** $\epsilon_x = 500$,     $\epsilon_y = -1000$,     $\gamma_{xy} = -3000$

**2-22.** $\epsilon_x = 1200$,     $\epsilon_y = 1200$,     $\gamma_{xy} = 2400$

**2-23.** $\epsilon_x = 0$,     $\epsilon_y = -1800$,     $\gamma_{xy} = 1350$

**2-24.** $\epsilon_x = -736$,     $\epsilon_y = 1487$,     $\gamma_{xy} = 0$

**2-25.** The strain readings from a delta rosette strain gage applied to a submarine hull were recorded at one instant during a severe maneuver with the follow-

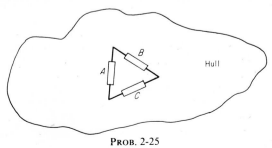

PROB. 2-25

ing results: $\epsilon_A$ was 4360 $\mu$ in./in. compressive; $\epsilon_B$ was 780 $\mu$ in./in. compressive; and $\epsilon_C$ was 2410 $\mu$ in./in. tensile. Did the maximum strain at the point exceed the design strain of 5000 $\mu$ in./in. compressive?

NOTE: For the rosette data in each of the problems for 2-26 to 2-32, determine the values and directions of the principal strains. The values of the angle $\theta$ are shown in the illustration, and all strains are in (in./in.) $\times$ $10^6$.

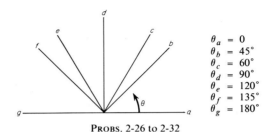

$$\theta_a = 0$$
$$\theta_b = 45°$$
$$\theta_c = 60°$$
$$\theta_d = 90°$$
$$\theta_e = 120°$$
$$\theta_f = 135°$$
$$\theta_g = 180°$$

PROBS. 2-26 to 2-32

| | | | |
|---|---|---|---|
| **2-26.** | $\epsilon_a = 1100,$ | $\epsilon_b = 4200,$ | $\epsilon_d = -550$ |
| **2-27.** | $\epsilon_b = 2500,$ | $\epsilon_d = -1600,$ | $\epsilon_f = -500$ |
| **2-28.** | $\epsilon_c = 5640,$ | $\epsilon_e = -240,$ | $\epsilon_g = -1260$ |
| **2-29.** | $\epsilon_a = 0,$ | $\epsilon_c = 900,$ | $\epsilon_e = 600$ |
| **2-30.** | $\epsilon_a = -600,$ | $\epsilon_b = -100,$ | $\epsilon_d = +1800$ |
| **2-31.** | $\epsilon_c = -2010,$ | $\epsilon_e = -1430,$ | $\epsilon_g = +270$ |
| **2-32.** | $\epsilon_b = +100,$ | $\epsilon_d = 0,$ | $\epsilon_f = -600$ |

## chapter 3

# Stress-Strain Relationships

## 3-1. Introduction

Having studied stress and strain separately, we will now consider how they are related to each other. The definitions of stress and strain in Chapters 1 and 2 were independent of each other and independent of the material of the body; we assumed only that the body was continuous and deformable. However, stress and strain *are* related to each other. Also, relationships between stress and strain depend directly on the material of the body, and it will be shown that these stress-strain relations are functions of several parameters whose values are dependent on the specific material. In this chapter these parameters will be defined, and their physical significance discussed.

Before proceeding with the main theme of this chapter, we feel it appropriate to mention something which is often overlooked or taken for granted. It is often said that a body acted on by a balanced force system (one for which the resultant force and the resultant moment are zero) is in equilibrium. But is this true? A body is in equilibrium if the body as a whole *or any part* of the body has no acceleration. Hence, if the body is originally at rest and is then acted on by a balanced force system, neither it nor any part of it can move, since it never has any acceleration and thus never gains velocity. But we know that every body deforms when forces are applied to it and that the deformation results in movement. Thus we seem to have a contradiction. The answer to the riddle is simply this. If a balanced force system is applied to a body, various parts of the body accelerate and move while the deformation is occurring. This deformation process will ultimately cease if and when the internal resistive forces become sufficiently large. When this state is reached, the body and every part of the body will once again be in equilibrium, and the equations of equilibrium will now be applicable. Throughout Chapter 1 it was assumed that the deformation process had ceased. Of course, if the internal resistive forces never become sufficiently large to stop the deformation, the body will ultimately rupture.

A static load or force is one which is applied in such a manner that the deformation occurs slowly enough to permit us to assume that the acceleration effects are negligible during the deformation process. Such a process is often called *quasi-static*. Hereafter we shall use the term "static loads" to mean loads which produce a *quasi-static deformation process*.

## 3-2. The Stress-Strain Curve

All strains at a point in a loaded body may be influenced by the stress in *any* direction at that point. However, the simplest and most convenient way to relate a stress to a strain is to relate the principal tensile stress in one direction to the resulting axial strain in the same direction in a case where the other two principal stresses are zero. This case is called a uniaxial state of stress and can be most easily obtained in a member loaded as in Fig. 3-1(*a*). The uniaxial forces acting on the element *A* are shown in Fig. 3-1(*b*), and the free-body diagram for a section of the member is shown in (*c*). The very common procedure in which a slowly increasing load is applied to such a member at room temperature is called a uniaxial tensile test. During the test the load and the resulting elongation of a selected gage length (previously marked off on the member) are recorded simultaneously. Most modern materials-testing laboratories have equipment which will plot the load-elongation data automatically as the test is performed. In order to make this a plot of stress vs. strain, the rela-

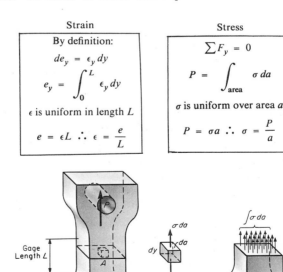

| Strain | Stress |
|---|---|
| By definition: | $\sum F_y = 0$ |
| $de_y = \epsilon_y \, dy$ | |
| $e_y = \int_0^L \epsilon_y \, dy$ | $P = \int_{\text{area}} \sigma \, da$ |
| $\epsilon$ is uniform in length $L$ | $\sigma$ is uniform over area $a$ |
| $e = \epsilon L \therefore \epsilon = \dfrac{e}{L}$ | $P = \sigma a \therefore \sigma = \dfrac{P}{a}$ |

Gage Length $L$

Element *A*
(b)

(a)

(c)

FIG. 3-1

tionships of Fig. 3-1 show that we can convert the load scale to a stress scale by dividing the loads by the original cross-sectional area of the member, and we can convert the elongation scale to a strain scale by dividing the elongations by the original gage length.

Many of the succeeding chapters of this book are concerned with relating loads and dimension changes to stresses and strains. Here we will only say that the stress and strain can be considered uniform throughout the gage length of a specimen such as that in Fig. 3-1 under the following conditions:

*a*) The load passes through the centroid of the cross section considered.

*b*) The member is of constant cross section in the gage length and somewhat beyond.

*c*) The material is homogeneous and isotropic.

In Fig. 3-2 is shown a plot of stress vs. strain for a tensile test on an extruded rod of a common aluminum alloy. As is customary in engineer-

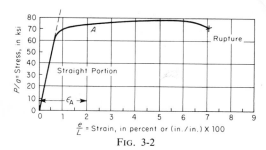

FIG. 3-2

ing, stress is plotted as the ordinate. This curve extends from zero strain to rupture. From an engineering-design viewpoint, the first part of the curve, or that up to about point *A*, is the most important portion since a material of this kind is usually unusable if the strain is much greater than $\epsilon_A$. The more important initial portion of the curve is shown to a larger scale in Fig. 3-3.

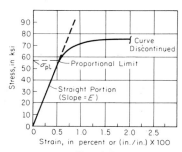

FIG. 3-3

It is generally true that for most metallic engineering materials the stress is proportional to strain for the first portion of the curve. The stress corresponding to the first point at which the curve deviates from the initial straight, or linear, portion is called the *proportional limit.* Up to the proportional limit the stress change $\Delta\sigma$ is equal to the corresponding strain change $\Delta\epsilon$ multiplied by some constant. Thus,

$$\Delta\sigma \propto \Delta\epsilon \qquad \Delta\sigma = (\Delta\epsilon) \times \text{constant}$$

This relationship is a simple manifestation of the familiar Hooke's law in which this constant of proportionality is called the modulus of elasticity and is denoted by $E$.

$$E = \frac{\Delta\sigma}{\Delta\epsilon} \qquad \text{for} \quad \begin{matrix} \sigma \leq \sigma_{PL} \\ \epsilon \leq \epsilon_{PL} \end{matrix} \qquad (3\text{-}1)$$

The modulus of elasticity can, and should, be thought of as the slope of the initial linear portion of the uniaxial stress-strain curve. It is a constant, which is unique for a particular material, and is a measure of the *stiffness* of the material. Since $\Delta\epsilon$ in in./in. is usually many orders of magnitude smaller than $\Delta\sigma$ in psi, the modulus $E$ is usually a relatively large number; for example, it is 30,000,000 psi for steel and 10,000,000 psi for aluminum (see Table I in the Appendix). For most metallic engineering materials the initial portion of the stress-strain curve for uniaxial compressive stress has the same slope as that for tensile stress. Hence, the tensile modulus of elasticity usually has the same value as the compressive modulus (this is fortunate, as we shall see later).

You must realize that the results obtained from any one uniaxial stress-strain test would only represent the response of that one particular sample of one particular material loaded in one particular direction. Thus, by themselves, the data for *one* test are of little use. However, if the material is homogeneous, then one test of one sample should be indicative of all such similar material loaded in the manner of the sample. Also, if the response of the material is independent of the orientation of the load axis of the sample, we say the material is *isotropic.* On the other hand, if the response is orientation *dependent*, the material is *anisotropic.* Most engineering materials are homogeneous and isotropic to a considerable degree. In this text we shall assume the materials under consideration to be homogeneous and isotropic unless specifically stated otherwise.

The relationship between the shear stress $\tau$ and the shear strain $\gamma$ can be obtained from a plot of torque on a circular cylinder vs. the resulting angle of twist. The conversion of the torque-twist curve to a shear stress-strain curve is more involved than is the conversion in the previous axial case. This conversion can be found in various references.[1]

[1] *Theory of Flow and Fracture of Solids*, by A. Nadai, McGraw-Hill Book Company, Inc., 1950, Chapter 21.

As in the case of uniaxial stress, the shear stress generally is proportional to the shear strain for the first portion of the curve. Thus,

$$\Delta \tau \propto \Delta \gamma \qquad \Delta \tau = (\Delta \gamma) \times \text{constant}$$

This constant is called the modulus of elasticity in shear and is denoted by $G$. It is also commonly called the modulus of rigidity or simply the shear modulus. For a shear stress not greater than the proportional limit in shear,

$$G = \frac{\Delta \tau}{\Delta \gamma} \quad \begin{array}{l} \tau \leq \tau_{PL} \\ \gamma \leq \gamma_{PL} \end{array} \qquad (3\text{-}2)$$

*Caution:* Our entire discussion of modulus of elasticity is confined to the initial linear portion of the stress-strain curve, i.e., the part below the proportional limit. The words "elasticity" and "elastic" are often used ambiguously in engineering. A more detailed discussion of these words is presented in Chapter 4. Until then, the word "elastic" will be used when referring to quantities associated with, or derived from, the modulus of elasticity.

## 3-3. Mathematical Modeling of Idealized Stress-Strain Curves

Each individual material exhibits its own unique response to the application of loads. In fact, it is this unique response which mechanically distinguishes one material from another. In this respect the quasi-static uniaxial stress-strain diagram plays an important role in *qualitatively* distinguishing one type of material from another. On the other hand, we can think of the stress-strain diagram as a graphical representation of a *quantitative* relationship between the stress and strain; that is, a mathematical function relating the stress to the strain, or vice versa. Such a mathematical expression which reflects the characteristic features of a particular mechanical response is called a *mathematical model* for that particular mechanical response. For example, the mathematical model for a linear response is the algebraic equation

$$y = mx$$

where $y$ is the output and $x$ the input. Thus, mathematical modeling is not simply curve-fitting of data, but rather is an attempt to analytically describe a type or class of response which may be exhibited by a great number of individual materials, e.g., many materials exhibit linear response to uniaxial loading.

Usually mathematical models contain parameters whose values are dependent upon the particular material exhibiting the response. These values are calculated from the experimental data and serve to distinguish one material from another even though both materials have the same mathematical model. If, for a material, a mathematical expression can be found which relates the stress to the strain for *all* possible types of loading, then that expression (or expressions) is called the *constitutive equation* for that material. In this section we concern ourselves with the construction of mathematical models for uniaxial stress-strain curves while in Sec. 3-5 we will construct a set of constitutive equations for one particular *class of materials*, namely, linear homogeneous isotropic materials.

**Uniaxial Stress-Strain Curves.** The initial portion of the uniaxial tensile or compressive stress-strain curve for most engineering materials usually takes a qualitative form approaching one of the five types in Fig. 3-4. Glass and some other brittle materials have curves that are practically linear to rupture, as illustrated by type I. The first part of the curve for

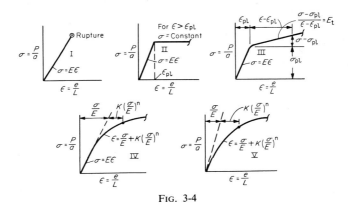

Fig. 3-4

structural or mild steel, which is one of the very common engineering materials, is like that of type II. A commonly used aluminum alloy, 2024-T4, has a curve similar to that of type III. Stress-strain curves for cast iron and concrete sometimes do not have any initial linear range, as illustrated by type V. The curves for many other engineering materials approximate that of type IV. Each specific material mentioned has a different value of $E$. Therefore, the slope of the initial part of the curve is different for each.

Experimental plots of stress-strain curves for uniaxial tension or compression can ordinarily be obtained quite easily, and it is usually not difficult to identify the initial linear portion, if one exists. Although experimental plots usually do not pass through the origin of the load and deformation axes, all analytical expressions discussed here are based on the assumption that the stress-strain curve passes through the origin of the coordinates. It is assumed that any necessary adjustment for a displaced origin is made when converting the load scale to stress and the deformation scale to strain.

Before the mathematical models for each of the five types of curves are discussed, two other quantities sometimes used in the expressions will be defined. The first is the *tangent modulus*, denoted by $E_t$, which is the *slope* of the stress-strain curve at any point *beyond* the proportional limit.

$$E_t = \lim_{\Delta\epsilon \to 0} \frac{\Delta\sigma}{\Delta\epsilon} = \frac{d\sigma}{d\epsilon} \quad \text{for} \quad \begin{array}{l} \sigma > \sigma_{PL} \\ \epsilon > \epsilon_{PL} \end{array} \qquad (3\text{-}3)$$

Generally the curve beyond the proportional limit is nonlinear, and the tangent modulus is a variable as for curves of types IV and V. The tangent modulus may be a constant, as for a curve of type III; may be zero, as for a curve of type II; or may not exist, as for a curve of type I.

The second quantity is the *secant modulus*, denoted by $E_s$, which is the ratio of the stress to the strain at any point beyond the proportional limit.

$$E_s = \frac{\sigma}{\epsilon} \quad \text{for} \quad \begin{array}{l} \sigma > \sigma_{PL} \\ \epsilon > \epsilon_{PL} \end{array} \qquad (3\text{-}4)$$

Figure 3-5 shows the difference between $E$ and $E_t$ and $E_s$.

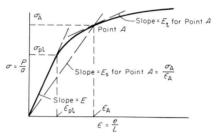

FIG. 3-5

We will now discuss the analytical expression or mathematical model and the method of obtaining it for each type of curve in Fig. 3-4. All curves are assumed to be adjusted so that they pass through the origin.

Type I: The analytical expression for the simple linear curve is $\sigma = E\epsilon$, where the constant $E$ is simply the measured slope $\Delta\sigma/\Delta\epsilon$ of the experimental curve.

Type II: The expression for the initial linear portion is the same as that for a curve of type I, or $\sigma = E\epsilon$, where $\epsilon \leq \epsilon_{PL}$. For the "flat topped" portion, where $\epsilon > \epsilon_{PL}$, the expression is $\sigma = $ constant. The constant stress is measured directly from the curve.

Type III: The expression for the initial linear portion is the same as that for a curve of type I or type II; thus, $\sigma = E\epsilon$, where $\epsilon \leq \epsilon_{PL}$. For the second linear portion, where $\epsilon > \epsilon_{PL}$, the expression is $(\sigma - \sigma_{PL}) = $

$E_t(\epsilon - \epsilon_{PL})$. The constants $E_t$, $\sigma_{PL}$, and $\epsilon_{PL}$ are measured directly from the curve, as indicated in the figure.

Types IV and V: An analytical expression for a part of a stress-strain curve which is nonlinear cannot be obtained so easily as one for a linear part. An expression for the linear part of a curve of type IV is like that for a curve of type I, II, or III. For a curve which is partially or wholly nonlinear, a commonly used approximation is the following expression given by Ramberg and Osgood:[2]

$$\epsilon = \frac{\sigma}{E} + K\left(\frac{\sigma}{E}\right)^n \qquad (3\text{-}5)$$

The first term $\sigma/E$ represents a straight line with slope $E$. For a curve of type V, which has no linear part, $E$ is the slope of the tangent to the curve at the origin. The second term $K(\sigma/E)^n$ is the deviation of the stress-strain curve from the straight line with slope $E$. For a curve of type IV the constants $K$ and $n$ are such that this second term is negligible below the proportional limit. To obtain a Ramberg-Osgood expression from experimental data, it is necessary to obtain suitable values of $K$ and $n$. Ramberg and Osgood show how to do this in their publication.

**Shear Stress-Strain Curves.** An analytical expression for a stress-strain curve for shear can be obtained by adopting much the same procedure as that used for uniaxial stress, *provided of course that the stress-strain curve for shear can be obtained from some experimental data.* As was pointed out earlier, the procedure for obtaining a stress-strain curve for shear is more involved than that for uniaxial stress.

### 3-4. Poisson's Ratio

You have no doubt noticed that when you stretch a rubber band, the cross section gets smaller; and if you squeeze a solid rubber ball, the diameter in a transverse plane (one at right angles to the axial squeezing forces) gets larger. Engineering materials also behave in this manner, but to a lesser degree and the changes in dimensions are not easily discernible to the naked eye. Getting to a more rigorous discussion, it can be shown experimentally that in a homogeneous and isotropic material subjected to a *uniaxial* state of stress, the resulting transverse strain is directly proportional to the axial strain in the direction of the load for stresses below the proportional limit. Thus,

$$\epsilon_{transverse} \propto \epsilon_{axial} \quad \text{or} \quad \epsilon_{transverse} = \epsilon_{axial} \times \text{constant}$$

[2] *Description of Stress-Strain Curves by Three Parameters*, by Walter Ramberg and William Osgood, *NASA-TN* 902, July 1943.

The ratio of the magnitude of the transverse strain to the magnitude of the axial strain is called Poisson's ratio.[3] It will here be denoted by $\mu$. Therefore,

$$\mu = \frac{|\epsilon_{transverse}|}{|\epsilon_{axial}|} \tag{3-6}$$

Poisson concluded that the theoretical value for the ratio $\mu$ was $\frac{1}{4}$ for any isotropic body. For most metallic engineering materials the actual value is in the vicinity of $\frac{1}{3}$ for strains less than the proportional limit (see Table I in the Appendix). Poisson's ratio can be measured experimentally rather easily by the simultaneous use of two strain gages on a test specimen subjected to uniaxial tensile or compressive stress; one gage is parallel to the load, and the other is perpendicular to it.

This discussion of Poisson's ratio is confined to strains in the elastic range below the proportional limit. For strains beyond the proportional limit, the ratio increases and in some cases approaches the limiting value of $\frac{1}{2}$.[4]

## 3-5. Generalized Hooke's Law

As we have seen, for stresses not greater than the proportional limit,

$$\epsilon = \frac{\sigma}{E}$$

This equation expresses the relationship between stress and strain (Hooke's law) for a *uniaxial* state of stress only when the stress is not greater than the proportional limit. Let us now consider the general triaxial state of stress shown in Fig. 3-6(a), which is Fig. 1-4(b) redrawn.

In order to analyze the deformational effects produced by all the forces, we will consider the effects of one axial force at a time. Since we presumably are dealing with strains of the order of one percent or less, these effects can be superposed arbitrarily. Figures 3-6(b), (c), and (d) show these effects separately. The undeformed element is represented by dotted lines, and the deformed element by full lines. The deformations in these figures are greatly exaggerated. Figure 3-6(b) shows that the force $\sigma_x \, da$ causes an increase in the x dimension and a decrease in the y and z dimensions because of the Poisson effect. Since this figure repre-

---

[3] The French mathematician and scientist S. D. Poisson (1781–1840) first defined this ratio as an elastic constant.

[4] For a discussion of the Poisson effect beyond the proportional limit, see *Strength of Materials*, by F. R. Shanley, McGraw-Hill Book Company, Inc., 1957, pp. 171–3.

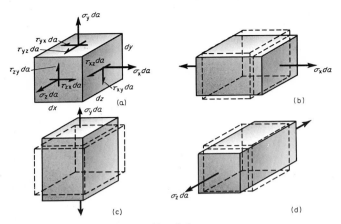

FIG. 3-6

sents a uniaxial state of stress in the $x$ direction, the strains in the $x$, $y$, and $z$ directions, respectively, would be as follows:

$$\epsilon_x = \frac{\sigma_x}{E} \qquad \epsilon_y = -\mu\epsilon_x \qquad \epsilon_z = -\mu\epsilon_x \qquad (3\text{-}7)$$

Then the resulting deformations would be

$$de_x = \epsilon_x \, dx \qquad de_y = -\mu\epsilon_x \, dy \qquad de_z = -\mu\epsilon_x \, dz \qquad (3\text{-}7a)$$

Here the minus signs indicate contractions.

In a similar manner, Figs. 3-6(*c*) and (*d*) illustrate the deformations produced by the forces $\sigma_y \, da$ and $\sigma_z \, da$, respectively. By applying the previous analysis to Fig. 3-6(*c*), we obtain

$$\epsilon_y = \frac{\sigma_y}{E} \qquad \epsilon_x = -\mu\epsilon_y \qquad \epsilon_z = -\mu\epsilon_y \qquad (3\text{-}8)$$

$$de_y = \epsilon_y \, dy \qquad de_x = -\mu\epsilon_y \, dx \qquad de_z = -\mu\epsilon_y \, dz \qquad (3\text{-}8a)$$

For Fig. 3-6(*d*),

$$\epsilon_z = \frac{\sigma_z}{E} \qquad \epsilon_x = -\mu\epsilon_z \qquad \epsilon_y = -\mu\epsilon_z \qquad (3\text{-}9)$$

$$de_z = \epsilon_z \, dz \qquad de_x = -\mu\epsilon_z \, dx \qquad de_y = -\mu\epsilon_z \, dy \qquad (3\text{-}9a)$$

The *total* deformation in any one direction would be the sum of the deformations in that direction resulting from all the individual forces. Thus

$$\overbrace{de_x}^{\text{total}} = \overbrace{\frac{\sigma_x}{E} \, dx}^{\text{from } \sigma_x} - \overbrace{\mu\frac{\sigma_y}{E} \, dx}^{\text{from } \sigma_y} - \overbrace{\mu\frac{\sigma_z}{E} \, dx}^{\text{from } \sigma_z} \qquad (3\text{-}10)$$

If we divide through by $dx$, we obtain

$$\frac{de_x}{dx} = \epsilon_{x(\text{total})} = \frac{\sigma_x}{E} - \mu \frac{\sigma_y}{E} - \mu \frac{\sigma_z}{E} \tag{3-11a}$$

In a similar manner, the total strains in the $y$ and $z$ directions become

$$\epsilon_{y(\text{total})} = \frac{\sigma_y}{E} - \mu \frac{\sigma_x}{E} - \mu \frac{\sigma_z}{E} \tag{3-11b}$$

$$\epsilon_{z(\text{total})} = \frac{\sigma_z}{E} - \mu \frac{\sigma_x}{E} - \mu \frac{\sigma_y}{E} \tag{3-11c}$$

The shear forces were not considered in the foregoing analysis. It can be shown that for *isotropic* materials a shear stress will produce only its corresponding shear strain and will not influence the axial strains.[5] We can, however, write Hooke's law for the individual shear strains and stresses in the following manner:

$$\gamma_{xy} = \frac{\tau_{xy}}{G} \tag{3-12a}$$

$$\gamma_{xz} = \frac{\tau_{xz}}{G} \tag{3-12b}$$

$$\gamma_{yz} = \frac{\tau_{yz}}{G} \tag{3-12c}$$

Equations 3-11 and 3-12 are usually called the *Generalized Hooke's Law* and are the constitutive equations for *linear elastic isotropic* materials. When these equations are used as written, the strains can be completely determined from known values of the stresses. The constitutive equations can also be written explicitly in strains (see Prob. 3-28). You should remember that $\sigma_x$, $\sigma_y$, and $\sigma_x$ are not necessarily principal stresses, but are merely the normal stresses in three mutually perpendicular directions.

We have remarked in Chapter 1 that the case of plane stress ($\sigma_z = \tau_{xz} = \tau_{yz} = 0$) is of particular importance to the engineer. For this special situation, Eqs. 3-11 and 3-12 reduce to

$$\epsilon_x = \frac{\sigma_x}{E} - \mu \frac{\sigma_y}{E}$$

$$\epsilon_y = \frac{\sigma_y}{E} - \mu \frac{\sigma_x}{E} \tag{3-13}$$

$$\gamma_{xy} = \frac{\tau_{xy}}{G}$$

[5]For a discussion of this, see *An Introduction to Mechanics of Solids*, by Crandall and Dahl, McGraw-Hill Company, Inc., 1959, pp. 185–6.

along with their inverse relations

$$\sigma_x = \frac{E}{1 - \mu^2} (\epsilon_x + \mu\epsilon_y)$$

$$\sigma_y = \frac{E}{1 - \mu^2} (\epsilon_y + \mu\epsilon_x) \tag{3-14}$$

$$\tau_{xy} = G\gamma_{xy}$$

## 3-6. Relationship Between Elastic Constants $E$, $G$, and $\mu$

The constitutive equations for a linear elastic isotropic material given by Eqs. 3-11 and 3-12 contain three parameters, namely, $E$, $G$, and $\mu$. It can be shown by various analytical and physical arguments that such a material has at most two *independent* parameters and that any other parameters can be expressed in terms of two known independent ones. To demonstrate this we will utilize the biaxial case of pure shear stress (which is not uncommon in engineering members). Figure 3-7(*a*) shows an element in a state of pure shear and labeled with the existing shear stress. Figure 3-7(*b*) is Mohr's stress circle for pure shear. This circle shows the following facts:

1. For pure shear the center of the circle must be at the origin.

2. The maximum principal stress, minimum principal stress, and maximum shear stress all have the same magnitude.

3. The algebraic sum of the normal stresses for any orientation of the element is always zero.

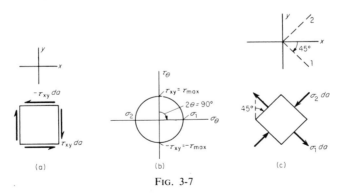

(a)                    (b)                    (c)

Fɪɢ. 3-7

Figure 3-7(*c*) shows the element oriented with its faces parallel to the principal planes. So $\sigma_1$ and $\sigma_2$ are the maximum and minimum principal stresses, respectively.

The Generalized Hooke's Law for a biaxial state of stress may be expressed by using Eqs. 3-11 and 3-12 and letting $\sigma_3 = 0$. The results are:

$$\epsilon_1 = \frac{\sigma_1}{E} - \mu \frac{\sigma_2}{E} \qquad (3\text{-}15)$$

$$\epsilon_2 = \frac{\sigma_2}{E} - \mu \frac{\sigma_1}{E} \qquad (3\text{-}15a)$$

$$\epsilon_3 = -\mu \frac{\sigma_1}{E} - \mu \frac{\sigma_2}{E} \qquad (3\text{-}15b)$$

$$\gamma_{xy} = \frac{\tau_{xy}}{G} \qquad (3\text{-}15c)$$

For pure shear, $\sigma_1 = + |\tau_{xy}|$ and $\sigma_2 = - |\tau_{xy}|$; so the principal strains $\epsilon_1$ and $\epsilon_2$ may be expressed as follows:

$$\epsilon_1 = \frac{\tau_{xy}}{E} + \mu \frac{\tau_{xy}}{E} = \frac{\tau_{xy}}{E}(1 + \mu) \qquad (3\text{-}16)$$

$$\epsilon_2 = -\frac{\tau_{xy}}{E} - \mu \frac{\tau_{xy}}{E} = -\frac{\tau_{xy}}{E}(1 + \mu) \qquad (3\text{-}16a)$$

These are principal strains, and they therefore represent the right-hand and left-hand extremities of Mohr's *strain* circle, which is shown in Fig. 3-8.

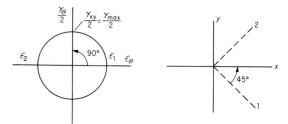

FIG. 3-8

From Fig. 3-8 we see that the magnitude of the maximum shear strain is the diameter of the circle, or twice $\epsilon_1$. So

$$\gamma_{max} = \gamma_{xy} = 2\epsilon_1 \qquad (3\text{-}17)$$

Substituting values from Eqs. 3-15c and 3-16 in Eq. 3-17, we get

$$\frac{\tau_{xy}}{G} = \frac{2\tau_{xy}(1 + \mu)}{E}$$

$$\frac{1}{G} = \frac{2(1 + \mu)}{E}$$

$$G = \frac{E}{2(1 + \mu)} \qquad (3\text{-}18)$$

Equation 3-18 relates the elastic constants $E$, $G$, and $\mu$. This equation shows that these elastic constants are not independent of each other and experimental tests have proved that this relationship is valid.

Our derivation of Eq. 3-18 was based on relationships which were obtained from Mohr's circle for plane strain and which were given in Chapter 2. It is possible also to derive the same relationship between $E$, $G$, and $\mu$ by using deformation relations for the element instead of Mohr's strain circle.[6] However, the method given here is sufficient.

Another defined elastic constant which is sometimes useful to the engineer is the bulk modulus. The bulk modulus is the average normal stress divided by the cubical dilatation. The cubical dilatation is the change in volume per original unit volume. Hence if an element originally has sides $dx$, $dy$, and $dz$, and deforms into an element with sides $dx(1 + \epsilon_x)$, $dy(1 + \epsilon_y)$, and $dz(1 + \epsilon_z)$, the dilatation becomes

$$\text{Dilatation} = \frac{V_f - V_o}{V_o}$$

$$= \frac{(1 + \epsilon_x)(1 + \epsilon_y)(1 + \epsilon_z)\,dx\,dy\,dz - dx\,dy\,dz}{dx\,dy\,dz}$$

$$\approx \epsilon_x + \epsilon_y + \epsilon_z = \epsilon_V$$

if products of strain are considered negligible compared with the strains themselves. Thus, by definition the bulk modulus $K$ becomes

$$K = \frac{\dfrac{\sigma_x + \sigma_y + \sigma_z}{3}}{\epsilon_x + \epsilon_y + \epsilon_z} = \frac{E}{3(1 - 2\mu)} \tag{3-19}$$

The intermediate details of this derivation are left as an exercise for the student (Prob. 3-29). Note that the bulk modulus is not an independent elastic constant but is dependent upon $E$ and $\mu$. Thus, while we have defined four elastic constants $E$, $G$, $\mu$, and $K$, only two (any two) are needed to completely know the elastic properties of an isotropic elastic material.

## 3-7. Closure

We close this chapter with a few important remarks. Hooke's Law is probably the most well known and widely used constitutive equation for

[6]See *Strength of Materials*, Part I, by S. Timoshenko, D. Van Nostrand Co., 1955, pp. 59–60.

engineering materials. However, we do not wish to leave you with the impression that all engineering materials are linear elastic isotropic ones, particularly in this age when new materials are being developed every day. Indeed, many useful materials exhibit non-linear response, are not elastic, and have a high degree of anisotropy. Unfortunately, mathematical models and constitutive equations for such materials are often extremely difficult to obtain and utilize. The modern engineer faces many interesting challenges in trying to make efficient and effective use of the structural capabilities of these new materials.

Remember, one of the purposes of this course is to provide you with some *basic ideas in the analysis of load resisting members*. In this regard the discussions in this chapter should be considered to be of fundamental importance but not the last word.

### PROBLEMS

**3-1.** Following are results from tensile test data of 2024-T4 aluminum alloy (extrusion) at room temperature. Plot the stress-strain curve. Use a straightedge for the linear portion and an irregular curve for the nonlinear portion. Take great care in the transition near the proportional limit stress.

| Stress, psi | Strain, in./in. | Stress, psi | Strain, in./in. |
|---|---|---|---|
| 0 | 0 | 30,000 | 0.00288 |
| 2000 | 0.00020 | 32,000 | 307 |
| 4000 | 0.00035 | 34,000 | 326 |
| 6000 | 55 | 36,000 | 345 |
| 8000 | 74 | 38,000 | 364 |
| 10,000 | 92 | 40,000 | 382 |
| 12,000 | 114 | 42,000 | 401 |
| 14,000 | 135 | 44,000 | 420 |
| 16,000 | 152 | 46,000 | 442 |
| 18,000 | 175 | 48,000 | 460 |
| 20,000 | 193 | 50,000 | 486 |
| 22,000 | 211 | 52,000 | 548 |
| 24,000 | 230 | 53,000 | 740 |
| 26,000 | 249 | 54,000 | 0.00940 |
| 28,000 | 0.00268 | | |

**3-2.** Find the proportional-limit stress and the elastic modulus for the material in Prob. 3-1.

**3-3.** For the material of Prob. 3-1, determine the $E_T$ and $E_S$ for a stress of 52,500 psi.

**3-4.** For the material of Prob. 3-1, plot the stress-tangent modulus relationship.

**3-5.** Following are data for a compression test of a 1.75 × 1.75 in. clear birch specimen 8 in. long. The gage length of the compressometer was 6.00 in. Plot the stress-strain curve.

| Stress, psi | Strain, in./in. | Stress, psi | Strain, in./in. |
|---|---|---|---|
| 0 | 0 | 7840 | 0.00606 |
| 654 | 0.00154 | 8500 | 654 |
| 1308 | 206 | 9150 | 717 |
| 1961 | 249 | 9810 | 801 |
| 2615 | 288 | 10110 | 873 |
| 3270 | 332 | 10460 | 0.00948 |
| 3925 | 373 | 10790 | 0.01065 |
| 4580 | 409 | 10950 | 1149 |
| 5230 | 456 | 11110 | 1256 |
| 5890 | 486 | 11280 | 1385 |
| 6540 | 518 | 11430 | 0.01591 (ultimate) |
| 7190 | 0.00563 | | |

**3-6.** Find the proportional limit stress and the elastic modulus for the material in Prob. 3-5.

**3-7.** For the material of Prob. 3-5 determine the $E_T$ and $E_s$ for a stress of 11,000 psi.

**3-8.** For the material of Prob. 3-5 plot the stress-tangent modulus relationship.

**3-9.** Following are data from the first part of a tensile test of a 0.505-in. diameter specimen of ZK60A magnesium alloy extrusion. A 2-in. gage length extensometer was used. Plot the stress-strain curve. Determine the elastic modulus and the proportional-limit stress.

| Load, lb | Extensometer Reading, in. | Load, lb | Extensometer Reading, in. |
|---|---|---|---|
| 0 | 0 | 7000 | 0.0120 |
| 700 | 0.0010 | 7500 | 0.0130 |
| 1300 | 0.0020 | 7700 | 0.0140 |
| 1960 | 0.0030 | 8070 | 0.0150 |
| 2560 | 0.0040 | 8260 | 0.0160 |
| 3300 | 0.0050 | 8370 | 0.0170 |
| 3780 | 0.0060 | 8400 | 0.0190 |
| 4440 | 0.0070 | 8420 | 0.0210 |
| 5000 | 0.0080 | 8440 | 0.0230 |
| 5540 | 0.0090 | 8490 | 0.0250 |
| 6080 | 0.0100 | 8500 | 0.0270 |
| 6560 | 0.0110 | Data stopped—No rupture | |

**3-10.** For the material of Prob. 3-9, determine the tangent modulus and the secant modulus for a stress of 37,500 psi.

**3-11.** For the material of Prob. 3-9, plot the stress-tangent modulus relationship.

**3-12.** If Poisson's ratio for the magnesium alloy of Prob. 3-9 is 0.34, what is the modulus of rigidity?

**3-13.** Following are data taken from a compressive test of a 1.00 in. diameter specimen of nuclear grade graphite. The strain readings are average of two electrical resistance strain gages which were on opposite sides. The gage axes were parallel to the load. Note that these gages read directly in microinches per inch.

Plot the stress-strain curve. Do you notice anything unique about the elastic modulus and the proportional limit stress for this material?

| Load, lb | Strain, $\mu$ in./in. | Load, lb | Strain, $\mu$ in./in. |
|---|---|---|---|
| 0 | 0 | 1500 | 3100 |
| 250 | 350 | 1750 | 4300 |
| 500 | 750 | 2000 | 6100 |
| 750 | 1160 | 2250 | 8600 |
| 1000 | 1700 | 2425 | 11,000 rupture |
| 1250 | 2400 | | |

**3-14.** For the material of Prob. 3-13, determine the $E_T$ and $E_s$ for a stress of 2500 psi.

**3-15.** For the material of Prob. 3-13, plot the stress-tangent modulus relationship.

**3-16.** A material has a stress-strain relationship which can be approximated by the equation

$$\epsilon = \frac{\sigma}{2 \times 10^7} + 180 \left( \frac{\sigma}{2 \times 10^6} \right)^3$$

Find the secant modulus and the tangent modulus for a stress level of 25,000 psi.

**3-17.** An aluminum alloy has a stress-strain relationship which can be approximated by the equation

$$\epsilon = \frac{\sigma}{10^7} + \left( \frac{\sigma}{10^5} \right)^{20}$$

Find the secant modulus and the tangent modulus for a stress level of 65,000 psi.

**3-18.** At some point in the elastic range of the specimen shown in the illustration, the elongation of $a$ is $2500 \times 10^{-6}$ in. and the contraction of $b$ is $375 \times 10^{-6}$ in. The original lengths were $a = 2.50$ in. and $b = 1.25$ in. What is Poisson's ratio for the material of the specimen?

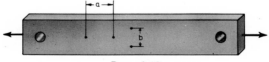

PROB. 3-18

**3-19.** The original 2-in. longitudinal gage length of a cylindrical compression specimen decreases 0.0024 in. During the same time the circumference, which was originally 7.16 in., increased 0.0021 in. Assuming that the action was elastic, find Poisson's ratio for the material.

**3-20.** Write Hooke's law for plane strain, explicit in stress and explicit in strain.

NOTE: For Probs. 3-21 to 3-27 assume that $\sigma_z = 0$ and determine the values and directions of the principle stresses and the maximum shear stress in each case.

Notice that these strains are the same as in Probs. 2-18 to 2-24. Strains are in $\mu$ in./in. and $E$, $K$, and $G$ in psi $\times$ $10^6$.

| | | | | | |
|---|---|---|---|---|---|
| **3-21.** $\epsilon_x = 400,$ | $\epsilon_y = 2000,$ | $\gamma_{xy} = 1800,$ | $E = 24,$ | $\mu = \frac{1}{3}$ |
| **3-22.** $\epsilon_x = -600,$ | $\epsilon_y = 600,$ | $\gamma_{xy} = -800,$ | $E = 28,$ | $\mu = 0.4$ |
| **3-23.** $\epsilon_x = -2800,$ | $\epsilon_y = 2000,$ | $\gamma_{xy} = -1400,$ | $E = 8.0,$ | $G = 3.2$ |
| **3-24.** $\epsilon_x = 500,$ | $\epsilon_y = -1,000,$ | $\gamma_{xy} = -3000,$ | $E = 20,$ | $G = 8.0$ |
| **3-25.** $\epsilon_x = 1200,$ | $\epsilon_y = 1200,$ | $\gamma_{xy} = 2400,$ | $G = 5.0,$ | $\mu = 0.3$ |
| **3-26.** $\epsilon_x = 0,$ | $\epsilon_y = -1800,$ | $\gamma_{xy} = 1350,$ | $G = 10,$ | $\mu = \frac{1}{4}$ |
| **3-27.** $\epsilon_x = -736,$ | $\epsilon_y = 1487,$ | $\gamma_{xy} = 0,$ | $E = 30,$ | $K = 20$ |

**3-28.** Express generalized Hooke's law for stresses in terms of the strains $\epsilon_x$, $\epsilon_y$, $\epsilon_z$, $\gamma_{xy}$, $\gamma_{xz}$, and $\gamma_{yz}$.

**3-29.** Carry out the details in the derivation of Eq. 3-19.

**3-30.** Prove rigorously that for plane stress in an isotropic material obeying Hooke's Law the principal directions of stress coincide with the principal directions of strain.

**3-31.** Show that for a linearly elastic isotropic material the normal stress $\sigma_i$ in any direction can be expressed as $\sigma_i = C_1 \epsilon_i + C_2 D$, where $\epsilon_i$ is the axial strain in the same direction as $\sigma_i$ and $D$ is the dilatation. Express $C_1$ and $C_2$ in terms of $E$ and $\mu$.

**3-32.** Give an argument that the value of Poisson's ratio for a material obeying Hooke's law must lie between $-1$ and $+\frac{1}{2}$.

## chapter 4

# Experimental Mechanical Properties of Engineering Materials

## 4-1. Elastic and Plastic Deformations

Materials are classified or grouped in an almost limitless variety of ways. First of all, they are classified as solids, liquids, fluids, gases, vapors, etc. We will concern ourselves only with solids, although many of our definitions will apply to materials of the other types. Solid materials are further classified according to their molecular or crystalline structure, physical and chemical properties, thermal properties, electrical properties, mechanical properties, etc. In this chapter, we shall discuss some of the mechanical and thermal properties which are of particular importance to the engineer. For the most part, the mechanical properties will be those obtained experimentally from a *uniaxial* tension or compression test. A typical screw-type universal test machine commonly used for such uniaxial tensile or compressive tests is shown in Plate I. The machine as pictured is set up for a uniaxial tensile test.

Before we discuss in detail specific experimental mechanical properties, it will be helpful to define some common general terms relating to material behavior. Several of these terms, such as elastic, plastic, stiffness, and brittle, are often used too loosely, incorrectly, and ambiguously.

**Elastic Action.** *Elastic* is an adjective meaning "capable of recovering size and shape after deformation." If a material is subjected to load, deformation will result. If, upon release of the load, the material returns to its original size and shape, it has undergone *elastic* action or *elastic* deformation. The stress was an *elastic* stress within the *elastic* range. *Elastic limit* is the maximum *uniaxial* stress that can be applied to a material without causing any permanent deformation. *Elastic range* is the range of stress below the elastic limit.

Nowhere in the discussion of the word elastic have we said or implied that the load and the deformation must be proportional or linearly related in the elastic range. They usually are, as we saw from the linear first por-

65

PLATE I. A typical screw-type universal test machine.

tions of most stress-strain curves in Chapter 3, but linearity is not a necessary condition for a material to be elastic. Many engineering materials behave as indicated in Fig. 4-1(*a*); however, some behave as in (*b*) or (*c*) while in the elastic range. When a material behaves as in (*c*), the stress-strain relationship is *not* single valued, since the strain corresponding to any particular stress will depend on the loading history.

In engineering literature the term elastic action or elastic range is used quite commonly to mean a linear stress-strain behavior. It is common for

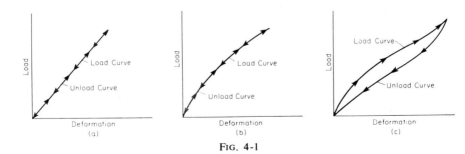

FIG. 4-1

an engineer to say elastic limit when he means proportional limit. This practice has come about naturally, because for most engineering materials the proportional-limit stress and the elastic-limit stress are approximately the same. It should be noted that, by definition, the proportional limit must be obtained from a load-deformation or stress-strain curve, but the elastic limit *cannot* be obtained from such a single curve. If desired, the elastic limit can be obtained from a laboratory test in which increasing loads are applied and released, and then a check for residual deformation is made. The value of the elastic limit determined experimentally in this way is dependent on the sensitivity of the measuring instrument used and on other variables such as time effect.

**Plastic Action.** *Plastic* deformation, or *permanent set*, is any deformation that remains in the material *after* the load has been removed. All deformation is composed of plastic and elastic deformation. However, when the plastic deformation is negligible compared with the elastic deformation, the material is said to be elastic, and vice versa.

The average person often confuses brittleness with stiffness. Stiffness is the ability of a material to resist deformation. Thus, the modulus of elasticity and the tangent modulus are measures of stiffness. A material which undergoes very little plastic deformation before rupture is said to be *brittle*. A material which undergoes a great amount of plastic deformation before rupture is said to be *ductile*. Since elastic deformations are generally small, the usual measure of ductility (or brittleness) is the total percent elongation up to rupture of a 2-in. gage length tensile specimen. Sometimes the percent reduction of cross-sectional area of a tensile specimen after rupture is also used as a measure of ductility. A very ductile material such as structural steel, may have an elongation of 30 percent at rupture; whereas a brittle material, such as gray cast iron or glass, will have relatively little elongation at rupture. Compare the ductilities of several engineering materials in Table I of the Appendix. Figure 4-2 illustrates the difference in stress-strain curves for a ductile material and a brittle material.

### 4-2. Ultimate Strength and Rupture Stress

The ultimate tensile strength of a material is the *maximum* tensile stress the material can withstand before rupture in a tensile test in which the load is applied *slowly*. It is obtained by dividing the ultimate load by the original cross-sectional area of the test specimen (measured before loading). The ultimate tensile strength is the stress corresponding to the uppermost point on the stress-strain curve.

The ultimate compressive strength of a material is obtained in a similar manner from a compressive test. It is an important property for a brittle material. Many ductile materials do not exhibit a clearly defined rupture in a compression test. For such materials the load becomes indefi-

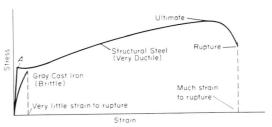

Fig. 4-2

Fig. 4-2

nitely large as the compression specimen "flattens out," and the term ultimate compressive strength is therefore meaningless. The usefulness of a material of this kind in compression is measured by some other property, such as the compressive yield strength or by its resistance to buckling.

The rupture stress is, as the name implies, the stress *at the time of rupture*. It is obtained by dividing the load at rupture by the original cross-sectional area of the test specimen. For a brittle material the ultimate stress and the rupture stress are likely to have the same value. If you look at several of the stress-strain diagrams in this chapter and in Chapter 3 for ductile materials, however, you will see that the rupture stress is not necessarily the same as the ultimate stress.

In a tensile test of most ductile engineering materials there is a very pronounced reduction in cross-sectional area of the specimen in the vicinity where rupture occurs. This phenomenon, which occurs shortly before rupture, is called "necking." Referring to the stress-strain curve for the ductile material in Fig. 4-2, necking occurs after the load reaches the ultimate strength. During this necking process, the test specimen becomes considerably weakened. Hence, the applied load drops rapidly as rupture is approached. Rupture stress is not usually an important quantity from a design standpoint, and it is seldom given in tables of mechanical properties for materials. However, for a ductile material the difference between the ultimate stress and the rupture stress is indicative of the degree of deformation that occurs during the necking process.

### 4-3. Yield Point and Yield Strength

Mild steel (for example, SAE 1020 or ASTM A-36), which is one of the most widely used engineering materials, has an unusual feature in its static stress-strain relationship. At the end of its linear elastic range there is a sudden and complete loss of stiffness, which is characterized by a sharp "knee" in the curve, as at point *A* in Fig. 4-2. At this point the material undergoes a rather rapid and extensive *plastic* deformation with *no* accompanying *increase* in load. Such a deformation process is called *yielding*.

The yield-point stress is specifically defined as the stress corresponding to the first point on the stress-strain curve at which there is a large increase in strain for no increase in stress. Although the yield point is easily obtained from the curve (the first place where the slope is zero), it is also commonly obtained simply by watching the load-indicator dial on the testing machine during a uniaxial tensile or compressive test. If the material has a yield point, and if the specimen is being loaded at a uniformly slow rate, the dial will halt temporarily at the yield point, indicating that the load remains constant even though the specimen is still being stretched (strained). This point is referred to as the yield point by "halt of dial" (or "drop of beam" for an old lever-type testing machine). The flaking off of the brittle mill scale on the surface of a specimen of ductile *hot rolled* mild steel is also a simple, but effective, indication that the yield point has been reached.

Although few materials exhibit a yield point, it is of great interest because it occurs in mild steel, which is such an important engineering material. For most purposes, mild steel is used only in the elastic range of stress. Since the yield point marks the end of the elastic range, the yield-point stress is commonly used as a measure of the "usable strength" of the material.

For a material with no yield point, there is no simple measure of usable strength. Hence, a somewhat arbitrary measure has been developed. From experience, it was decided that a measure of the usable strength would be the stress causing a certain permissible permanent strain (set) after release of the load in a uniaxial test. For most metallic materials, a strain of 0.2 percent (or 0.002 in./in.) is commonly used for the permissible set, although 0.1 percent is sometimes used for ferrous metals.

As we now know, the stiffness of a material decreases when it is loaded above the proportional limit into the inelastic range. If a material with no yield point is loaded into the inelastic range and is then unloaded, the unloading curve will be essentially parallel to the initial linear portion of the loading curve.[1] See Fig. 4-3(a). The usual method of determining the yield strength corresponding to a specific permanent set is as follows: First, offset the stress-strain curve by the desired amount, such as 0.2 percent in Fig. 4-3(b), to locate point A. Then through point A draw a straight line AB parallel to the original straight portion of the stress-strain curve. The stress level at B, at which this construction line crosses the stress-strain curve, is the *yield strength*.

The yield strength determined in the manner just outlined is approximately the stress which will cause the specified permissible set. This stress is commonly used as a measure, or at least as an index, of the usable

---

[1] There will be a slight curving off toward the origin at the bottom of the unloading curve. This residual action is known as the Bauschinger effect. See *Strength of Materials*, by F. R. Shanley, McGraw-Hill Book Company, Inc., 1957, pp. 118, 157.

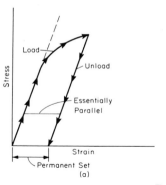
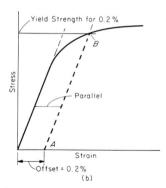

FIG. 4-3

strength of a metallic engineering material that does not exhibit a yield point. The yield strength should always be associated with a specific offset or permanent set. Engineers and technical writers are often careless about this requirement. If no offset is given, 0.2 percent is usually implied, since this is by far the most common offset.

Yield point and yield strength are very important mechanical properties for the classification of engineering materials, because they are easily reproducible. Properties such as proportional limit and elastic limit cannot be reproduced so easily. For example, suppose that a group of twenty students (on second thought, let us say twenty experienced engineers) was given a set of numerical values for plotting a stress-strain curve, and each one was asked to determine the proportional limit and the yield strength for a 0.2 percent offset. It would not be surprising to find that their values of the proportional limit would vary by 15 to 25 percent, while the variation in their values of the yield strength might probably be less than 5 percent. This illustration is indicative of the difference in reproducibility.

## 4-4. Resistance to Energy Loads

Work is produced by a force acting through a distance. The loads acting on an engineering material move through certain distances as the material undergoes deformations. Therefore, when loads are applied to a body, work is being done on the body, and energy is absorbed by the body. Energy is the capacity or ability to do work. The energy that a body absorbs as the result of its deformation under load is called *strain energy*. Strain energy which can be recovered as the loading is released is called *elastic strain energy*. In fact, a more sophisticated definition of an elastic body is one for which the entire strain energy is recoverable. In an elastic body, no energy is dissipated during the deformation process. According to this definition, no ideally elastic body exists. However, practically speaking, an elastic body is one for which the dissipated energy is negligible.

Consider a load-deformation curve for a uniaxial tensile test, such as that shown in Fig. 4-4(*a*). By definition, the total work done on the body by loading up to some value of deformation *e* is

$$W = \int_0^e P \, de$$

which in turn equals the area under the *P* vs. *e* curve up to the point corresponding to the particular deformation *e*. The percentage of the strain

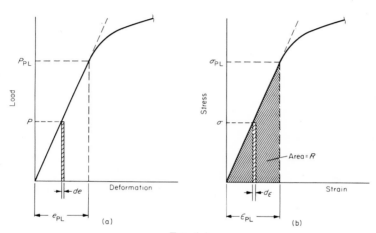

Fig. 4-4

energy which is recoverable elastic energy will depend on the shape and path of the unloading curve (see Fig. 4-1). The amount of energy dissipated will be the difference between the total energy (the area under the loading curve) and the recovered energy (the area under the unloading curve). For most common engineering materials, the loading and unloading curves will be very nearly the same as long as the load-deformation relationship is linear.

The total strain energy determined from a load-deformation curve is not really indicative of the material, since the results will depend on the size of the test specimen. In order to eliminate the specimen size as a factor, we consider the strain energy per unit volume (sometimes called the strain energy density). This quantity can be obtained from a uniaxial test simply by dividing the total strain energy by the volume of the specimen. However, since both the stress and the strain are uniformly distributed in the uniaxial test, the strain energy per unit volume becomes

$$U = \frac{W}{V} = \int_0^e \frac{P \, de}{AL} = \int_0^\epsilon \sigma \, d\epsilon$$

which in turn equals the area under the stress-strain curve; see Fig. 4-4(*b*).

The units of strain energy per unit volume usually are inch-pounds of energy per cubic inch of volume.

$$U = \frac{\text{in-lb}}{\text{cu in.}}$$

The area under the linear portion of the uniaxial stress-strain curve is a measure of a material's ability to store elastic energy. This measure is called the *modulus of resilience* and will be denoted by $R$. Thus,

$$R = \int_0^{\epsilon_{PL}} \sigma\, d\epsilon = \frac{1}{2}\, \sigma_{PL}\, \epsilon_{PL} = \frac{\sigma_{PL}^2}{2E}$$

As illustrated in Fig. 4-5, it is possible for a material with low strength and a low elastic modulus to have a higher modulus of resilience than a material with high strength and a high elastic modulus.

The area under the entire stress-strain curve is a measure of the strain energy per unit volume required to rupture a material. This measure is called the *toughness* of the material. Toughness is an important engineering property, since it indicates a material's resistance to energy loads before rupture. For example, a high degree of toughness would be important for the material in an automobile bumper, a highway guard rail, or an airplane landing gear.

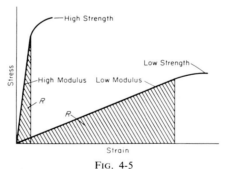

FIG. 4-5

The general expression for toughness is

$$T = \int_0^{\epsilon_{\text{rup}}} \sigma\, d\epsilon$$

However, there is seldom any simple analytical way of expressing $\sigma$ as a function of $\epsilon$. This integral should be thought of as the area under the entire stress-strain curve (see Fig. 4-6). The area can usually be obtained with sufficient accuracy by visually approximating an over-all average stress

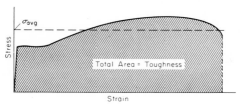

FIG. 4-6

from the diagram and multiplying this average stress by the strain at rupture from the diagram.

## 4-5. True Stress

During a uniaxial tensile or compressive test, the cross section of the test member changes because of the Poisson effect. This lateral change is greater in the plastic range, since Poisson's ratio increases in that range. For a ductile material in tension the change in cross section is quite drastic just before rupture, because of the necking effect. For the ordinary stress-strain curve discussed so far in this text, the stress at any point is based on the *original* cross-sectional area.

The *true stress* in a tension or compression test is found by dividing the load by the *actual* cross-sectional area. To obtain data for plotting

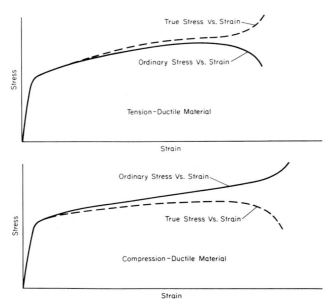

FIG. 4-7

true stress vs. strain, the actual cross section would have to be measured for each plotted point. The additional precise measurements required make it more difficult to perform this type of test. As seen from Fig. 4-7, there is no appreciable difference between true stress and ordinary stress in the range of stress usually encountered in engineering members. Ordinary stress is much more commonly used by the engineer, since in the design of a member the engineer is concerned with finding the *original* cross-sectional area required to support specific forces.

## 4-6. The Nature of Solids

Engineering solids are made up of numerous particles of various sizes. The particles may be atoms, crystals, molecules, fibers, pieces of gravel, grains of sand, or even voids. Some of these particles are macroscopic (visible to the unaided eye); some are microscopic (for example, crystals); and some are submicroscopic (for example, atoms). The cohesive forces between the particles, and the energies associated with them, are the major factors which influence the mechanical properties of a solid that are of interest to the engineer.

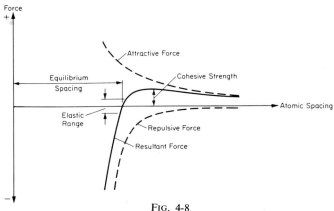

FIG. 4-8

Starting with the submicroscopic scale, we realize that all materials are composed of atoms. Bonds which are formed between atoms enable them to combine in very large groups to form solids. Adjacent atoms in a solid exert attractive and repulsive forces, which tend to maintain the atoms at an equilibrium distance from one another. The forces making up the bonds are balanced at this equilibrium spacing. Any tendency to change this spacing causes an unbalance which tends to pull (or push) the atoms back to the equilibrium spacing. The curves in Fig. 4-8 are plots of attractive force, repulsive force, and resultant force versus the distance between the atoms. When there is a tendency, due to some external cause, to increase the spacing of the atoms, a positive or pulling force is built up between the atoms, and this force tends to cause the atoms to return to the equilibrium spacing. If an external cause tends to decrease the spacing, there is set up between the atoms a negative or compressive force which also tends to cause a return to the equilibrium spacing.

The atoms are not completely stationary at the equilibrium spacing. There is continual movement in the form of vibration because of the heat present. The amplitude of these oscillations due to thermal agitation is of the order of one-tenth of the equilibrium spacing. The equilibrium spacing is of the order of $10^{-8}$ cm. We see then that the atoms are continually oscillating back and forth about the equilibrium spacing.

Notice in Fig. 4-8 that beyond the spacing at which the cohesive strength is developed, the resultant force decreases as the spacing increases. The strength of any engineering solid on the macroscopic level is only a fraction of the theoretical cohesive strength considered here on a submicroscopic level. How are groups of atoms built up to form solids? Usually the atoms combine to form crystals or molecules. Solids are generally aggregates of these crystals or molecules. These structural "building blocks" are extremely important in determining the mechanical behavior of engineering solids.

Most metallic engineering solids are crystalline, that is, they are composed of many crystals. A crystal is made up of an *orderly* array of atoms. As a crystal is formed, the atoms arrange themselves in three-dimensional rows so that each atom is at the equilibrium distance from the adjacent atoms. This structural arrangement is called a space lattice. A small portion of a simple cubic lattice is represented in Fig. 4-9. Such a lattice can be thought of as a three-dimensional array of cubic blocks with an atom at each corner where four cubes come together.

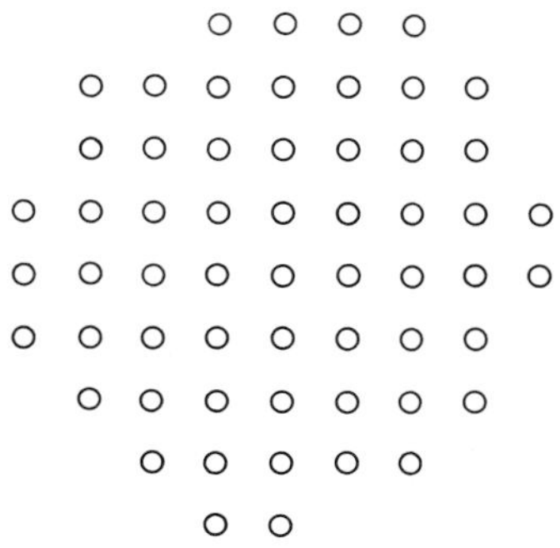

Side, Top and Bottom View
of a Simple Cubic Lattice

FIG. 4-9

One crystal may consist of millions of atoms, the number depending on the crystal size. Crystal size (or grain size, as it is often referred to) can vary widely. Although crystals usually are of microscopic size, they can sometimes be quite large; and under special conditions they can be made to grow to several inches. Engineering tests on single crystals are not uncommon in research laboratories.

In some materials atoms arrange themselves into chain-like or sheet-like structural units called molecules. A molecule is the smallest particle which retains the chemical properties of the original material. As molecules are united to form solids, molecular bonds are created between them. These bonds are not nearly so strong as are the atomic bonds discussed earlier. Chain-like molecules often intertwine in a solid. Such an arrangement has an appreciable effect on the mechanical behavior of the material.

Solids are usually classified as crystalline, amorphous, or a combination of these two. A crystalline material is usually made up of a number of crystals, in which case it may be called polycrystalline, whereas an amorphous material is usually made up of molecules. Crystalline materials have a much more orderly atomic arrangement, and as a result generally have greater density and strength than amorphous materials. Most engineering metals are crystalline. Some important amorphous engineering materials are wood, plastics, glass, and rubber. Other materials, such as concrete and some ceramics are combinations of crystals and molecules.

A brief general discussion of this kind may be misleading, since it may give the impression that the make-up of solids and the factors influencing their properties are relatively simple. Of course, such is not the case. It is outside the scope of this text to consider other details of atomic physics and chemistry influencing atomic bonds. It is important for the engineering student to study elsewhere such things as the major influence of imperfections on crystalline materials and the chemistry of polymerization in influencing properties of some amorphous materials.[2]

## 4-7. The Nature of Elastic and Plastic Action

**Elastic Action.** Elastic action in a polycrystalline material, such as almost any engineering metal, consists primarily of distortions of the atomic space lattice of the crystals. In the case of tension, there are increases in the distances between the atoms in the direction of external loading. These increases develop attractive forces between the atoms to balance the external applied forces. For compression the distances are decreased, and repulsive forces are built up. Accumulations of these small

[2]See Chapter 2 of *Engineering Materials Science,* by C. W. Richards, Wadsworth Publishing Company, 1961.

increases (for tension) result in over-all elastic elongations. Upon release of the external load, the atoms return to their equilibrium spacing, and essentially each atom maintains the same position relative to all other atoms. In Fig. 4-8 the part of the curve for resultant force versus distance just to either side of the equilibrium spacing can be considered approximately straight for the small changes in spacing considered here. Therefore, for most metals important in engineering, elastic action is essentially linear.

In an amorphous material or in a material that is a combination of amorphous and crystalline portions, elastic action is often approximately linear, but it is just as likely to be nonlinear. Although the same type of action as described in the preceding paragraph may occur with atoms of the molecules of an amorphous material, the configurations and arrangements of the molecules may contribute to a nonlinear elastic action. This nonlinear action is likely in such materials as rubber and thermoplastic polymers. "Combination" type materials that are important in engineering, such as concrete and plastic-fiberglass laminates, often exhibit this nonlinear characteristic.

Elastic action is essentially an atomic and/or molecular distortion caused by the external loads, wherein the atoms and molecules maintain similar relative positions while the loads are being applied or removed. When the load is entirely removed, the material on a macroscopic, microscopic, or submicroscopic scale essentially goes back to its original configuration.

**Nature of Yielding.** When external loads reach a sufficient magnitude, the elastic range ends, and either one of two effects is possible. The material could rupture, or it could yield. If a material were to rupture at the end of the elastic range, it would be said to be perfectly brittle. Although glass and a few other common engineering materials are very brittle, they yield slightly before rupture.

What is the nature of yielding and how is it different from elastic action? Yielding occurs in a material when the distances between atoms increase to the extent that atomic bonds are broken and place changes occur in the atomic lattice. The cumulative effect of such place changes is a greater over-all elongation. When elongation with yielding is plotted against load, the curve becomes nonlinear. These place changes are changes of positions of atoms relative to each other. The behavior is therefore *not reversible* upon release of the load. Yielding or initial plastic action, as it is often called, is a *permanent* action.

The atomic place changes associated with yielding are transcrystalline; that is, they occur on planes through the crystals. Yielding is a *slipping* action, associated with a tendency to shear, and it occurs on planes of weakness through the crystal. The locations of these planes of weakness depend on the specific arrangement of atoms in the space lattice, and this arrangement varies for different metals. Generally the weakest planes are those in which the atoms are most closely spaced. An example of slip in a portion of a crystal subjected to uniaxial tension is illustrated in Fig. 4-10. Only a microscopic portion of a crystal is represented, and this figure is, of course, somewhat idealized. Elastic lattice distortion is not shown in the figure. Notice that place changes are shown in view (*b*), and that new bonds have been established between atoms along the slip plane and different neighboring atoms. If a steel engineering member were to yield when in uniaxial tension, the initial yield planes on a macroscopic level would show that the statistical accumulation of thousands of submicroscopic slips is a shear phenomenon and is associated with maximum shear stress because the over-all yields would develop on planes making angles of approximately 45° with the external load.

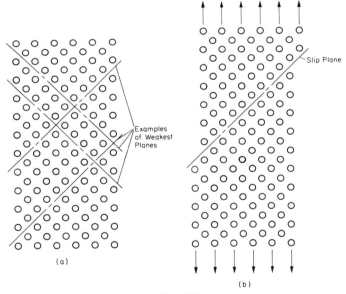

(a)

(b)

FIG. 4-10

Where and how does yielding start in a loaded body? From a macroscopic viewpoint, it is reasonable to assume that yielding will start in a region of high stress. Regardless of attempts to distribute stress uniformly in any real member of an engineering material, there will always be some regions of relatively higher stress concentration and therefore higher strain concentration. A concentration may be caused by the method of loading, the geometry of the member, or a material imperfection such as a flaw (due perhaps to inclusions of foreign matter or voids). Now consider the region of high stress submicroscopically. Although the arrangement in an atomic lattice of metallic crystals has a high degree of order, it is not absolutely perfect. There are different types of imperfections, called dislocations, which provide starting places for the slip planes.

**Yielding of Crystalline Materials.** In a polycrystalline material, yielding or plastic strain is primarily a rearranging of atoms along certain planes of slip with little or no change in the final atomic distance. The new atomic bonds formed are just as strong as the old ones. Therefore, yielding in a metal is not generally associated with any weakening or loss of strength. For example, if a mild-steel bar is stressed in tension or compression somewhat beyond the yield point, is then unstressed, and is again stressed to the yield point, it will be found that the yield-point stress is just as high for the second time or possibly even higher.

A combination of elastic action and plastic action results in an overall deformation. Since elastic action involves changes of the atomic spacing, it is accompanied by a volume change, which is an increase for tension and a decrease for compression. Remember that Poisson's ratio for most materials is only about $\frac{1}{3}$ for the elastic range. Since plastic action involves primarily place changes, rather than atomic spacing changes, it is *not* accompanied by any significant volume change. Can you now give a valid physical argument to show that Poisson's ratio in the plastic range should be $\frac{1}{2}$ for metals?

To get a correct mental picture of the yielding process in an engineering metal, you must have an appreciation of the relative scales on which the slipping actions occur. The over-all observable behavior is the cumulative statistical effect of a great many atomic, microscopic, and macroscopic actions. For example, in a single crystal many dislocations may cause the presence of many slip planes, and the length of each slip may be more than a thousand times the distance between atoms. Also, across a typical section of an engineering member there may be hundreds of crystals. Hence, the observable effect may be the result of a very great number of unobservable actions.

In a static tensile test of a flat specimen of mild steel, yielding will start across some section, perhaps at a grip. Once this region of the material yields initially, it is strengthened and resists further yielding; but other portions, usually adjacent, then begin to yield. This progression of yielding eventually covers the entire specimen between grips before the specimen as a whole will sustain additional load. As a result there is a rather flat region on the stress-strain curve for this material. In Fig. 4-11 the flat region is between points $A$ and $B$.

Notice that the strain at $B$ in Fig. 4-11 is approximately 20 times the limiting elastic strain at $A$. When point $B$ on the curve has been reached, all the material in the specimen will have yielded initially. Further yielding or slipping in the crystals will require more stress, mainly because of the interference to further slip provided by the crystal boundaries and adjacent crystals. There are other rather complex factors which contribute to this strengthening, but a discussion of such factors is beyond the scope of this text. This strengthening phenomenon, which takes place approximately between $B$ and $C$ in Fig. 4-11, is called strain hardening or work hardening. Most engineering metals exhibit some degree of strain hardening.

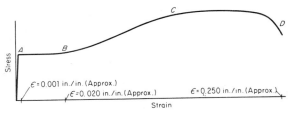

FIG. 4-11

**Yielding of Amorphous Materials.** In an amorphous material, or in a combination amorphous-crystalline material, yielding is also a manifestation of atomic and molecular bond failure. Usually, however, the number of new bonds formed is much lower than the number of bonds broken. As a result, there is a general weakening of the material, rather than a strengthening like that which may occur when a polycrystalline material yields. Other rather complex internal behaviors may take place in amorphous or "combination" type materials. Fragmentation of, and friction between, different materials may occur in a combination type material such as concrete. In some polymer plastics and similar materials long chain molecules often continue to slip past each other as molecular bonds are broken, even though new bonds are being formed continually. This action actually causes a kind of viscous flow, and does not permit a static or quasi-static condition of the type considered in this text.

The science of rheology is devoted to the study of materials which flow to some extent. In fact, one basic law of rheology says that all materials flow to some degree. However, the most interesting and difficult to describe are those materials that are solid but border on being fluid-like. These are referred to as viscoelastic materials. Many new load carrying plastics are viscoelastic as are load carrying biological materials such as bone, skin, tendon, and muscle tissue. The essential feature of a viscoelastic material is that the behavior under load is time dependent. The influence of time and other variables on the load-deformation behavior is discussed briefly in the next article.

## 4-8. Effects of Other Variables on the Stress-Strain Relationship

In this elementary text we are primarily concerned with deformable bodies subjected to static or slowly applied (quasi-static) loads at approximately room or ambient temperatures. Disregarding time, temperature, and type of load is in many cases a great simplification, because these items are major parameters influencing the stress-strain relationship.

In our usual room-temperature tests of materials, we apply the load slowly and assume that the time rates of change of stress and strain are zero. That is, we assume that

$$\frac{d\sigma}{dt} = 0 \qquad \frac{d\epsilon}{dt} = 0$$

This assumption is also made for most engineering materials in use at room temperature. An increased rate of loading usually results in an increase in the tensile proportional limit and ultimate strength of the material.

Under constant load (constant stress) the strain may change with time. This behavior is called *creep*. That is,

$$\frac{d\epsilon}{dt} \neq 0 \qquad (\sigma = \text{const.}) \qquad \text{Creep}$$

Creep is usually more pronounced, and often becomes a critical problem, at elevated temperatures. Some materials, such as metals with low melting points (for example, lead) and many plastics, exhibit appreciable creep at room temperatures. Engineering members designed and stressed "elastically" on the basis of an ordinary static stress-strain relationship may rupture by creep after a long period, perhaps years, at a constant stress level.

Sometimes the strain is kept constant, but the stress changes (usually decreases) with time. This behavior is called *relaxation*. That is,

$$\frac{d\sigma}{dt} \neq 0 \qquad (\epsilon = \text{const.}) \qquad \text{Relaxation}$$

Relaxation is also more prevalent at higher temperatures. This phenomenon often occurs in a bolted connection where the stress relaxes with time while the strain remains relatively constant. Creep and relaxation are discussed further in Chapter 15.

The stress in an engineering member often changes continuously with time because the loading is applied and removed continually. This action is referred to as repeated loading, or fatigue loading, and is discussed in more detail in Chapter 15.

Temperature has a considerable influence on the stress-strain relationship and properties of practically every engineering material. In the case of an engineering metal the ultimate strength, yield strength, and stiffness decrease appreciably with increasing temperatures, while the percent elongation (ductility) increases as the temperature rises. The reverse is generally true as the temperature is lowered. The loss of ductility in steels at very low temperatures became a major engineering problem in World War II, and as a result a great research effort was undertaken in this area.

In recent years the behavior of materials at extremely high temperatures has remained one of engineering's biggest problems. Operating temperatures in gas turbines, rocket motors, hypersonic aircraft, space structures, and nuclear devices, for example, have been limited to lower, less thermally efficient values because of the unavailability of materials which will maintain their strength properties at extremely high temperatures. Today, sustained repeated loading at well above 2000° F is possible with some materials. Many new temperature-resistant materials are under development for use at higher temperatures.

## PROBLEMS

**4-1.** Find the yield strength for a 0.2 percent offset for the aluminum alloy material of Prob. 3-1.

**4-2.** From the curves in the accompanying illustration, find the proportional limits for the materials of (a), (b), (c), and (d). Also find the yield strengths for a 0.2 percent offset for the materials of (b), (c), and (d) if possible.

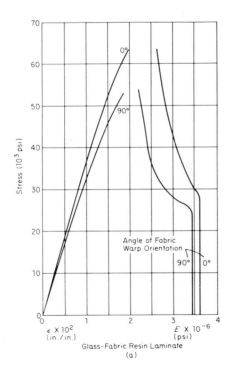

Glass-Fabric Resin Laminate
(a)

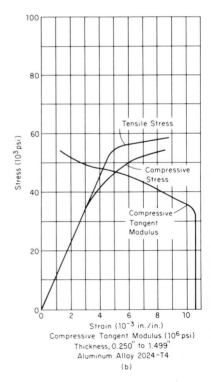

Compressive Tangent Modulus ($10^6$ psi)
Thickness, 0.250″ to 1.499″
Aluminum Alloy 2024-T4
(b)

PROB. 4-2

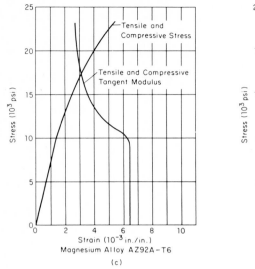

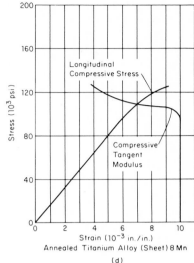

PROB. 4-2 (*continued*)

**4-3.** Find the modulus of resilience for each of the materials in Prob. 4-2.

**4-4.** Find the modulus of resilience for the material of Prob. 3-1.

**4-5.** For the material of Prob. 3-5, determine the yield strength for an offset of 0.05 percent.

**4-6.** Find the modulus of resilience for the material of Prob. 3-5.

**4-7.** Find the yield strength for an offset of 0.2 percent and the modulus of resilience for the material of Prob. 3-9.

**4-8.** How much strain energy was absorbed by the material of Prob. 3-1 up to the point where the data was discontinued?

**4-9.** How much strain energy was absorbed by the materials of (*b*) and (*d*) in Prob. 4-2 up to the points at which the curves were discontinued?

**4-10.** For the stress-strain relationship of Prob. 3-16, what is the strain energy absorbed up to a stress level of 25,000 psi?

**4-11.** For the aluminum alloy whose stress-strain relationship is given in Prob. 3-17, find the strain energy absorbed up to a stress level of 60,000 psi.

*part* **II**

# Load Analysis

## chapter 5

# Types of Loads

## 5-1. Introduction

In previous chapters we stated that the function of an engineering member is to resist loads. It is now in order to discuss the various types of loads and the ways in which they are commonly classified in engineering.

In your course in statics you were concerned primarily with rigid bodies in which there was little concern about the exact manner in which the specific forces were transmitted to the body and about their influence on the deformation of the body. In such rigid-body analysis you could even use the laws of transmissibility and have a force act on the body at any convenient position along its line of action. Only the over-all action was important. In deformable-body mechanics, of course, the manner in which the forces are transmitted to the body can have a great influence on the deformations, strains, and stresses in the body, particularly in the neighborhood of the areas of applications of surface forces. Different degrees of idealization of load systems may be justified for simplification of engineering design and analysis. The next section discusses the idealization of actual loads.

## 5-2. Representation of Loads

Loads on engineering bodies are usually represented by force vectors. The representation of these forces as vectors is an idealization for convenience of analysis. In Fig. 5-1 are illustrated examples of such idealized representations. You will note that Fig. 5-1 appears to be simply an elementary exercise in drawing free-body diagrams. It is certainly true that sketching free-body diagrams involves representing forces as vectors. Notice, however, the degree of idealization in each case. For example, the "concentrated" force on the beam in Fig. 5-1($a$) is not really concentrated at a point but is actually distributed over a finite area having the dimensions $w$ and $d$. If the distance $d$ were appreciable and a more refined analysis were required, the representation might be as shown in Fig. 5-2($a$). If $d = l$, we would have the case represented in Fig. 5-2($b$). If the weight of a member is one or more orders of magnitude smaller than the other loads,

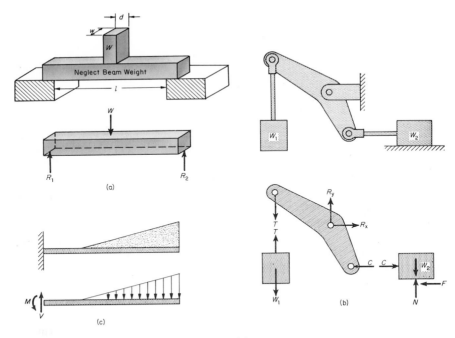

FIG. 5-1

it can usually be neglected. This is often the case in the design of a metal member; but in the design of a concrete member, for example, its weight must often be considered.

The more refined the analysis, the less is the permissible degree of idealization of load representation. For a first approximation a great amount of idealization may be justified. For more accurate analyses it may be necessary to know the effect of the loads on the members exactly as they are actually applied. This effect may become of utmost importance if an analysis of the localized behavior in the vicinity of the load is necessary.

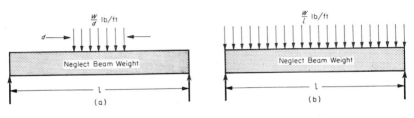

FIG. 5-2

## 5-3. Types of Loads

The adjectives used in all phases of engineering to modify the noun "load" are perhaps as numerous as those used with the word engineer. Also, they are often as misleading and ambiguous. We will try to define and discuss some of the most commonly used terms, and will try to outline some logical categories for classifying loads. Following are some types of loads classified generally on the basis of their causes.

**Body-Contact Loads.** Any force that is transmitted to a body from another body by means of direct contact over an area on the surface of the first body is a load due to body contact and is often called a surface traction. If the area of contact is relatively small, the load may be considered to act at a point and is usually called a *concentrated load* or a *point load*. If the area of contact is very narrow relative to its width, the load is called a *line load*. If the area of contact is too large to allow the load to be considered concentrated, it is then referred to as a *distributed load*. Line loads and distributed loads may be uniformly distributed or nonuniformly distributed; the loading in a particular case depends on how the force is distributed over the area of contact.

If the other "body" applying the load to the body under consideration is a liquid, a gas, or a vapor, then the load is usually called a *pressure*. Fluid-pressure loads may be uniform or nonuniform. For example, the tank of an air compressor may be subjected to a uniform load caused by a positive gage pressure, while the leeward side of a tall structure in a hurricane may be subjected to a nonuniform load caused by a negative air pressure. Here, negative means less than atmospheric.

Another important type of body-contact load common in engineering is the *friction load,* or *friction force*. This load, as you learned in statics, acts parallel to the contact area and is a function of the normal load, the materials in contact, and the relative velocity of the contact surfaces.

**Other Causes of Loads.** Many types of loads are not transmitted by body contact. The more important of these types are *gravitational* (weight), *inertial,* and *magnetic*. *Thermal loads* will also be considered here.

The gravitational load, or gravity load, on any body is usually considered to be a uniformly distributed downward load. It may, in some cases, be idealized as a concentrated load acting through the mass center of the body. Although a gravity load is really the attractive force between the earth and a body, it may be necessary to consider a "gravity" force exerted by some other planet or large body in space on some relatively smaller body in space. For example, as a space vehicle approaches the moon, the attractive force exerted on the vehicle by the moon will become larger than the attractive force (gravity) exerted on it by the earth.

Inertial forces are produced when a body is accelerated. The internal organs of a human being can withstand only a limited number of "*G*'s" (a common measure of inertial loads). This condition limits the maximum acceleration of a manned rocket. If the lines of a control-line model plane failed, the failure would no doubt be due to excessive inertial loading. Rotating machinery of any kind is subjected to inertial loads. Even if a part is rotating at a constant speed, the radial acceleration can cause critical inertial loads (centrifugal force) and therefore stresses.

Magnetic loads are the result of electromagnetic attractive and repulsive forces acting on bodies of ferrous metallic materials (those in the iron-cobalt-nickel group). Although it is possible for magnetic loads to cause critical stresses in some instances, they are not commonly significant in a great majority of engineering bodies or structures.

Thermal loads are mentioned here for lack of a better category. A thermal load may be said to be the internal load in a body resulting from the resistance to free thermal deformation (expansion or contraction). If the temperature of an isolated body is changed uniformly and slowly, then there will be free thermal deformation. If the body is restrained so as to prevent this free thermal deformation, then so-called thermal loads are produced. If complete restraint were assumed, the thermal load could be determined as follows: Hypothetically allow free thermal deformation to take place; determine the amount of this deformation from the coefficient of thermal expansion for the material and the temperature change; and then determine the restraining loads needed to return the body to its restrained dimensions. This procedure would require the use of load-deformation relationships which are to be studied in later chapters.

Much more complex thermal loads are sometimes induced in structures by severe nonuniform temperature changes. Large temperature gradients in a structure can produce critical stresses, and can even cause rupture of a brittle material. For example, consider what happens if you take an ordinary glass bottle from the refrigerator and immediately put it under the hot-water faucet. Such a suddenly induced temperature gradient is called a thermal shock.

## 5-4. Other Load Classifications

Loads are often classified on the basis of the primary type of deformation they produce, such as *axial, shearing, bending,* or *torsional.* Each of these types of loading can be produced by any of the various forces discussed in Sec. 5-3, namely, surface tractions, inertial forces, gravitational forces, and thermal loads. Special attention will be devoted to these classifications in succeeding chapters.

*Dynamic loads* are generally those that vary with time, whereas *static loads* do not change significantly in a relatively short time period. Any of the types of loads discussed so far could be static or dynamic. Some dynamic loads common in engineering are *repeated loads, impact loads,* and *energy loads.*

In the case of a repeated load, often referred to as *fatigue loading,* the load either is continually applied and removed or is varied periodically between two values. As an example, the loads on a connecting rod in an auto engine vary periodically with the rotation of the crankshaft.

In the case of an impact load, a large force is applied in a relatively short time period.

An energy load is a "load" which is, for convenience, expressed in terms of the amount of energy (for example, foot-pounds) it imparts to the body. For example, some impact loads can be expressed much more easily in terms of the energy transmitted during the impact period than in terms of applied force.

Each phase of engineering has its own jargon, and as a result has its own specialized adjectives that are used to classify loads. Structural engineers, for example, call the weight of the structure the *dead load* and the additional load applied in use the *live load*. Also, the aero-space industry today is developing its own vocabulary so fast that dictionaries cannot keep up with the terminology.

## PROBLEMS

**5-1 to 5-5.** For each of the accompanying illustrations, draw a free-body diagram of the member *A* with the loads thereon idealized as necessary.

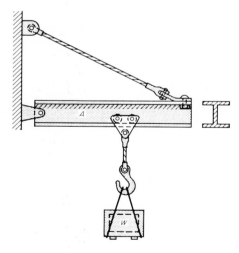

PROB. 5-1

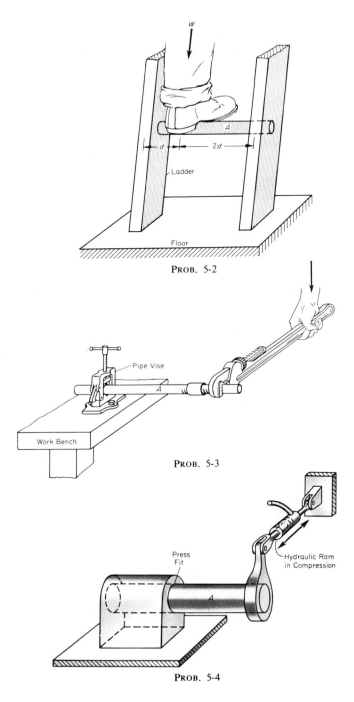

PROB. 5-2

PROB. 5-3

PROB. 5-4

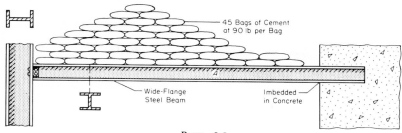

45 Bags of Cement
at 90 lb per Bag

Wide-Flange
Steel Beam

Imbedded
in Concrete

PROB. 5-5

NOTE: In Probs. 5-6 to 5-13, sketch a free-body diagram (FBD) whenever desirable or possible.

**5-6.** Discuss and classify as well as possible the design loads on the following:
*a*) auto steering wheel
*b*) sailboat mast
*c*) propeller blade
*d*) motorcycle wheel spoke

**5-7.** Discuss and classify as well as possible the design loads on the following:
*a*) screwdriver
*b*) bridge girder
*c*) golf ball
*d*) paper clip

**5-8.** Discuss and classify as well as possible the design loads on the following:
*a*) drill bit
*b*) fishhook
*c*) four-prong auto lug wrench
*d*) support brackets for astronaut's couch

**5-9.** Discuss and classify as well as possible the design loads on the following:
*a*) air compressor tank
*b*) surgical hemostat
*c*) human femur (thigh bone)
*d*) strapless bra

**5-10.** Discuss and classify as well as possible the design loads on the following:
*a*) coat hanger
*b*) floor joist in a residence
*c*) ski pole
*d*) piston in an auto engine

**5-11.** Discuss and classify as well as possible the design loads on the following:
*a*) pencil lead
*b*) wing lift strut on light airplane
*c*) guitar string
*d*) beer can

**5-12.** Discuss and classify as well as possible the design loads on the following:
*a*) cable in a highway suspension bridge

*b*) safety pin

*c*) plate glass store window

*d*) fireman's suspenders (red)

**5-13.** Discuss and classify as well as possible the design loads on the following:

*a*) automobile rear axle

*b*) door hinge

*c*) outdoor TV antenna mast

*d*) the chair you're sitting in

**5-14.** Give an example or two (other than those in the text) of a body, member, structure, or material subjected to the following types of loads:

*a*) uniformly distributed repeated

*b*) acoustical

*c*) moving line

*d*) uniformly distributed dead

**5-15.** Give an example or two (other than those in the text) of a body, member, structure, or material subjected to the following types of loads:

*a*) thermal shock

*b*) high-velocity impact

*c*) inertial fatigue

*d*) nonuniformly distributed static pressure

**5-16.** Be prepared to discuss in class the meanings, as applied to loads, of the following adjectives:

*a*) point

*b*) transient

*c*) wind

*d*) shear

*e*) line

*f*) live

*g*) moving

*h*) inertial

*i* ) blast

*j* ) buoyancy

*k* ) vibratory

*l* ) aerodynamic

*m*) shock

*n* ) transverse

*o* ) acoustical

*p* ) dead

## chapter 6

# Static Analysis of Load-Resisting Members

## 6-1. Introduction

In our study of the response of engineering bodies or members to external loads, we must necessarily be concerned with the resulting internal reactions, since, as we have already seen in Chapters 1 and 2, these internal reactions are related to the stresses and strains. Proper design of members and structures is dependent upon the ability to relate the external loads to the resulting stresses and strains. Future design courses will deal with the problem of actually selecting the proper shapes and sizes of structural elements so that the members will fulfill their specific functions. The main purpose of this course, however, is to provide a firm foundation on which to base the design theories.

## 6-2. Method of Sections

In the usual statics course, much of the time is spent in analyzing force systems acting on complete bodies or structures as a whole. When necessary, an individual member of a complex structure is "taken out," and the force system on it is analyzed separately. In every case the primary concern is to determine the magnitudes, locations, and lines of action of the external forces on the body or member under consideration. Little or no attempt is made to find out how the body *responds* to the loads, or even if the body is capable of supporting the loads. In the present course the emphasis is quite different, although free-body analysis and equilibrium equations are still used. Here our primary concern is to investigate how deformable bodies respond to various loads.

To investigate this response, it is usually necessary to "go inside the body" and find out how the internal reactions are related to the externally applied loads. The process of "going inside," usually called the *method of sections*, is very similar to the procedure you have used in isolating an *individual* member of a structure composed of several members. The basic principle is simply this: If an entire body or structure is in equilibrium when acted upon by some force system, then *any* portion of the body or

structure must itself be in equilibrium in the deformed position, provided
that the body has stopped deforming. When an individual member of a
complex structure is isolated, the equations of equilibrium can be used to
determine the unknown forces on that member; these unknown forces are
generally in the form of pin reactions. However, when a *portion* of a body
is isolated from the remainder of the body, the unknowns are usually in the
form of internal *resultant* normal and shear forces and possibly *resultant*
bending and twisting moments.

When a framed structure was analyzed in a statics course, the usual
procedure was to draw a free-body diagram of the whole structure or an
entire individual member of the structure. For instance, in the case of a
framed structure like that in Fig. 6-1(*a*), a free-body diagram of the entire
member *BDF* was drawn, as in Fig. 6-1(*b*). You probably did not consider
the *internal* forces in different parts of the member, but analyzed only the
external forces, shown in Fig. 6-1(*b*), as if they were acting on pins at the
ends of the member and at that intermediate joint *D*. Since we are now

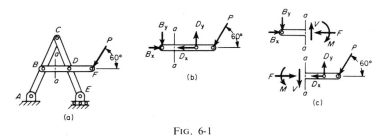

FIG. 6-1

interested in evaluating the internal forces on various sections of the body,
we may pass imaginary cutting planes through members or bodies as we did
in Chapter 1 in Figs. 1-1, 1-2, and 1-3. A free-body diagram for that part
of the member or body to either side of the cutting plane can then be considered. For example, see the common two-dimensional diagrams in Fig.
6-1(*c*). If the load *P* and the various distances were known, the internal reactions *F*, *V*, and *M* could be determined. (Another supplemental free-
body diagram would first be required to find $B_x$ and/or $D_x$.)

Each internal reaction on a section represents a *resultant* reaction of a
very large number of small fiber forces. In Fig. 6-1(*c*), *F* represents the resultant normal force on section *a-a*; *V*, the resultant shear force on the section; and *M*, a resultant moment of the fiber forces. The value of *M* will depend on the location chosen for the line of action of *F*. The actual distribution of the fiber forces will be discussed later; now we are concerned only
with their cumulative effects.

The example in Fig. 6-1 involves only coplanar forces, but the method
of sections can be applied to a three-dimensional member subjected to any

variety and combination of the loads discussed in Chapter 5. As you will recall from your statics course, not all structures are statically determinate; that is, in many structures the equations of equilibrium are not sufficient to completely determine the external and internal reactions. From time to time we shall encounter such structures, and in subsequent chapters you will see how some of them can be solved. For the present, however, it is only necessary that you be able to recognize when a structure or member is statically indeterminate and know how to establish its degree of inde-terminateness.

One additional comment is worth mentioning here. When equilibrium equations are written, the dimensions of the body often enter into the prob-lem. These dimensions should be those of the *deformed* body. In most cases, the allowable deformations will be so small in comparison with the original dimensions of the body that using the undeformed dimensions will not result in significant error. However, there is a very important excep-tion, buckling, in which the deformations greatly affect the equilibrium equations. Buckling will be discussed in greater detail in Chapter 11.

EXAMPLE 6-1. The body shown in Fig. 6-2($a$) is fixed at the left-hand end and subjected to the given loads. Determine the internal resultant reactions at section *a-a*.

*Solution:* The free-body diagram for the portion of the body to the right of section *a-a* is drawn in Fig. 6-2 ($b$). Here the equivalent force and couple on the right-hand end replace the original eccentric 700-lb force. The internal (external to this free body) resultant reactions $F$ and $M_y$ are determined by using equilibrium equations. Note that $F$ has been placed collinear with the longitudinal $x$ axis. Con-sidering only horizontal forces, we have

$$\overset{+}{\underset{\longleftarrow}{}}\sum F_x = 0$$

$$+ 700 - 2(600) + F = 0$$

$$F = +500 \text{ lb (Plus means as assumed in Fig. 6-2.)}$$

Considering moments about the $y$ axis,

$$\sum M_y = 0$$

$$- (700)(2) + M_y = 0$$

$$M_y = +1400 \text{ in-lb (Plus means as assumed in Fig. 6-2.)}$$

Force equations in the $y$ and $z$ directions show that there is no internal re-sultant shear at section *a-a* in either of these directions. A moment equation with

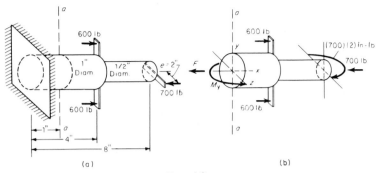

FIG. 6-2

respect to the $z$ axis indicates that there is no resultant internal moment about this axis. A moment equation with respect to the $x$ axis indicates that there is no internal resultant twisting moment on section $a$-$a$.

### 6-3. Variation of Internal Reactions

As we have already seen, the internal reactions in a member depend upon where and how the cutting plane is passed through the member. For example, consider the internal resultant axial force for the member in Fig. 6-3($a$) at various cutting planes perpendicular to the length of the member. In Fig. 6-3($b$) is a typical free-body diagram obtained by passing a cutting

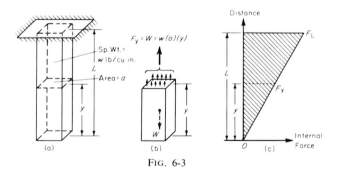

FIG. 6-3

plane at an arbitrary distance $y$ from the bottom end. In Fig. 6-3($c$) is a plot of the variation of the internal resultant reaction with the length below the section. Notice that $F$ is a linear function of $y$ in this case.

In the example in Fig. 6-4, the variation of the internal resultant force is nonlinear, since $F = f(y^2)$. Here the member is shown in ($a$); a free-body diagram of the lower portion of length $y$ is shown in ($b$); and a plot of the variation of the internal resultant reaction with the length $y$ is shown in ($c$).

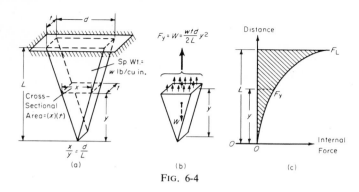

FIG. 6-4

These examples are given merely to illustrate how an internal reaction may vary with some parameter, such as the length of the member. An internal reaction may be a function of some parameter other than length, but in most of the problems we will consider, length will be the parameter.

In order to make a *complete* static analysis of a load-carrying member, it is usually necessary to determine how the internal reactions vary with the length of the member. An internal-reaction diagram is merely a graphical representation to show how *one* reaction varies with the length. Such diagrams are quite useful in helping you to understand how and why a member behaves as it does.

In writing equilibrium equations and in plotting the reaction diagrams, it is usually necessary to select some direction for each reaction and also to select an algebraic sign convention. Both selections are purely arbitrary and have absolutely no influence on the member. Thus, although two different assumptions in regard to the directions of the forces or moments may result in *apparently different* equilibrium equations and reaction diagrams, the two different assumptions must actually yield the *same result* for the types, magnitudes, and directions of the internal reactions. The same statement also applies to the arbitrary choice of the sign convention. To illustrate this point, consider the examples in Fig. 6-5. In each example the weight of the member is neglected. In Example I, $M_x$ is equal to plus $T$. Thus, the direction of $M_x$ assumed in the free-body diagram is correct. In Example II, $V_y$ is equal to minus $P$. The minus sign indicates that since $V_y$ was assumed to be acting downward, it is actually acting upward. However, as long as a sign is correctly interpreted, there is no need to go back and change the direction of the force on the free-body diagram. In fact, changing the direction of a force *after* the equilibrium equations have been written can often result in confusion rather than clarification.

In each example in Fig. 6-6 there are combinations of various internal reactions, and in some cases, the external loads have abrupt changes or are concentrated at points. In such cases separate equations of equilibrium

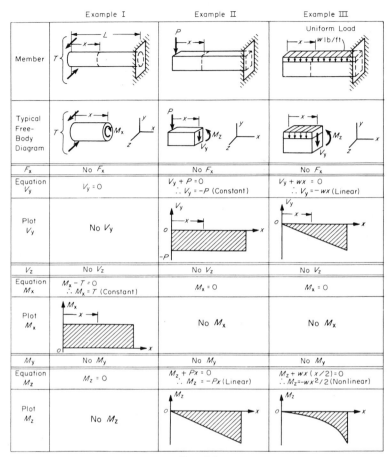

FIG. 6-5

must be written for each region of the member in which the loading conditions are continuous. Study each example carefully. Notice that the typical free-body diagrams have been omitted. You should draw a typical one for each region of each member, in order to verify and interpret the obtained results. Also, in example III, you should derive the equation for $M_z$ for $2L/3 \leq x \leq L$. Be aware of the origin of coordinates for each equation.

## 6-4. Conventional Shear-Force and Bending-Moment Diagrams for Beams

**Types of Beams.** Of particular importance to the engineer are the so-called shear and moment diagrams for beams. A beam may be defined as a

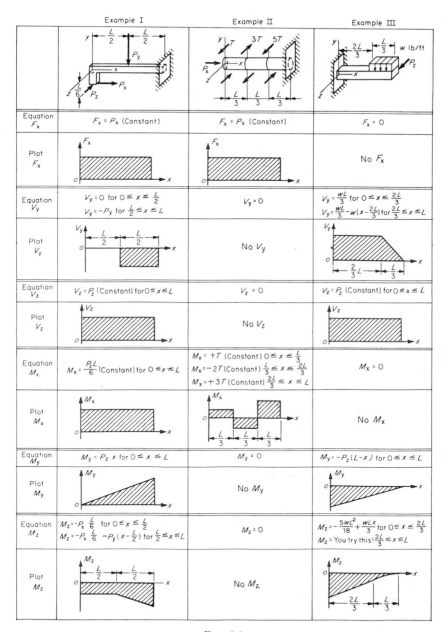

FIG. 6-6

member whose length is large in comparison with its thickness and depth, and which is loaded with couples and/or transverse loads that produce significant bending effects. Beams are so common in engineering structures that their importance cannot be overemphasized, and Chapter 10 is devoted to a detailed investigation of their behavior. In actual structures, beams can be found in an infinite variety of orientations. For purposes of discussion and analysis, however, we will usually orient the beam so that its length is horizontal.

Beams are generally classified according to their geometry and the manner in which they are supported. Geometrical classification includes such features as the shape of the cross section, whether the beam is straight or curved, whether the beam is tapered or has a constant cross section and other features which will be considered in more detail in Chapter 10. For the present, we will confine our discussion to straight beams. They may be readily classified according to the manner in which they are supported. Some types that occur in ordinary practice are shown in Fig. 6-7, the names of some of these being fairly obvious from direct observation. Note that the beams in $(d)$, $(e)$, and $(f)$ are statically indeterminate.

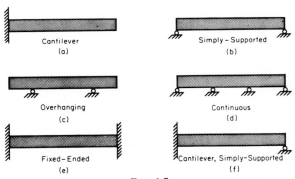

FIG. 6-7

A beam can be further classified according to the type of load or loads it is carrying. For instance, a cantilever beam may be carrying a uniformly distributed load. In that case the beam might be classified as a uniformly loaded cantilever beam. Further extension of this method of classification is possible and if carried to extremes the names become rather complex and lengthy. In this text, however, we shall merely classify beams according to their geometry and support.

**Sign Conventions for Beams.** Consider now a simply supported beam, such as that shown in Fig. 6-8($a$). This beam carries some arbitrary continuously distributed load expressed as $w$ pounds per unit length, where $w$ is a function of $x$.

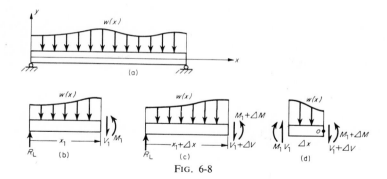

FIG. 6-8

A free-body diagram of a portion of the beam extending $x_1$ units from the left-hand end is shown in Fig. 6-8(*b*), while a similar diagram for a portion having a length of $x_1 + \Delta x$ units is shown in (*c*). In Fig. 6-8(*d*) is a free-body diagram of the $\Delta x$ portion of the beam. Note that the moments and shears in (*d*) are consistent with those in (*b*) and (*c*).

In American engineering practice it is customary to assume that a *positive* bending moment is one which causes a beam to bend *concave upward*. Another way of saying the same thing is as follows: A positive bending moment produces tension in the lower fibers of the beam and compression in the upper fibers. A negative moment produces the opposite effects. Hence, *all* the bending moments in Fig. 6-8(*b*), (*c*), and (*d*) are considered positive.

The directions of the shear forces in Fig. 6-8(*b*), (*c*), and (*d*) are also customarily assumed to be positive; that is, a positive shear is one which acts downward on the right-hand side of the section or upward on the left-hand side. We will now see that such positive shear forces are directly related to positive bending moments and the loading *w*.

**Basic Equations for Beam.** For equilibrium in the *y* direction in Fig. 6-8(*d*),

$$\downarrow +$$

$$\sum F_y = 0$$

$$(V_1 + \Delta V) + w(x') \Delta x - V_1 = 0$$

where $x'$ is some value of $x$ between $x_1$ and $x_1 + \Delta x$. Hence,

$$\frac{\Delta V}{\Delta x} = -w(x')$$

Taking the limit as $\Delta x$ approaches zero, we have

$$\frac{dV}{dx} = -w(x) \qquad (6\text{-}1)$$

Thus, at any point in a beam *where the loading is continuous*, the rate of change of the shear force is equal to the negative loading $w(x)$; the negative sign indicates that the shear decreases when the loading acts downward, as assumed here.

Taking moments about point $o$ in Fig. 6-8($d$), we have

$$\overset{+\curvearrowright}{\sum M_o} = 0$$

$$(M_1 + \Delta M) + w(x') \Delta x \, \alpha \, \Delta x - M_1 - V_1 \Delta x = 0$$

$$V_1 = \frac{\Delta M_1}{\Delta x} + w(x') \, \alpha \, \Delta x$$

where $\alpha$ is some number between 0 and 1. Now, taking the limit as $\Delta x$ approaches zero and omitting the subscript, we obtain

$$V = \frac{dM}{dx} \tag{6-2}$$

Thus, at any point along a beam *where the loading is continuous*, the shear is equal to the derivative of the moment at the corresponding point.

Writing Eq. 6-2 in another form and integrating between two points *1* and *2* at distances $x_1$ and $x_2$ from the left-hand end of the beam, we have

$$V \, dx = dM$$

$$\int_{x_1}^{x_2} V \, dx = \int_{x_1}^{x_2} dM = M_2 - M_1 = \Delta M_{1,2} \tag{6-2a}$$

Thus, the integral $\int_{x_1}^{x_2} V \, dx$ represents the *change* in moment between points *1* and *2* (if there are no jump discontinuities in the moment due to concentrated bending couples).

Finally, from Eq. 6-1 we see that wherever the loading is continuous, $V$ will be differentiable and thus

$$\frac{d^2 M}{dx^2} = \frac{dV}{dx} = -w(x) \tag{6-3}$$

**Shear and Moment Diagrams.** Equations 6-2 and 6-2a are quite useful in drawing shear and moment diagrams, since Eq. 6-2 relates the slope of the moment diagram to the value of the shear, and Eq. 6-2a relates an area on the shear diagram to the *change* in the value of the moment. The following example illustrates the use of these relationships.

EXAMPLE 6-2. For the beam in Fig. 6-9($a$), write shear and moment equations for each region. Then sketch the complete shear and moment diagrams, labeling the values at points of discontinuity. Also, determine the value of the maximum moment and locate the section at which it occurs.

*Solution:* From the free-body diagram of the entire beam in Fig. 6-10(a), we first find the reactions $R_1$ and $R_2$. Using the left-hand end as the origin of the $x$-$y$ coordinate system, the following shear equations are obtained by referring to the free-body diagrams in Fig. 6-9(b), (c), and (d):

For $0 \leq x < 4$, $V = -2000x$
For $4 < x \leq 15$, $V = -2000x + 23{,}440$
For $15 \leq x < 20$, $V = (-2000)(15) + 23{,}440 = -6560$

Notice that the first two equations represent straight lines each with a negative slope of 2000. The last equation is straight line with a zero slope. Observe

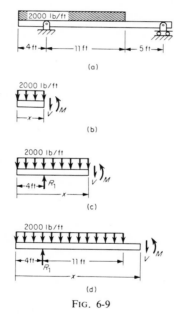

FIG. 6-9

how these three lines are plotted in Fig. 6-10(b) to form the shear diagram. The vertical line in the shear diagram is a jump discontinuity caused by the concentrated reaction.

Again using the same origin and free-body diagrams, we get the following moment equations:

For $0 \leq x \leq 4$, $M = -2000x(x/2) = -2000x^2/2$
For $4 \leq x \leq 15$, $M = -2000x^2/2 + 23{,}440(x-4)$
For $15 \leq x \leq 20$, $M = -2000(15)[x - (15/2)] + 23{,}440(x-4)$

Notice that the first two equations are those of parabolas, and the third is that of a straight line with a negative slope. Verify the fact that $dM/dx = V$ by

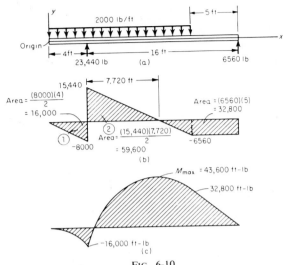

FIG. 6-10

differentiating each moment equation and comparing the result with the appropriate shear equation.

To obtain the proper shape of each curve in the moment diagram, we can utilize Eq. 6-2. In the region where $0 \leq x < 4$, the shear is negative and its value decreases as $x$ increases. Consequently, the slope of the moment diagram should be negative and *decreasing*, and the parabola, therefore, opens concave downward. A similar argument shows that the parabola in the region where $4 \leq x \leq 15$ is also concave downward, and that the straight line in the region where $15 \leq x \leq 20$ has a constant negative slope.

The location of the maximum moment will be either at a discontinuity or at a point of zero slope on the moment diagram. In the latter case, since $dM/dx = V$, *the maximum moment occurs where the shear is zero*. Therefore, equating the second shear equation to zero and solving for $x$ will give the distance to the maximum moment from the left-hand end. This distance is 11.720 ft. This location can be found just as easily from the geometry of the shear diagram in Fig. 6-10(b) in this way: Divide the ordinate (15,440) by the slope of the line (2000) to get the distance from the left-hand reaction to the point of zero shear (maximum moment). The distance is 7.720 ft.

The value of the shear or moment at any location in the beam can be found by substituting the distance from the origin in the appropriate equation. For instance, the maximum moment can be found by substituting 11.720 ft for $x$ in the second moment equation. The result is

$$M_{max} = -2000 (11.720)^2/2 + 23,440 (11.720 - 4)$$
$$= 43,600 \text{ ft-lb}$$

The values at various points on the moment diagram can also be obtained by utilizing Eq. 6-2a, which relates areas on the shear diagram to *changes* in the value of the moment. For example, area ① in Fig. 6-10(b) is 16,000 *negative* (below the $x$ axis). So the moment *decreased* from 0 to $-16,000$ between $x = 0$ and $x = 4$. Also, since area ② is 59,600 *positive*, the moment *increased* from $-16,000$ for $x = 4$ to $+43,600$ for $x = (7.720 + 4) = 11.720$.

In working problems and drawing free-body diagrams, it is usually convenient to assume that the shears and moments on the free body diagrams are positive, without worrying about whether they are actually positive or negative. If you initially assume a quantity to be positive, then a positive answer means that it is positive and a negative answer means that it is negative. Thus, the interpretation of the sign is automatic. However, if you initially assume that a shear or moment is negative, then you must be very careful to interpret your answer correctly.

## 6-5. Singularity Functions

Recall relations 6-1, 6-2, and 6-3 between the loading $w(x)$, shear $V$, and moment $M$;

$$w(x) = -\frac{dV}{dx} \qquad [6\text{-}1]$$

$$V = \frac{dM}{dx} \qquad [6\text{-}2]$$

$$w(x) = -\frac{d^2M}{dx^2} \qquad [6\text{-}3]$$

which are valid so long as the loading function $w(x)$ is continuous. However, from our example we see that it is often necessary to deal with discontinuous loadings such as point loadings (concentrated forces such as pin reactions), jump discontinuities (such as at $x = 15'$ in Example 6-2, and concentrated external couples (for example, Probs. 6-32 and 6-33). We got around this difficulty in Example 6-2 by separately considering intervals of the beam for which the loading was continuous. This procedure will always work but a more sophisticated and less cumbersome treatment can be made by the use of singularity functions.

Briefly, we seek to define some loading functions which can physically represent such things as point loads, jump loading conditions, and point couples but still satisfy Eqs. 6-1, 6-2, and 6-3. To this end we introduce the Dirac delta function, step function, and doublet function, respectively.

**Dirac Delta Function.** The basic idea is to replace the concentrated load $P$ by a high intensity distributed load $P/\epsilon$ distributed over a very small length $\epsilon$, so that even in the limit as $\epsilon \to 0$ the product of $P/\epsilon$ with the length $\epsilon$ remains the finite value $P$, as illustrated in Fig. 6-11(a) and (b). For convenience we let $P$ equal unity and define the Dirac delta function as

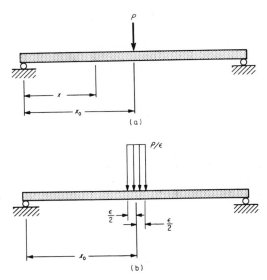

FIG. 6-11

$$\delta(x - x_o) = \lim_{\epsilon \to 0} \begin{cases} 0 & \text{when } x < \left(x_0 - \dfrac{\epsilon}{2}\right) \\[2ex] \dfrac{1}{\epsilon} & \text{when } \left(x_o - \dfrac{\epsilon}{2}\right) < x < \left(x + \dfrac{\epsilon}{2}\right) \\[2ex] 0 & \text{when } x > \left(x_o + \dfrac{\epsilon}{2}\right) \end{cases} \qquad (6\text{-}4)$$

Then, using this delta function we see that the distribution

$$w(x) = P\delta(x - x_o)$$

integrated over the length of the beam $l$ gives

$$\int_0^l P\delta(x - x_o)\,dx = \lim_{\epsilon \to 0} \int_{x_o - (\epsilon/2)}^{x_o + (\epsilon/2)} \frac{P}{\epsilon}\,dx = P$$

Hence the distribution

$$w(x) = P\delta(x - x_o)$$

can be used to represent a concentrated load $P$ applied at point $x_o$.

We note that the delta function is not a function in the usual mathematical sense of continuity and differentiability. However, we justify our introduction of it simply because its formal use leads to results which we can interpret physically, such as the concentrated load discussed above.

**Step Function.** We now introduce the second singularity function, the unit step function, defined by

$$u(x - x_o) = \int_o^x \delta(t - x_o) \, dt \tag{6-5}$$

which, from the definition of $\delta(t - x_o)$ has the property

$$u(x - x_o) = \begin{cases} 0 & x < x_o \\ 1 & x > x_o \end{cases} \tag{6-6}$$

and physically can be interpreted as a unit step or jump beginning at the point $x_o$. With this function we are now able to formally handle the initiation and termination of distributed loads. For example the distributed load of Fig. 6-12(a) is given by

$$w(x) = w_o u(x - x_o) - w_o u(x - x_1) = \begin{cases} 0 & x < x_o \\ w_o & x_o < x < x_1 \\ 0 & x > x_1 \end{cases}$$

while that for Fig. 6-12(b) is given by

$$w(x) = 100 \, \frac{x}{5} - 100 \, \frac{x}{5} \, u(x - 5) = \begin{cases} 20x & x < 5 \\ 0 & x > 5 \end{cases}$$

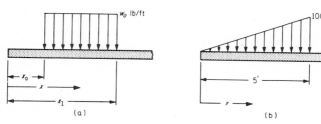

(a) $w_o$ lb/ft

(b) 100 lb/ft

FIG. 6-12

We note for later use that the definite integral of the product of a continuous function $f(x)$ with the unit step function $u(x - x_o)$ is

$$\int_o^x f(t) \, u(t - x_o) \, dt = \int_{x_o^+}^x f(t) \, u(t - x_o) \, dt = u(x - x_o) \int_{x_o}^x f(t) \, dt \tag{6-7}$$

where $x_o^+$ means $x_o + \epsilon$ as $\epsilon \to 0$, since $u(t - x_o)$ is not defined for $t = x_o$. In particular, if $f(x) = (x - x_o)^n$ we have the useful results

$$\int_o^x (t - x_o)^n u(t - x_o)\, dt = u(x - x_o) \int_{x_o}^x (t - x_o)^n\, dt$$

$$= u(x - x_o) \frac{(x - x_o)^{n+1}}{n + 1} \tag{6-8}$$

for $n > -1$.

**Doublet Function.** The delta function enables us to represent a concentrated load as a distribution $P\delta(x - x_o)$ and the unit step function enables us to introduce or terminate distributed loads. We now introduce the third singularity function, the doublet function, in order to express concentrated couples in terms of a load distribution function. To this end we visualize the concentrated couple $C$ as a pair of very large concentrated forces $C/\epsilon$ separated by a very small distance $\epsilon$, as illustrated in Fig. 6-13,

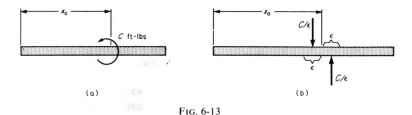

(a)                                             (b)

Fig. 6-13

so that the product of $C/\epsilon$ with $\epsilon$ equals the finite number $C$. For convenience we let $C$ equal unity. Now, if we think of the loads $1/\epsilon$ as being distributed over the lengths $\epsilon$, we define the doublet function by

$$\psi(x - x_o) = \lim_{\epsilon \to 0} \begin{cases} 0 & \text{when} \quad x < (x_o - \epsilon) \\[2mm] \dfrac{1}{\epsilon^2} & \text{when} \quad (x_o - \epsilon) < x < x_o \\[2mm] -\dfrac{1}{\epsilon^2} & \text{when} \quad x_o < x < (x_o + \epsilon) \\[2mm] 0 & \text{when} \quad (x_o + \epsilon) < x \end{cases}$$

Hence, the distribution corresponding to a concentrated counterclockwise couple $C$ applied at $x = x_o$ is given by

$$w(x) = C\,\psi(x - x_o) \tag{6-9}$$

Integration of the doublet function yields

$$\int_o^x \psi(t - x_o)\, dt = \lim_{\epsilon \to 0} \begin{cases} 0 & \text{when} \quad x < (x_o - \epsilon) \\[2mm] \dfrac{1}{\epsilon} & \text{when} \quad x = x_o \\[2mm] 0 & \text{when} \quad x > x_o + \epsilon \end{cases}$$

which we recognize as the delta function. Hence

$$\int_o^x C\psi(t - x_o)\, dt = C\delta(x - x_o) \tag{6-10}$$

Thus with the doublet function we can represent a loading distribution corresponding to the effect of a concentrated couple; with a delta function we can represent a distribution corresponding to a concentrated load; and with the step function we can introduce or terminate ordinary distributed loads. We now give an illustration of the application of these functions for determining shear and moment equations for beams.

EXAMPLE 6-3. Using singularity functions, obtain the shear and moment equations for the beam of Example 6-2.

*Solution:* Refer to Fig. 6-10(a):

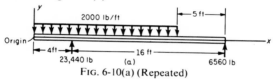

FIG. 6-10(a) (Repeated)

Recalling that $w(x)$ is positive when directed downward, we write the loading functions for each of the loadings.

For the uniformly distributed load of 2000 lb/ft which terminates at $x = 15$

$$w(x) = 2000 - 2000\, u(x - 15) \tag{a}$$

For the left pin reaction located at $x = 4$

$$w(x) = -23{,}440\, \delta(x - 4) \tag{b}$$

For the right pin reaction located at $x = 20$

$$w(x) = -6560\, \delta(x - 20) \tag{c}$$

Hence for the whole beam

$$w(x) = 2000 - 2000\, u(x - 15) - 23{,}440\, \delta(x - 4) - 6560\, \delta(x - 20) \tag{d}$$

The boundary conditions are

$$V = 0 \quad \text{at} \quad x = 0, x = 20^+$$
$$M = 0 \quad \text{at} \quad x = 0, x = 20^+$$

(the $20^+$ means that we consider the values just to the right of the pin reaction).

Now, Eq. 6-1 can be written as

$$dV = -w\,dx$$

whose definite integral is

$$\int_0^{V(x)} dV = -\int_0^x w(t)\,dt$$

$$V(x) = -\int_0^x [2000 - 2000\,u(t - 15) - 23{,}440\,\delta(t - 4) - 6560\,\delta(t - 20)]\,dt$$

$$= -2000x + 2000\,(x - 15)\,u(x - 15) + 23{,}440\,u(x - 4)$$
$$+ 6560\,u(x - 20) \tag{e}$$

where we have utilized Eq. 6-5 and Eq. 6-8 for $n = 0$. We see that

$$V(0) = 0,\; V(20^+) = 0$$

as required. Now, Eq. 6-2 can be rewritten as

$$dM = V\,dx$$

whose definite integral is

$$\int_0^{M(x)} dM = \int_0^x V(t)\,dt$$

Thus, using equation (e),

$$M(x) = \int_0^x [-2000t + 2000\,(t - 15)\,u(t - 15) + 23{,}440\,u(t - 4)$$

$$+ 6560\,u(t - 20)]\,dt$$

$$= -2000\,\frac{x^2}{2} + 2000\,\frac{(x - 15)^2}{2}\,u(x - 15) + 23{,}440\,(x - 4)\,u(x - 4)$$

$$+ 6560\,(x - 20)\,u(x - 20) \tag{f}$$

when we have again used Eq. 6-8. We see that

$$M(0) = 0, \quad M(20^+) = 0$$

as required.
As a check, let us evaluate $M$ at $x = 11.720$ as in Example 6-2.

$$M(11.72) = -2000\,\frac{(11.720)^2}{2} + 0 + 23{,}440(11.720 - 4)\,(1) + 0 = 43{,}600 \text{ ft-lb}$$

## PROBLEMS

**6-1 to 6-18.** For each of the illustrations of an engineering member, determine the internal resultant reactions at sections *a-a*, *b-b*, and *c-c*.

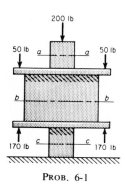

PROB. 6-1

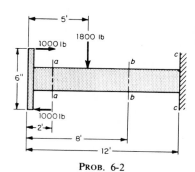

PROB. 6-2

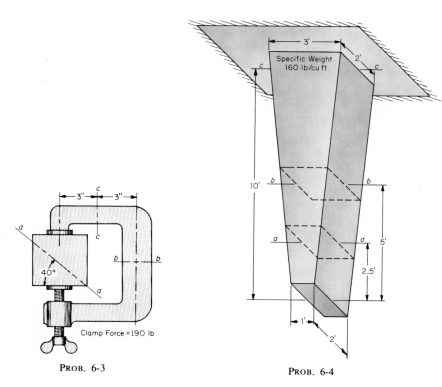

PROB. 6-3

PROB. 6-4

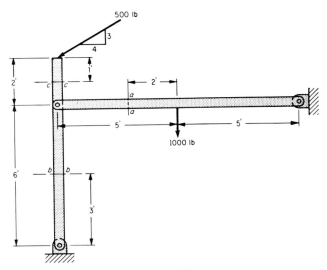

PROB. 6-5

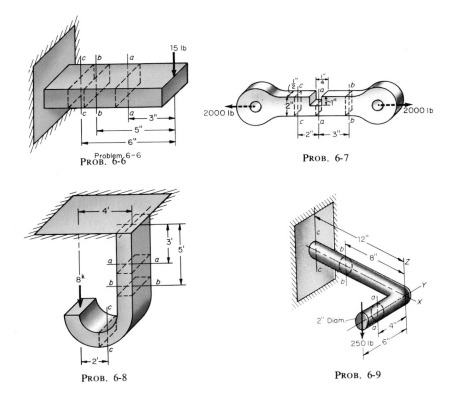

Problem 6-6
PROB. 6-6

PROB. 6-7

PROB. 6-8

PROB. 6-9

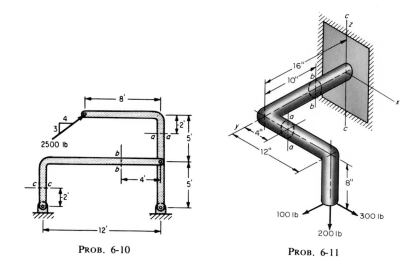

PROB. 6-10

PROB. 6-11

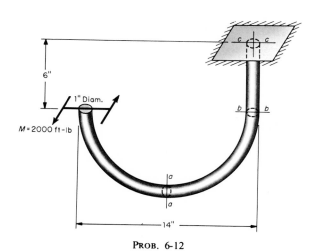

PROB. 6-12

PROB. 6-13

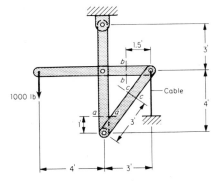

PROB. 6-14

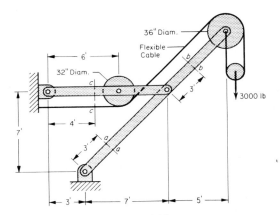

PROB. 6-15

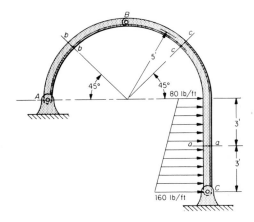

PROB. 6-16

**6-17 to 6-35.** For each of the accompanying illustrations, write the general equations for the internal reactions for each continuous region over the length of the member. Choose a convenient coordinate system, and sketch the variation of each reaction.

PROB. 6-17

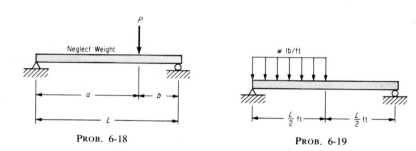

PROB. 6-18                    PROB. 6-19

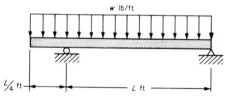

PROB. 6-20

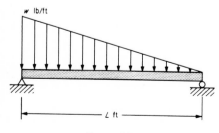

PROB. 6-21

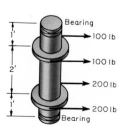

PROB. 6-22

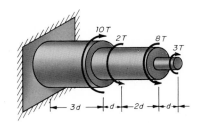

PROB. 6-23

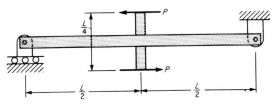

PROB. 6-24

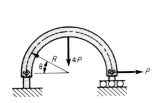

PROB. 6-25

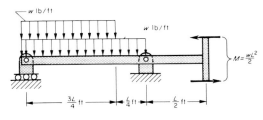

PROB. 6-26

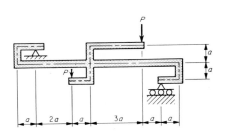

PROB. 6-27

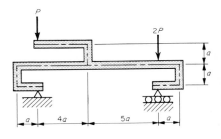

PROB. 6-28

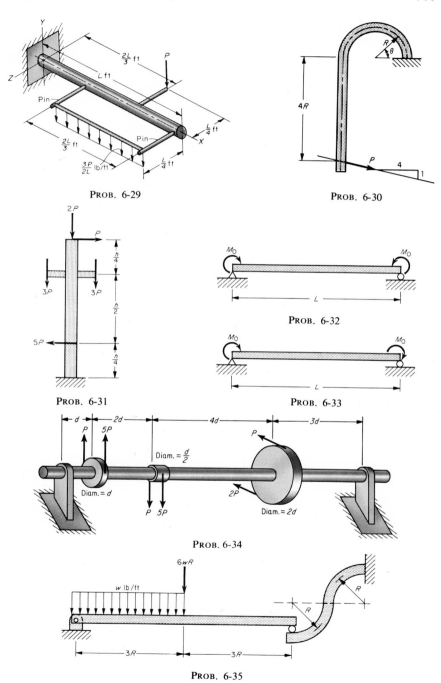

PROB. 6-29

PROB. 6-30

PROB. 6-31

PROB. 6-32

PROB. 6-33

PROB. 6-34

PROB. 6-35

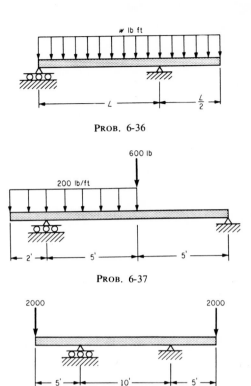

PROB. 6-36

PROB. 6-37

PROB. 6-38

**6-39 to 6-60.** Draw complete shear and moment diagrams for the members shown in the illustrations. Neglect the weights of the members.

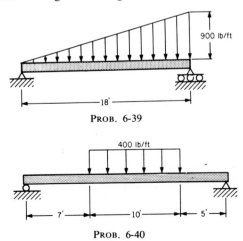

PROB. 6-39

PROB. 6-40

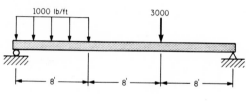

PROB. 6-41

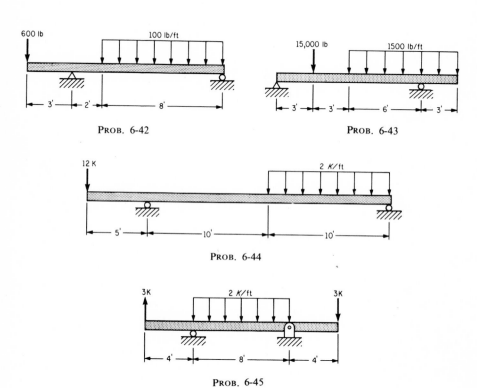

PROB. 6-42

PROB. 6-43

PROB. 6-44

PROB. 6-45

PROB. 6-46. See illustration for Prob. 6-2.
PROB. 6-47. See illustration for Prob. 6-6.

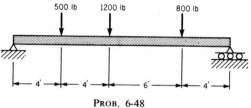

PROB. 6-48

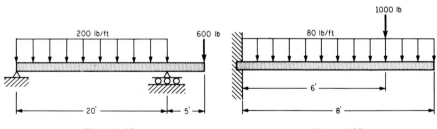

PROB. 6-49                                            PROB. 6-50

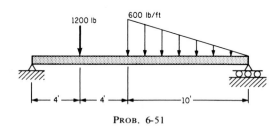

PROB. 6-51

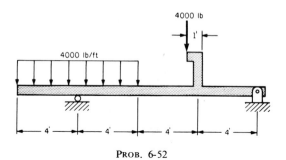

PROB. 6-52

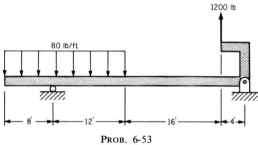

PROB. 6-53

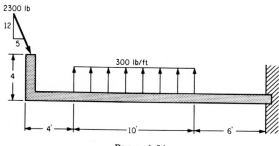

PROB. 6-54

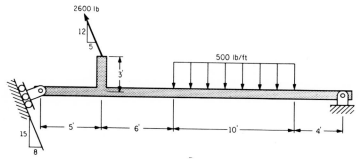

PROB. 6-55

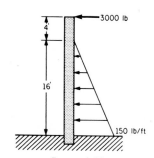

PROB. 6-56

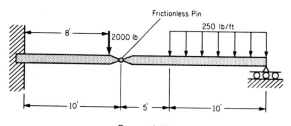

PROB. 6-57

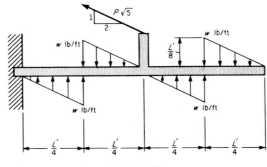

PROB. 6-58

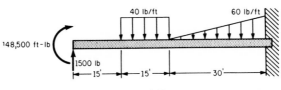

PROB. 6-59

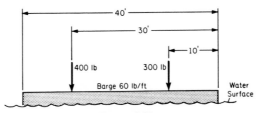

PROB. 6-60

*part* **III**

# Relationships Between Loads and Factors Important in Design

## chapter 7

# Design Procedure

## 7-1. Introduction

The engineer generally uses mechanics of deformable bodies in two ways. One use is to *investigate* the stresses and/or the deformations in a structure or member that has already been designed. The other use is in the design of a proposed structure. The first requires only that the relations between loads, dimensions, and stresses or deformations be known. The development of a design is more involved. We will now concentrate on some of the aspects of this latter use.

According to the dictionary, the word "design" means "to fashion according to a plan." We shall interpret the verb "fashion" to mean "to select the materials and to determine the dimensions." This chapter will organize and set forth a rational or reasonable "plan" for design.

The first step in the design of a member is for the engineer to gather and examine all general information, special requirements, and specifications relating to the member. This discussion will be confined to load-resisting members, although quite often these load-resisting members have special requirements placed on them. Among such requirements are those placed by architects for appearance purposes, by mechanical engineers for insulating purposes, by chemical engineers for environmental purposes, by electrical engineers for conduction purposes, and by aero-space engineers for minimum-weight purposes and others you can no doubt think of. Engineering load-resisting members occur in an almost infinite variety, from a test stand for a rocket with a multimillion-pound thrust to a sub-miniature mainspring in a fine watch.

After all information and possible special stipulations about the proposed design are known to the engineer, he should then proceed with the additional steps in his plan or design. These additional steps depend on the concepts of "failure" and the relation between working load and failure load. The concepts of failure will be discussed first before we take up a design procedure. These concepts are very important and are commonly misinterpreted. You should read the following explanations carefully.

## 7-2. Failure and Safety

Failure occurs when a member or structure ceases to perform the function for which it was designed. A member can cease to perform its function, or fail, for any one or more of a number of reasons. These "modes of failure" will be discussed in the next section. The word "failure" is often mistakenly used to mean fracture or break (separation). Fracture is a common and important type of failure, but every failure does not result in a fracture. Some failures do not even result in inelastic action or a permanent deformation of the member. It is possible for a member or structure to cease to perform its function because of excessive *elastic* deformation. As an example, a machine lathe that is made too flimsy (is not stiff enough) will not hold the close tolerances necessary for performing its function properly. This excessive elastic deformation may occur at a very low stress level. Thus you must remember that failure of a member is defined with reference to the function of the member, and not necessarily to its degree of destruction.

Since the primary function of an engineering member is to resist loads, we must relate failure to load; that is, failure will occur when the load reaches a value, called the failure load, at which the member ceases to perform its function. The failure load will be denoted by $P_f$. The margin of safety of a member depends on how close the working load $P_w$ is to the failure load. The working load, or design load, is the load that the member supports in normal use. The "factor of safety" is defined as the ratio of the failure load to the working load. If $N$ denotes the factor of safety

$$N = \frac{P_f}{P_w} \tag{7-1}$$

$$P_f = N P_w \tag{7-1a}$$

The factor of safety should be greater than unity. If it is unity, the working load equals the failure load, and there is no margin of safety.

Many handbooks and design books define the factor of safety as a ratio of stresses. This definition is valid only if the stresses can be directly related to the failure load; this can be done in many design problems. In some instances, however, the failure load cannot be directly related to any particular stress, and in such cases it is incorrect to apply the factor of safety to a stress. The best way to stay out of trouble in this respect is to base the factor of safety on the load.

The intelligent choice of a proper factor of safety is often difficult, to say the least. It requires a thorough and intimate knowledge of all aspects of the use and function of the member. Textbooks on design cover the selection of safety factors in detail. By measurements and theoretical calculations, an engineer can generally predict with good accuracy at what load a member will fail, but the decision as to how close to this failure load the member should operate must be based on many tangible, intangible, and interrelated factors. Often the factor of safety is increased because of uncertainties or ignorance. A few of the questions to be answered are:

1. What would an unpredictable failure cost in lives, in dollars, and in time?

2. How reliable are the material properties used in the calculations of the design?

3. How reliable is the material itself?

4. To what extent are the assumptions used in the design relationships valid?

5. Is the fabrication or manufacture exactly as specified in the design?

6. Can the extra weight of the structure resulting from a high factor of safety be tolerated?

7. Who will use the product or member—an experienced expert or an unskilled, careless person who is likely to misuse (overload) the member?

## 7-3. Modes of Failure and Design Criteria

A few of the more common modes of failure will now be discussed briefly.

**Fracture by Static Load.** When a fracture occurs in a *brittle* material, it is usually sudden and complete in nature and is likely to begin with a crack in an area of high stress concentration. However, failure by fracture under static loads is not so common in ductile materials as in brittle materials. In a ductile member, failure usually occurs as a result of excessive inelastic action which leads to very large over-all deformations long before fracture.

Breaking a piece of chalk by bending or twisting it between the fingers is an example of fracture of a brittle material. A ductile steel leader (the wire between the hook and the line) in a deep-sea fishing rig will fail by fracture when subjected to the pull imposed by a large fish, if it is designed only for smaller fishes. It does not cease to perform its function until fracture occurs. The same is true of a thin, ductile nonferrous blowout diaphragm in a safety valve.

**Fracture by Repeated Load.** Fracture caused by a repeated load is commonly referred to as a "fatigue" failure. Regardless of whether the material is brittle or ductile, no appreciable inelastic deformation is associated with this mode of failure. This mode is responsible for a large number of the failures in engineering. Such failures are often catastrophic in nature. The failure load for this mode cannot be predicted easily. The nature of the fatigue failures is discussed in more detail in Sec. 15-3. Let it suffice here to say that this type of fracture usually starts at a microscopic imperfection in a highly stressed area, and that the resulting crack progresses as the repetitions of loading are continued. Finally the crack grows to such an extent that one more application of the load results in a sudden and complete fracture of the member. Failure of an internal moving part of an engine or turbine would most likely be due to fatigue. Similarly, many aircraft structural failures can be traced to fatigue resulting from vibration.

**General Yielding.** When a structure or member fails by general yielding, it loses it ability to support the load. General yielding can occur only in a member or structure of a ductile material. It is necessary to distinguish general yielding from localized yielding, which often occurs in a ductile member at a point of stress concentration but is not widespread enough to significantly effect the over-all structural integrity of the member. General yielding must be widespread enough throughout one or more complete sections of the structure to allow the cumulative deformations to render the structure unfit for performing its function. These deformations usually lead to total collapse. This mode of failure is not associated with fracture, although in some cases of total collapse, localized ruptures may occur as secondary effects.

If you try to hang an extra-heavy overcoat on a common wire coat hanger, it ceases to perform its function because of general yielding. A heavy snow often causes a thin and poorly supported metal roof to collapse as the result of general yielding.

**Excessive Elastic Deformation.** At what stage elastic deformation becomes excessive depends, of course, on the function of the member. If the deformation is great enough to make the member cease to perform its function, the member can be said to have failed. For example, if you walk on a long, thin plank to get across a creek, and it deflects down into the water and lets you get your feet wet, then it has failed. It doesn't have to "break" to fail. In fact, in this case the deflection is elastic and the plank itself would not in any way be damaged. The flimsy lathe mentioned in the first paragraph of Sec. 7-2 is another example of failure due to excessive elastic deformation.

**Excessive Inelastic Deformation.** Failure due to excessive inelastic deformation is similar to the mode of failure just described except the deformations required to cause failure result in stresses beyond the elastic limit. This failure is distinguished from general yielding in that the *amount of deformation* is the quantity that prevents the member from functioning properly. On the other hand, in general yielding the member or structure loses its ability to support the load.

**Buckling.** Failure by buckling occurs when a member or structure becomes unstable. Failures resulting from instability usually occur rapidly and without warning, and such failures are generally catastrophic in nature.

Secondary effects may be general yielding and/or fracture. As an example of failure by buckling, roll a sheet of paper into a tube about an inch or two in diameter. Then put one flat end on the desk, and slowly apply a vertical load downward at the other flat end with the palm of your hand. The tube is surprisingly strong, but when it collapses, it does so with little or no warning. Can you think of any specific engineering member or structure that would probably buckle if its design load were drastically exceeded?

Many other types of failure may occur in various engineering structures, but these are usually peculiar to the specific structure. For example, a pressure vessel might spring a leak as a result of any one of many possible actions, and a leak would certainly constitute failure. However, in most structures, failure usually results from one or more of the six modes previously discussed.

## 7-4. Steps in Design Procedure

The procedures in the design of different types of structures vary widely. Following is a general step-by-step procedure for the design of engineering load-resisting members. It is by no means exhaustive or rigid. It must be tailored somewhat to each individual case. Some steps may not be applicable for a certain member. Also, although it is usually desirable to perform each step in the given order, it may sometimes be impossible to do so. Therefore, the following outline should be used only as a guide in the design of a particular member.

1. Determine the use and the special requirements *not directly related to load.* Some of the more important of these requirements are:
   *a*) Environment requirements
   *b*) Temperature requirements
   *c*) Special material requirements
   *d*) Expected life (total time and/or load repetitions)
   *e*) Cost requirements
   *f*) Finish requirements
   *g*) Size and weight limitations
   *h*) Effect of failure on human life

2. Determine or estimate the working load $P_w$ from all the various possible types of loads that the member may be required to withstand.
   Some of the possible types follow (see Chapter 5):
   *a*) Static
   *b*) Steady-state dynamic (frequency)
   *c*) Transient
   *d*) Impact (velocity)
   *e*) Shock
   *f*) Acoustical

It is necessary to consider all likely combinations of loads and, if possible, to determine the relationship between load and time.

   3. Determine the mode of failure of the member. Some possible modes are:

   *a*) General yielding (over-all inelastic behavior)
   *b*) Rupture or fracture
      1) Sudden—caused by static or dynamic load on brittle material
      2) Slow—caused by static load on ductile material
      3) Progressive—caused by repeated load (fatigue)
   *c*) Excessive deformation
      1) Elastic
      2) Inelastic
   *d*) Buckling
      1) Elastic
      2) Inelastic
   *e*) Creep (deformation under constant stress)
   *f*) Relaxation (changing stress with constant strain)
   *g*) Abrasion (wear)
   *h*) Corrosion

   4. Determine the quantity that is significant in the failure. Possible quantities are yield point, yield strength, ultimate strength, endurance limit, fatigue strength, some specific elastic stress, strain, deformation, buckling load, creep stress, energy absorbing ability (toughness), and stiffness.

   5. Obtain a reliable value which the failure quantity may have. This step is often difficult, but a suitable value can usually be determined by experimental tests of the material.

   6. Determine the factor of safety $N$ by analyzing the requirements in step 1 and all other pertinent information.

   7. Determine the failure load $P_f$ by multiplying the working load $P_w$ selected in step 2 by the factor of safety.

   8. Relate the failure load $P_f$ to the failure quantity selected in step 4 and the dimensions of the member. This is usually done by use of some type of load-stress or load-deformation relationship. A major portion of this course deals with the derivation and use of these relationships.

   9. After substituting suitable values in the relationship in step 8, carefully examine and interpret the results obtained to be certain that they are reasonable and compatible with the over-all structural requirements.

   Any design is practically always a compromise. Often a preliminary design must be made. This may be based on crude assumptions or only on the experience of the design engineer. The preliminary design is then refined and modified, sometimes many times. Modifications are often based on the results of an experimental stress analysis of a prototype design.

## PROBLEMS

**7-1.** What is the probable mode of structural failure of each of the following:
*a*) cable on a construction crane
*b*) pre-cast concrete bridge girder
*c*) construction gin pole
*d*) wood floor joist in a residence

**7-2.** What is the probable mode of structural failure of each of the following:
*a*) turbine blade in jet engine
*b*) aircraft wing spar
*c*) windshield in commercial aircraft
*d*) aircraft control cable

**7-3.** What is the probable mode of structural failure of each of the following:
*a*) connecting rod in auto engine
*b*) automobile bumper
*c*) automobile tire
*d*) automobile suspension torsion bar

**7-4.** What is the probable mode of structural failure of each of the following:
*a*) high pressure $O_2$ tank in spacecraft
*b*) aerosol spray can
*c*) electric motor-generator coupling shaft
*d*) plate glass window

**7-5.** What is the probable mode of structural failure of each of the following:
*a*) automobile fender
*b*) door key
*c*) lower limb prostheses (artificial leg)
*d*) dentures (false teeth)

**7-6.** What is the probable mode of structural failure of each of the following:
*a*) outboard motor drive shaft
*b*) astronauts EVA tether
*c*) boiler tube in steam plant
*d*) wheel bearing

**7-7.** What is the probable mode of structural failure of each of the following:
*a*) vaulting pole (fiberglas, metal)
*b*) golf club shaft
*c*) fishing rod
*d*) baseball bat

**7-8.** What is the probable mode of structural failure of each of the following:
*a*) paper clip
*b*) umbrella frame
*c*) guitar string
*d*) beer can

**7-9.** What is the probable mode of structural failure of each of the following:
*a*) sailboat mast
*b*) garden hose
*c*) watch balance spring
*d*) toothpick

**7-10 to 7-18.** Where possible or applicable, discuss design-procedure steps numbered 1, 2, 3, 4, and 6 for each of the load resisting members in Probs. 7-1 to 7-9.

## chapter 8

# Axially Loaded Members

## 8-1. Introduction

The primary function of an engineering member is to resist loads. In Chapter 6 we saw that a complex force system could be resolved into axial, shear, bending, and torsion components. In this and in the succeeding three chapters, we will analyze members subjected to component force systems and will gradually work into the more difficult analysis of members subjected to combined force systems.

This chapter deals primarily with members which transmit only axial loads, i.e., tensile or compressive forces parallel to the major axis of the member. The basic analysis covers both centroidal and noncentroidal loading, and also includes a treatment of thin-walled pressure vessels. The last section is devoted to a brief discussion of some simple connections.

As you study the succeeding sections, bear in mind that the primary purpose of this presentation is to illustrate the *procedure* for relating applied loads to resulting stresses and deformations.

## 8-2. Basic Load-Stress Relationship

Consider a member of *constant* cross section subjected to a resultant tensile force $P$ parallel to the $y$ axis, such as that in Fig. 8-1(a). By passing a cutting plane *mm* normal to the line of action of the resultant external load, we see that the only *resultant* internal reaction is a normal force equal to $P$ and having a line of action which is collinear with that of $P$. Equilib-

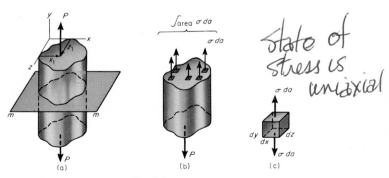

FIG. 8-1

**134**

rium also requires that there be no resultant shear force parallel to the cutting plane and no resultant twisting moment about a vertical axis. From these considerations we might then expect the state of stress in the member to be uniaxial [Fig. 8-1(*c*)], and the resultant reaction on plane *mm* to be composed of many minute cohesive normal forces which in turn produce normal stresses as indicated in Fig. 8-1(*b*). Hence, the equilibrium requirement for Fig. 8-1(*b*) gives

$$\overset{\uparrow+}{\sum} F_y = 0$$

$$\int_{\text{area}} \sigma \, da - P = 0$$

$$P = \int_{\text{area}} \sigma \, da \qquad (8\text{-}1)$$

where the integral $\int_{\text{area}} (\sigma \, da)$ represents the summation of all the infinitesimal cohesive forces $(\sigma \, da)$ over the cross-sectional area. Equation 8-1 represents the *basic* load-stress relationship for every axially loaded member.[1]

When Eq. 8-1 is to be applied to any axially loaded member, an expression for the stress $\sigma$ must be determined *before* the integral $\int_{\text{area}} (\sigma \, da)$ can be evaluated. The expression for stress is a mathematical function indicating how the stress is distributed throughout the body. In general, the stress distribution is dependent on two major factors.

*a*) The strain distribution in the member. This distribution indicates how the strain varies from point to point throughout the body. *In most cases, the strain distribution is governed by the geometry of the member and the location of the line of action of the applied load.*

*b*) The existing stress-strain relationship for the material of the member. This relationship can usually be obtained from the stress-strain diagram and a knowledge of the loading history of the member.

Thus, before any statement can be made about the distribution of the stress throughout a loaded member, two questions must be answered: 1) How is the strain distributed or what is the kinematic response of the member to the applied load? 2) How is the stress related to the strain? Throughout this and the succeeding chapters we will continually be asking ourselves these two vital questions.

EXAMPLE 8-1. The member shown in Fig. 8-2(*a*) is axially loaded. After loading, the strain does not *vary* in the *z* and *x* directions. After the load *P* was applied, the strain varied linearly in the *y* direction from 0.0030 in./in. tensile along *AA* to

[1]Slender compression members will not be considered in this chapter because of their buckling mode of failure. This mode was mentioned in Chapter 7 and will be considered in more detail in Chapter 11.

0.0010 in./in. compressive along $BB$, as in Fig. 8-2($b$). If the material has the stress-strain curve (tensile and compressive) of Fig. 8-2($c$), determine the magnitude and vertical location of $P$.

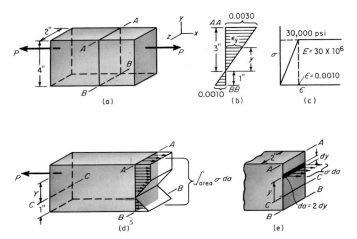

FIG. 8-2

*Solution:* From the stress-strain curve we see that as long as the axial strain is equal to or less than 0.0010 in./in., the stress will be directly proportional to the strain. For strains larger than 0.0010, the stress has a constant value of 30,000 psi. Hence, the stress distribution is shown in Fig. 8-2($d$), where the fibers below $CC$ are in compression and those above $CC$ are in tension. From Fig. 8-2($d$) and ($e$),

$$\overset{+}{\underset{}{\Sigma}} F_x = 0$$

$$+P - \int_{area} \sigma \, da = 0$$

$$P = \int_{-1}^{+3} \sigma \, 2 \, dy$$

To evaluate the integral, $\sigma$ must be expressed as a function of $y$. Because of the change in distribution which occurs at $y = +1$, the integral must be handled in two parts. Thus,

$$P = \int_{-1}^{+3} \sigma \, 2 \, dy = \int_{-1}^{+1} \sigma \, 2 \, dy + \int_{+1}^{+3} \sigma \, 2 \, dy$$

From Fig. 8-2($d$) and ($e$), we have

For $-1 \leq y \leq +1$      $\dfrac{\sigma}{30,000} = \dfrac{y}{1}$    or    $\sigma = 30,000 \, y$

For $+1 \leq y \leq +3$      $\sigma = 30,000$

Hence,

$$P = \int_{-1}^{+1} (30{,}000\,y)2\,dy + \int_{+1}^{+3} (30{,}000)2\,dy$$

$$= 60{,}000\,\frac{y^2}{2}\,\bigg]_{-1}^{+1} + 60{,}000\,y\,\bigg]_{+1}^{+3}$$

$$= 30{,}000 - 30{,}000 + 180{,}000 - 60{,}000 = 120{,}000\ \text{lb}$$

To determine the vertical location of the line of action of $P$, we must satisfy rotational equilibrium. For this purpose a moment equation with respect to $CC$ will be convenient. If the distance from $CC$ to the line of action of $P$ is denoted by $Y$,

$$\overset{\curvearrowright}{\underset{+}{}}\ \sum M_{CC} = 0$$

$$-PY + \int_{-1}^{+3} y\,(\sigma\,da) = 0$$

$$PY = \int_{-1}^{+3} y\,\sigma 2\,dy$$

Again, two integrals will be necessary. Using the expressions for stress previously obtained, we have

$$PY = \int_{-1}^{+1} y\,(30{,}000\,y)2\,dy + \int_{+1}^{+3} y\,(30{,}000)2\,dy$$

$$= 60{,}000\,\frac{y^3}{3}\,\bigg]_{-1}^{+1} + 60{,}000\,\frac{y^2}{2}\,\bigg]_{+1}^{+3}$$

$$= 20{,}000 + 20{,}000 + 270{,}000 - 30{,}000 = 280{,}000\ \text{in-lb}$$

Hence, the distance from $CC$ to $P$ is

$$Y = \frac{280{,}000}{120{,}000} = 2.33\ \text{in.}$$

and the distance from the bottom is 3.33 in.

*Alternate Solution:* The stress distribution of Fig. 8-2($d$) may be interpreted merely as a distributed load or, more conveniently, as three distributed loads, as shown in Fig. 8-3. Thus,

$$\overset{+}{\underset{\leftarrow}{}}\ \sum F_x = 0$$

$$P - 120{,}000 - 30{,}000 + 30{,}000 = 0$$

$$P = 120{,}000$$

Also,

$$\overset{\curvearrowleft}{\underset{+}{}}\ \sum M_{CC} = 0$$

$$PY - 120{,}000\,(2) - 30{,}000(2/3) - 30{,}000\,(2/3) = 0$$

$$Y = 2.33\ \text{in.}$$

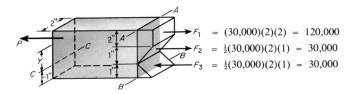

$$F_1 = (30,000)(2)(2) = 120,000$$
$$F_2 = \tfrac{1}{2}(30,000)(2)(1) = 30,000$$
$$F_3 = \tfrac{1}{2}(30,000)(2)(1) = 30,000$$

FIG. 8-3

## 8-3. Centroidal Loading

In a great many engineering problems the primary design criterion is the maximum normal stress produced by an axial load. The engineer is often called upon to determine the proper dimensions and material of an axially loaded member that is to resist a specified load, or to investigate the stress in a particular member for comparison with the critical stress for the particular material and use of the member. In either case he must make use of the relationship between load and stress. In order to determine the specific load-stress relationship for pure tension or compression, let us examine Eq. 8-1 more closely.

Figure 8-4(a) is a graphical representation of the integral $\int_{\text{area}} \sigma \, da$, where $\sigma$ varies over the area $a$. In this figure we can interpret the in-

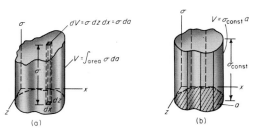

FIG. 8-4

tegral $\int_{\text{area}} \sigma \, da$ as a volume having a variable height represented by $\sigma$ and a cross sectional area in the x-z plane. Also, we can see that, for any desired maximum stress, the maximum obtainable volume would be one with uniform height $\sigma_{\text{const}}$, as in Fig. 8-4(b). Thus,

$$\text{Maximum volume} = \sigma_{\text{const}} \times \text{area}$$

Since the volume represents the integral $\int_{\text{area}} \sigma \, da$, which in turn is equal to the axial load $P$ (from Eq. 8-1), the *maximum* axial load for any desired maximum stress is obtained when the stress is uniform over the cross-sectional area. Hence,

$$P_{max} = \int_{area} \sigma_{const} \, da = \sigma_{const} \int_{area} da$$

$$P_{max} = \sigma_{const} \, a \tag{8-2}$$

When the normal stress in an axially loaded member is *uniformly* distributed over its cross-sectional area, the member is said to be in "pure" tension or "pure" compression, the nature of the stress depending on the type of the applied load. However, before we plunge ahead and apply Eq. 8-2 to engineering members, let us first investigate the conditions necessary to produce a uniform stress distribution.

**Axially Loaded Straight Members.** Let us first determine the location of the line of action of the externally applied force $P$ which is necessary to produce a uniform stress distribution. We first write an equilibrium moment equation for the free-body diagram in Fig. 8-5($c$) by taking the moments of the forces with respect to the $x$ axis.

$$\sum M_x = 0$$

$$P z_1 - \int_{area} (\sigma \, da) z = 0$$

$$P z_1 = \int_{area} (\sigma \, da) z$$

But for pure tension $\sigma$ is uniform over the cross-sectional area and $P = \sigma a$. So

$$(\sigma a) z_1 = \sigma \int_{area} z \, da$$

$$z_1 = \frac{\int z \, da}{a} = \bar{z} \tag{8-3}$$

Similarly,

$$x_1 = \bar{x} \tag{8-3a}$$

Here $\bar{x}$ and $\bar{z}$ are the $x$ and $z$ coordinates of the centroid of the cross section. These results show that for uniform stress distribution *the line of action of the resultant externally applied force must be coincident with the longitudinal centroidal axis of the member;* that is, the member must undergo centroidal loading. This statement is true whether the external load is considered concentrated or distributed.

It was pointed out in Chapters 3 and 4 that most engineering materials can be considered to be *homogeneous* and *isotropic*; so let us consider the material of the body in Fig. 8-5 to be homogeneous and isotropic. In Fig. 8-5(*a*) two parallel planes *aa* and *bb* are passed through an unloaded body so that they are perpendicular to the *y* axis and are separated by a distance $L_{ab}$. Let us assume now that the body is loaded centroidally so that the stress is uniform across section *aa*, section *bb*, and every parallel section between *aa* and *bb*, as indicated in Fig. 8-5(*b*). Thus, since the cross section is constant, the stress is uniform throughout every longitudinal fiber between sections *aa* and *bb*. The state of stress is uniaxial, and the corresponding strain in the *y* direction can be obtained directly from the appropriate stress-strain curve of the material.

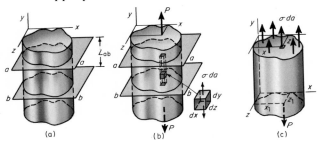

FIG. 8-5

Since the material is homogeneous, the *uniform* stress in the region between sections *aa* and *bb* must accompany a *uniform* strain throughout the length of each fiber. Hence, the elongation of a single fiber is given by

$$\frac{de_y}{dy} = \epsilon_y = \text{constant} \qquad e_y = \int_{\substack{\text{over} \\ L_{ab}}} \epsilon_y \, dy = \epsilon_y L_{ab} \qquad (8\text{-}4)$$

Since all fibers have the same strain and the same original length, all fibers will undergo the same deformation. *Thus sections aa and bb must remain parallel, if the member is to have a uniform stress distribution.* A more general statement is that *parallel cross sections must remain parallel.* This condition must be satisfied for any length, such as $L_{ab}$, of an axially loaded *homogeneous* member if it is to have uniform stress distribution within that length. The condition applies regardless of whether the stress is above or below the proportional limit of the material.

It should be mentioned that the requirement for uniform stress distribution in a homogeneous axially loaded straight member—*parallel sections must remain parallel*—generally will not be satisfied in the immediate vicinity of the applied external load. However, if the longitudinal dimension of the body is relatively large in comparison with the transverse dimensions, the localized effects near the applied load disappear in sections sufficiently distant from its point of application.[2]

[2]A French mathematician, Saint Venant, first showed that localized effects diminish rapidly as you consider sections further and further away from such locations. This is a topic usually covered in Theory of Elasticity.

Although the preceding analysis was concerned with a homogeneous member of constant cross section, Eq. 8-2 has been found to be applicable for homogeneous symmetrical members of gradually varying cross section. However, any abrupt change in cross section will produce a nonuniform strain distribution and parallel cross sections will not remain parallel. Figure 8-6 illustrates how the line of action of the load and the geometry of the member affect the strain distribution.

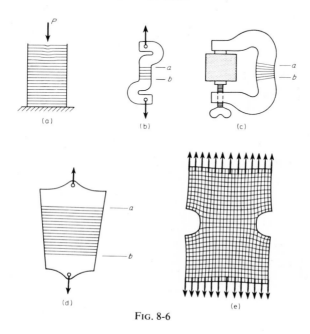

FIG. 8-6

In Fig. 8-6(*a*) the member is centrally loaded, straight, and has a uniform cross section. The strain is uniform, except for localized effects near the point of application of the load. Although the member in (*b*) is curved, it is straight and of uniform cross section in the length *ab*. Also, the line of action of the load coincides with the longitudinal centroidal axis in this length. So the load produces uniform strain in the length *ab*. In (*c*) the loading on the length *ab* is eccentric, and the strain distribution is nonuniform.

The cross section of the member in Fig. 8-6(*d*) varies *gradually*, but the line of action of the load coincides with the longitudinal centroidal axis of the member. Although the stress varies from cross section to cross section in the length *ab*, there is essentially uniform stress across any one particular cross section. There is an abrupt change in the cross section of the member in Fig. 8-6(*e*) at the notch. As a result, the strain distribution over the cross section is not uniform in the vicinity of the notch.

The strain distribution and the accompanying stresses produced by eccentric loading and bending will be studied in Chapter 10. Localized nonuniform strain distribution resulting from abrupt changes in geometry are called *strain concentrations*. Such *strain concentrations* are often very significant in determining the load-carrying capacity of structural members, particularly for certain types of materials and loading conditions. The importance of strain concentrations and their accompanying *stress concentrations* as a design consideration is discussed in Chapter 15.

EXAMPLE 8-2. The structural-steel link in Fig. 8-7(*a*) will continue to perform its function regardless of any dimension changes it may undergo. Its ends are overdesigned so that they do not influence failure of the link. What should be the diameter of the circular cross section of the link, if a factor of safety of 3 is recommended?

*Solution.* Referring to Sec. 7-4 for design procedure, the following facts can be stated:

*a*) The mode of failure is rupture, since only then will the member cease to perform its function.

*b*) The significant quantity in failure is the ultimate stress, which must be reached before rupture.

*c*) The average ultimate tensile stress for structural steel is 72,000 psi (see Appendix).

*d*) The factor of safety is given as 3. So

$$P_{\text{failure}} = P_w N = 3 P_w$$

*e*) The applicable relationship between stress, load, and dimensions is

$$P_{\text{failure}} = \sigma_{\text{failure}} a \qquad \text{(a)}$$

This equation is applicable since the link can be assumed to be homogeneous and centroidally loaded and thus has a uniform stress distribution between the overdesigned ends.

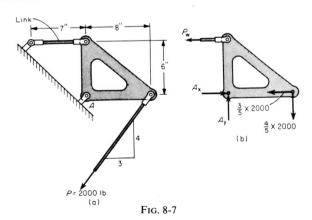

FIG. 8-7

One of the equilibrium requirements of the free-body diagram in Fig. 8-7(b) gives

$$\overset{\curvearrowright}{+} \\ \sum M_A = 0$$

$$+\left(\frac{4}{5} \times 2000\right)(8) - (P_w)(6) = 0$$

$$P_w = 2133 \text{ lb}$$

$$P_{\text{failure}} = (2133)(3) = 6399 \text{ lb}$$

From equation (a),

$$P_{\text{failure}} = \sigma_{\text{failure}} a$$

$$6399 = (72,000)(\pi d^2/4)$$

$$d = 0.336 \text{ in.}$$

The closest standard diameter larger than this is ⅜ in., which would probably be used. You might get by with ⁵⁄₁₆ in. if the specifications would allow a small sacrifice in the factor of safety.

## 8-4. Thin-Walled Cylindrical Pressure Vessels

Consider a section of a thin-walled pressure vessel carrying a fluid of negligible weight under a *gage* pressure $p$, as shown in Fig. 8-8(a). An end view of the unpressurized vessel with two radial planes $OA$ and $OD$ is shown in Fig. 8-8(b). Figure 8-8(c) illustrates the deformation of an infinitesimal section of the wall of the cylinder caused by the pressure $p$. It is assumed that the radial planes remain radial and that the thickness of the wall does not change due to the pressure. From this figure the circumferential strains in the inside and outside fibers are

$$\epsilon_{\text{cir}} = \frac{R_i d\theta - r_i d\theta}{r_i d\theta} = \frac{R_i - r_i}{r_i} \quad \text{(inside)} \tag{a}$$

$$\epsilon_{\text{cir}} = \frac{R_o d\theta - r_o d\theta}{r_o d\theta} = \frac{R_o - r_o}{r_o} \quad \text{(outside)} \tag{b}$$

Now if

$$R_o = R_i + t \quad \text{and} \quad r_o = r_i + t$$

Eq. (b) becomes

$$\epsilon_{\text{cir}} = \frac{R_i + t - (r_i + t)}{r_i + t} = \frac{R_i - r_i}{r_i + t} \approx \frac{R_i - r_i}{r_i}$$

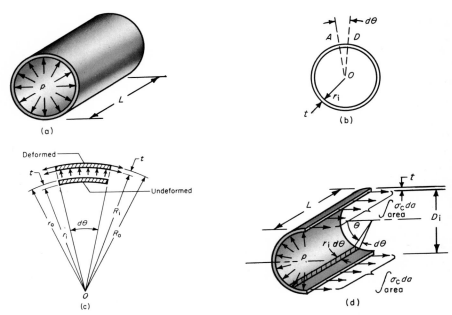

FIG. 8-8

*if t is small in comparison with $r_i$.* Hence, for these conditions the outside and inside circumferential strains are the same, and we assume that the strain is uniform throughout the thickness of the wall.

This assumption will be reasonably good if the wall thickness is not more than about $\frac{1}{10}$ of the radius. Such vessels are classified here as "thin-walled." A skin-diver's air tank would be considered a thin-walled vessel. A rifle barrel would be thick-walled, and there would be a nonuniform circumferential strain through the wall.[3]

If the circumferential strain is uniform and the material is homogeneous, the circumferential stress $\sigma_c$ in the wall of the pressure vessel will be uniform. An analysis of the free body in Fig. 8-8(d) will result in the desired relationship between gage pressure and circumferential stress. Thus,

$$\overset{+}{\underset{\rightharpoonup}{}}\ \sum F = 0$$

$$\int_{-\pi/2}^{+\pi/2} (pLr_i\,d\theta)\cos\theta - 2\int_{\text{area}} \sigma_c\,da = 0$$

[3]For an analysis of thick-walled cylinders see *Advanced Mechanics of Materials,* by Seely and Smith, John Wiley & Sons, Inc., Chapter 10, 1952.

$$pLr_i \sin \theta \left.\right]_{-\pi/2}^{+\pi/2} - 2\sigma_c Lt = 0$$

$$\sigma_c = \frac{2pLr_i}{2Lt} = \frac{pD_i}{2t} \tag{8-5}$$

The circumferential stress $\sigma_c$ is produced by the internal *gage* pressure only, and this pressure will produce no shear stress along a longitudinal section of the vessel. A circumferential stress is commonly called a *hoop* stress.

If a vessel has, or could be considered to have, closed ends and contains a fluid under a gage pressure $p$, as in Fig. 8-9(a), the wall of the vessel will have a longitudinal stress as well as a circumferential stress.

FIG. 8-9

If the deformation of the vessel does not alter its basic geometry, that is, if the vessel does not "bulge," then the longitudinal stress $\sigma_L$ in the wall of the vessel will be uniform, since parallel transverse sections would remain parallel. From Fig. 8-9(b)

$$\overset{+}{\overrightarrow{\sum F}} = 0$$

$$(p)\left(\frac{\pi D_i^2}{4}\right) - \int_{area} \sigma_L \, da = 0$$

$$\frac{p\pi D_i^2}{4} = \sigma_L a = \sigma_L (\pi D_{avg} t)$$

Since $D_{avg} \approx D_i$ when $t$ is very small in comparison with $D_i$,

$$\sigma_L = \frac{pD_i}{4t} \tag{8-6}[4]$$

Thus, the longitudinal stress produced in the wall of a vessel by an internal pressure is equal to one-half of the circumferential stress due to the same pressure. From this analysis we see that a biaxial state of stress exists

[4]If $D_i$ were replaced by $D_{avg}$ in Eqs. 8-5 and 8-6, the results would agree more closely with those obtained from the Theory of Elasticity.

in the wall of a thin-walled pressure vessel with closed ends. Actually there also exists a third normal stress in the wall called the radial stress which is due to the compressive effects of the pressure on the internal surface of the vessel. However, the radial stresses are usually neglected in the analysis of a thin-walled vessel.

Caution: *Circumferential stresses* exist on *longitudinal sections* [see Fig. 8-8(*d*)]. *Longitudinal* stresses exist on transverse sections [see Fig. 8-9(*b*)]. Do not confuse the two.

EXAMPLE 8-3. A cylindrical rocket motor case represented in Fig. 8-10(*a*) is made from a sheet of special steel 0.080 in. thick. The case is 70 in. in diameter and 8 ft long. It is subjected to 125,000 lb of tensile longitudinal thrust, in addition to internal pressure. What is the operating pressure, if the tensile stress in the wall must not exceed 210,000 psi?

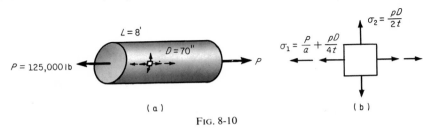

(a)                                                     (b)

FIG. 8-10

*Solution:* The stresses on an element along the longitudinal and transverse planes are shown in Fig. 8-10(*b*). No shear acts on these planes. The longitudinal stress $\sigma_1$ is the sum of the stress due to internal pressure and that due to the thrust. If $a$ represents the area of the wall in a transverse section, and $p$ is the internal pressure, the stress $\sigma_1$ in pounds per square inch is

$$\sigma_1 = +\frac{P}{a} + \frac{pD}{4t} = \frac{125,000}{\pi 70(0.080)} + \frac{p(70)}{4(0.080)}$$

$$= 7240 + 218.8p$$

The circumferential stress $\sigma_2$ is due to internal pressure alone, since the thrust does not cause any stress in this direction. This stress in pounds per square inch is

$$\sigma_2 = \frac{pD}{2t} = \frac{p(70)}{2(0.080)} = 437.5p$$

It is not obvious which is the larger stress. So each will be equated to the allowable stress to determine the allowable pressure. For $\sigma_1$,

$$218.8p + 7240 = 210,000$$

$$p = 927 \text{ psi}$$

For $\sigma_2$,

$$437.5p = 210,000$$

$$p = 480 \text{ psi}$$

The *smaller* of the two results, or 480 psi, is the maximum allowable pressure. If it is exceeded, the circumferential tensile stress on any longitudinal section will exceed 210,000 psi.

## 8-5. Deformation of Members in Pure Tension or Compression

In Chapter 2 we defined axial strain for an infinitesimal element as deformation per unit length. Thus,

$$\epsilon = de/dL \qquad (8\text{-}7)$$

where $dL$ is an infinitesimal length in the direction of the strain $\epsilon$, and $de$ is the elongation or contraction of that length. Consider the axially loaded member of Fig. 8-11. From Eq. 8-7 the axial deformation of the infinitesimal element becomes

$$de = \epsilon\, dL \qquad (8\text{-}8)$$

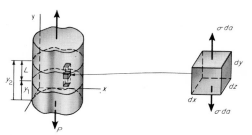

FIG. 8-11

Suppose that we desire to determine the deformation along the $y$ axis of a single fiber having a length $L$ and a cross-sectional area $dx\, dz$. This finite deformation can be obtained by summing the infinitesimal deformations of all the infinitesimal elements throughout the length $L$ of the fiber. From Eq. 8-8 the summation becomes

$$e_y = \int de_y = \int_{\text{over } L} \epsilon_y\, dy = \int_{y_1}^{y_2} \epsilon_y\, dy \qquad (8\text{-}9)$$

Equation 8-9 represents the deformation in the $y$ direction for a single fiber. In order to obtain the deformation of a particular fiber, we must know how its strain $\epsilon_y$ varies over the length $L$; that is, we must express the strain $\epsilon_y$ as a function of $y$. Once this strain function has been obtained, the integration process of Eq. 8-9 can be completed to obtain the desired deformation. The same basic procedure of summation of infinitesimal deformations can also be employed in determining the deformation of fibers in the $x$ and $z$ direction. Thus,

$$e_x = \int_{\text{over } L} \epsilon_x\, dx \qquad (8\text{-}9\text{a})$$

$$e_z = \int_{over\ L} \epsilon_z dz \tag{8-9b}$$

Since our primary objective is to relate *load* to deformation, the strain is usually expressed in terms of the applied load. This can be done if the load-stress and stress-strain relationships are known. The only specific load-stress relationship we have derived so far is that for the case of pure tension or pure compression. Therefore, our present applications will be limited to this case.

EXAMPLE 8-4. In Fig. 8-12(*a*) are given the dimensions of a homogeneous right circular cone resting on its base. The specific weight of the material is *w* lb/cu in. Assuming that the material acts elastically and that the height *h* is much greater than the diameter *b*, determine the longitudinal deformation of the axis of the cone due to its own weight.

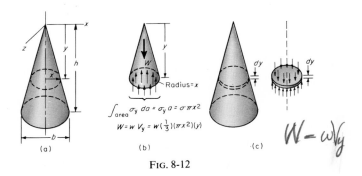

FIG. 8-12

*Solution:* Figure 8-12(*b*) shows a free-body diagram for the part of the cone above a section at a distance *y* from the apex. Since the cone is homogeneous, the loading is centroidal. Let $V_y$ denote the volume of the upper section of the cone, and $\sigma_y$ denote the stress at *y*. Here, $\sigma_y$ is assumed to be constant over the entire area of the section. Therefore from Fig. 8-12(*b*), we have

$$\overset{\downarrow\ +}{\sum F_y} = 0$$

$$wV_y - \int_{area} \sigma_y da = 0$$

$$wV_y - \sigma_y a = 0$$

Hence, the load-stress relationship is

$$\sigma_y = \frac{wV_y}{a} = \frac{w(\frac{1}{3})\pi x^2 y}{\pi x^2} = \frac{wy}{3}$$

Since the material acts elastically and the loading is uniaxial, the load-strain relationship is

$$\epsilon_y = \frac{\sigma_y}{E} = \frac{wy}{3E}$$

Then the load-deformation relationship becomes

$$e_y = \int_0^h \epsilon_y \, dy = \int_0^h \left(\frac{wy}{3E}\right) dy = \frac{w}{3E} \int_0^h y \, dy = \frac{wh^2}{6E}$$

where the deformation is in inches.

A quick dimensional check of units follows:

$$e_y = \left(\frac{\text{lb}}{\text{in.}^3}\right) \left(\frac{1}{\dfrac{\text{lb}}{\text{in.}^2}}\right) (\text{in.}^2) = \text{in.}$$

EXAMPLE 8-5. The dimensions in Fig. 8-13 are those before the loads were applied. In which direction and through what distance in inches will the plate on

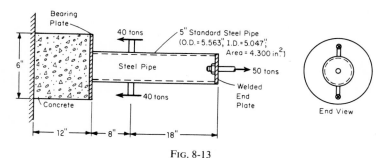

FIG. 8-13

the right end of the pipe move upon application of the loads? Neglect localized stress and strain concentrations and the weight of the structure. The compressive stress-strain curves for the steel and concrete are shown in Fig. 8-14(*a*).

*Solution:* From Eq. 8-9a,

$$e_x = \int_{\substack{\text{over} \\ \text{length}}} \epsilon_x \, dx$$

Since the strain in a member depends on the load, dimensions, and material, the strain $\epsilon_x$ will not be constant over the entire length of the structure. Therefore, it will be necessary to determine the strain in the concrete, in the left-hand portion of the steel pipe, and in the right-hand portion of the pipe separately. Thus,

$$e_{\text{total}} = \int_{\text{concrete}} \epsilon_c \, dx_c + \int_{\substack{\text{left} \\ \text{steel}}} \epsilon_{sL} \, dx_{sL} + \int_{\substack{\text{right} \\ \text{steel}}} \epsilon_{sR} \, dx_{sR} \qquad (a)$$

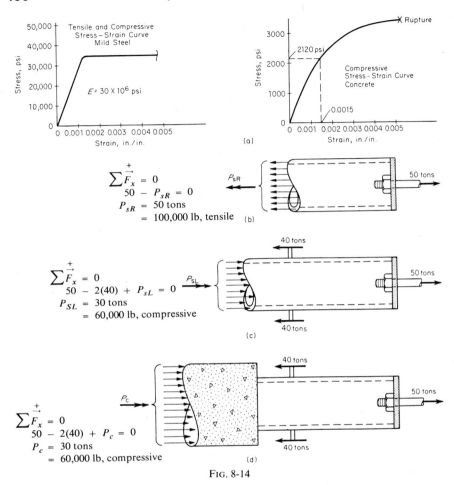

FIG. 8-14

Each individual section being considered meets the necessary conditions of loading and geometry for uniform stress and strain. So equation (a) becomes

$$e_{\text{total}} = \epsilon_c \int_0^{12} dx_c + \epsilon_{sL} \int_0^{8} dx_{sL} + \epsilon_{sR} \int_0^{18} dx_{sR}$$

$$= \epsilon_c(12) + \epsilon_{sL}(8) + \epsilon_{sR}(18) \qquad\qquad (b)$$

By static analyses of the free-body diagrams in Fig. 8-14(b), (c), and (d), the loads $P_c$, $P_{sL}$, and $P_{sR}$ are determined as shown. The stresses in the individual sections are now found as follows:

$$\sigma_c = \frac{P_c}{a_c} = \frac{60,000}{(\pi/4)(6^2)} = 2120 \text{ psi (compression)}$$

$$\sigma_{sL} = \frac{P_{sL}}{a_{sL}} = \frac{60,000}{4.300} = 13,950 \text{ psi (compression)}$$

$$\sigma_{sR} = \frac{P_{sR}}{a_{sR}} = \frac{100,000}{4.300} = 23,250 \text{ psi (tension)}$$

The strain in the concrete can be obtained directly from the stress-strain curve. Since the stresses in the steel are below the proportional limit, the strains in the steel can be obtained by dividing the stresses by the modulus of elasticity. The strains are as follows:

$$\epsilon_c = 0.0015 \text{ in./in. (compression)}$$

$$\epsilon_{sL} = 0.000465 \text{ in./in. (compression)}$$

$$\epsilon_{sR} = 0.000775 \text{ in./in. (tension)}$$

Substituting these values in equation (b), we obtain

$$e_{total} = -(0.0015)(12) - (0.000465)(8) + (0.000775)(18)$$

$$= -0.0180 - 0.00372 + 0.01392$$

$$= -0.00780 \text{ in.}$$

The end plate moves to the left, since the net deformation is compressive.

## 8-6. Statically Indeterminate Axially Loaded Members

In each of the previous example problems in this chapter, it was possible to determine the loads and stresses in the members by utilizing equations of equilibrium, i.e., the members were statically determinate. However, there are numerous structures for which the equilibrium equations *alone* are not sufficient to determine the loads. Such structures are *statically indeterminate*. In order to solve for the loads and stresses in members of such structures, it becomes necessary to supplement the equilibrium equations with additional relationships based on any conditions of restraint which may exist. These conditions of restraint usually govern the geometry of the deformation of the member.

Throughout Part III of this book, various types of indeterminate members and structures will be considered. The following examples merely illustrate some statically indeterminate structures which are loaded in pure tension or compression. The basic approach used in the solutions of these problems can be employed in the succeeding chapters where other types of loaded structures are considered.

EXAMPLE 8-6. A very stiff bar of negligible weight is suspended horizontally by two vertical rods, as in Fig. 8-15(a). One of the rods is of steel, and is $\frac{1}{2}$ in. in diameter and 4 ft long; the other is of brass and is $\frac{7}{8}$ in. in diameter and 8 ft long.

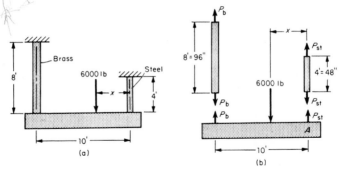

Fig. 8-15

If a vertical load of 6000 lb is applied to the bar, where must it be placed in order that the bar will remain horizontal? Assume that the rods behave elastically and that the bending of the bar is insignificant. Also, take $E_{st}$ as $30 \times 10^6$ psi and $E_b$ as $14 \times 10^6$ psi.

*Solution:* Two independent equations of static equilibrium may be written for the free-body diagram of Fig. 8-15(b). Two possible equations are

$$\overset{\uparrow+}{\sum F_y} = 0 \qquad \overset{\curvearrowright}{\underset{+}{\sum}} M_A = 0$$

$$P_{st} + P_b - 6000 = 0 \qquad P_b(10) - 6000x = 0$$

Since no more *independent* equations of equilibrium can be written and there are three unknown quantities, the structure is statically indeterminate. One additional independent equation is needed. The problem requires that the bar remain horizontal. Therefore, the rods must undergo equal elongations, and we have

$$e_b = e_{st}$$

$$\int_0^{96} \epsilon_b \, dL_b = \int_0^{48} \epsilon_{st} \, dL_{st}$$

If the strains are uniform in each rod and the rods act elastically, the last equation becomes

$$\epsilon_b (96) = \epsilon_{st} (48)$$

$$\frac{\sigma_b}{E_b} (96) = \frac{\sigma_{st}}{E_{st}} (48)$$

$$\frac{P_b}{a_b} \frac{1}{E_b} (96) = \frac{P_{st}}{a_{st}} \frac{1}{E_{st}} (48) \qquad (a)$$

Equation (a) is an *independent* equation relating the load in the brass to the load in the steel. When this equation is solved simultaneously with the equilibrium equations, the results are as follows:

$$P_b = 2510 \text{ lb} \qquad P_{st} = 3490 \text{ lb} \qquad x = 4.175 \text{ ft}$$

EXAMPLE 8-7. The aluminum rod in Fig. 8-16(a) is firmly welded to the ceiling and to the 150-lb block. The rod has no load at the temperature at which it is installed, and the entire weight of the block is carried by the floor. a) If the temperature of the rod is decreased 30° F at a uniform rate, what force will be exerted on the floor by the block? b) How much must the temperature decrease to cause the rod to pick up the 150-lb block so that it will be 0.10 in. above the floor? Neglect the weight of the rod.

*Solution:* a) Figure 8-16(b) shows the one available equilibrium equation when the temperature is decreased. Since this equation has two unknowns, $P_f$ and $P_r$, an additional relationship is necessary. A deformation relationship can best be visualized and derived by assuming that the rod is *detached* at the top and that free thermal contraction occurs, as indicated in Fig. 8-16(c). The shortening $e_t$ in the rod would then be as follows (see Appendix for the coefficient of thermal expansion):

$$e_t = \alpha(L)(\Delta t) = 13 \times 10^{-6}(20)(12)(-30) = -0.0936 \text{ in.}$$

We can now determine the load $P_r$ in the rod that would be necessary to pull the top of the rod up 0.0936 in. to where it is actually fastened. Thus,

$$e_t = \epsilon L = \left(\frac{\sigma}{E}\right)L = \frac{P_r L}{aE} = \frac{P_r(240)}{(0.02)\,10^7} = 0.0936 \text{ in.}$$

$$P_r = 77.8 \text{ lb}$$

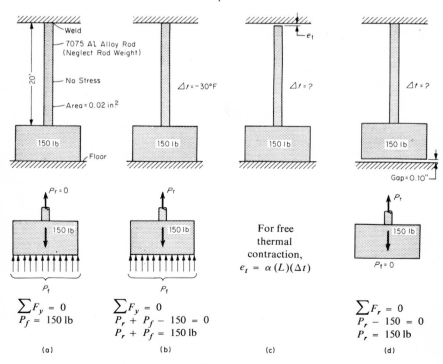

$$\sum F_y = 0$$
$$P_f = 150 \text{ lb}$$

(a)

$$\sum F_y = 0$$
$$P_r + P_f - 150 = 0$$
$$P_r + P_f = 150 \text{ lb}$$

(b)

For free thermal contraction,
$$e_t = \alpha(L)(\Delta t)$$

(c)

$$\sum F_r = 0$$
$$P_r - 150 = 0$$
$$P_r = 150 \text{ lb}$$

(d)

FIG. 8-16

The magnitude of $P_r$ is the amount by which the original load on the floor is reduced.  Substituting this load back in the equilibrium equation in Fig. 8-16($b$), we get

$$P_f = 150 - 77.8 = 72.2 \text{ lb}$$

*b*) If the rod undergoes sufficient contraction to raise the weight, the rod will then support the entire 150 lb, as shown in Fig. 8-16($d$).  Hence, the rod must first contract an amount $e_w$ to remove the weight from the floor, and then contract an additional 0.10 in.  The total contraction becomes

$$e_t = - e_w + (-0.10)$$

$$\alpha(L)(\Delta t) = - \frac{P_r L}{aE} + (-0.10)$$

$$13 (10^{-6})(240)(\Delta t) = - \frac{150(240)}{0.02(10^7)} - 0.10$$

$$\Delta t = -89.8°$$

It is very important that you be able to physically visualize all the deformation relationships.

## 8-7.  Simplified Shear and Bearing Stresses

In many engineering structures, loads are transmitted from one member to another member by means of connectors, such as bolts, pins, rivets, welds, and glue.  In most cases an analysis of the connecting joint would reveal that the load is transmitted *from* the first member *to* the connector over a contact surface or bearing surface, *through* the connector parallel to a shearing surface, and *to* the second member over a second contact surface or bearing surface.  Figure 8-17($a$) illustrates the critical tensile, shear, and bearing areas in a single riveted joint, as well as the path of the transmitted force through the joint.  Figure 8-17($b$) shows the important areas for another simple connection.

The force which a connector transmits may become large enough to cause the connecting joint to fail.  Failure of a joint may be due to the shear force in the connector, the bearing force between the connector and a member, and/or the tensile or shear force in a connected member.  The shear force in the connector could produce extensive distortion or even fracture of the connector, whereas the bearing force could cause either the connector or the member to crush at the area of contact.  The tensile or shear force in a connected member could result in a "tearing out" of the pin, rivet, or bolt.

The failure of a connecting joint can usually be related directly to a nominal or average force intensity.  The average intensity of the bearing

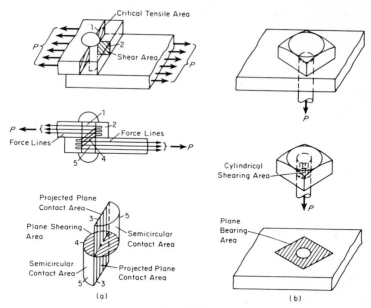

FIG. 8-17

force is obtained by dividing the bearing force by the projected contact area. The average intensity of the shear force in the connector is obtained by dividing the shear force in the connector by the cross-sectional area of the connector taken parallel to the shear force. The average intensity of the tensile force across the critical (minimum) cross section of a connected member is obtained by dividing the tensile force by the minimum tensile area. These average force intensities are often called simplified or average shear, bearing, and tensile stresses, although they are *not* actual stresses according to our definition. The actual shear, bearing, and tensile stresses vary over the corresponding areas. Some of the stresses are larger than the value of the average force intensity, while others are smaller. However, it has been found through experience that reliable designs can be based on the value of the average force intensity. This is particularly true for a connecting joint made of ductile material subjected to static loads, since a ductile material has the ability to deform to such an extent that the load becomes almost uniformly distributed across the shear, bearing, and tensile areas. We close this chapter with an illustration of a typical connector-type problem.

EXAMPLE 8-8. In the pin connected structure shown in Fig. 8-18, determine the largest bearing stress on the pin *B* and the shearing stress in the pin *B*.

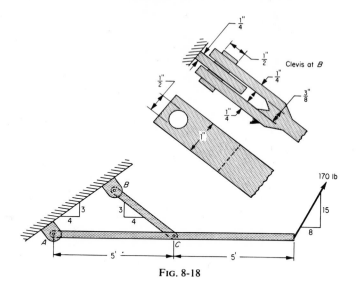

FIG. 8-18

*Solution:* Since member *BC* is a two-force member, the free body diagram of Fig. 8-19(*a*) shows that it carries a compressive load of 500 lb.

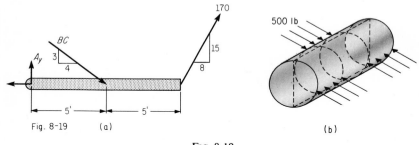

Fig. 8-19     (a)                                    (b)

FIG. 8-19

$$\sum M_a = 0$$

$$\frac{15}{17} 170 (10) - \frac{3}{5} BC (5) = 0$$

$$BC = 500 \text{ lb}$$

The pin bears against the wall fixture and the member *BC* in the manner indicated in Fig. 8-19(*b*). The *projected* bearing area against the fixture is (1/2)(1/4) = 1/8 sq in. while that against the member is 2(1/2)(1/4) = 1/4 sq in.

Hence the maximum bearing stress is

$$\sigma_B = \frac{500}{1/8} = 4000 \text{ psi}$$

The pin is double shear. Hence the shear area is $2(\pi/4)(1/2)^2 = \pi/8$ sq in. Hence the shear stress in the pin is

$$\tau = \frac{500}{\pi/8} = 1272 \text{ psi}$$

## PROBLEMS

NOTE: In each of the problems from 8-1 to 8-10, determine the magnitude and vertical location of the resultant load $P$.

**8-1.** The member in part (*a*) of the illustration is loaded parallel to its longitudinal axis in such a manner that the strain does not vary in the *z* direction. After the resultant load $P$ was applied, the strain varied linearly from 0.0008 in./in. tensile along $AA$ to 0 in./in. along $BB$, as indicated in part (*b*) of the illustration. The stress-strain curve (tensile and compressive) for the material is as shown in part (*c*).

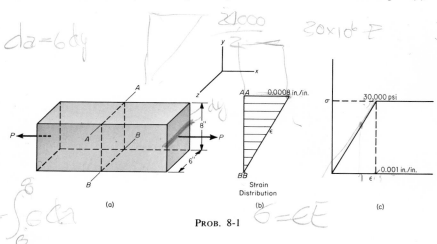

(a)         (b)         (c)

PROB. 8-1

**8-2.** The member in Prob. 8-1 is loaded so that the strain varies linearly from 0.0025 at $AA$ to zero at $BB$.

**8-3.** The member in Prob. 8-1 is loaded in compression so that the strain does not vary in the *z* direction. After the load was applied, the strain varied linearly from 0.0012 in./in. compressive along $BB$ to 0.0004 in./in. compressive along $AA$, as indicated in part (*a*) of the illustration for Prob. 8-3. The material has the stress-strain curve (tensile and compressive) shown in part (*b*).

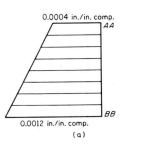

PROB. 8-3

**8-4.** The material for the member in Prob. 8-3 has the stress-strain curve shown in part (*c*) of the illustration for Prob. 8-1.

**8-5.** The member in Prob. 8-1 is loaded in such a manner that the strain does not vary in the *z* direction. After the load *P* was applied, the strain varied linearly from 0.0006 in./in. compression at *AA* to 0.0018 in./in. tensile at *BB*, as indicated in the illustration for Prob. 8-5. The material has the stress-strain curve (tensile and compressive) shown in part (*b*) of the illustration for Prob. 8-3.

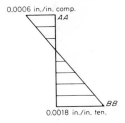

PROB. 8-5

**8-6.** The material for the member in Prob. 8-5 has the stress-strain curve shown in part (*c*) of the illustration for Prob. 8-1.

**8-7.** The member in Prob. 8-1 is loaded in such a manner that the strain does not vary in the *z* direction. After the load *P* was applied, the strain varied linearly from 0.006 in./in. tensile at *BB* to zero at *AA*. The stress-strain curve (tensile and compressive) for the material is shown in the illustration for Prob. 8-7.

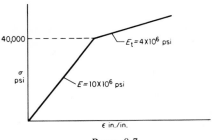

PROB. 8-7

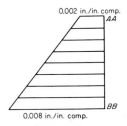

PROB. 8-8

**8-8.** The member in Prob. 8-7 is loaded so that the strain varies as shown in the illustration for Prob. 8-8.

**8-9.** The material for the member in Prob. 8-1 has a stress-strain relationship which can be approximated by $\sigma = k\epsilon^{1/2}$, in which $k = 400,000$.

**8-10.** The material for the member in Prob. 8-5 has a stress-strain relationship which can be approximated by $\sigma = k\epsilon^{1/3}$, in which $k = 8,000$.

**8-11.** Derive a relationship between the maximum tensile stress in the wall and the internal gage pressure for a thin-walled spherical pressure vessel.

**8-12.** A thin-walled circular tube carries a fluid stream at a gage pressure of 400 psi. Find the maximum tensile stress and the maximum shear stress developed in the wall of the tube. The inside diameter of the tube is 4 in. and the wall thickness is $\frac{1}{8}$ in.

**8-13.** The diameter of a spherical pressure vessel is 60 in., and its wall is 0.15 in. thick. What internal pressure is possible in the vessel if the maximum shear stress may not exceed 18,000 psi?

**8-14.** The tank in the illustration has a spiral seam as shown which is welded. If the tank is 34 in. in diameter, how much shearing and normal force will be carried per linear inch of weld when the tank is pressurized to 120 psi?

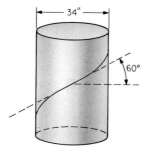

PROB. 8-14

**8-15.** Determine the percent increase in volume of a thin-walled cylindrical pressure tank expressed as a function of its dimensions, the internal pressure, and the mechanical properties for the material of which it is made. Assume elastic action and neglect products of very small quantities.

**8-16.** A closed-ended cylindrical pressure vessel of aluminum, for which $E = 10^7$ psi and $\mu = \frac{1}{3}$, has an inside diameter of 48 in. and a thickness of $\frac{1}{4}$ in. While the vessel is unpressurized, two electrical strain gages are bonded on the curved longitudinal surface at *right* angles to each other. The tank is then pressurized until the strain gages read 800 and 550 micro-inches per inch, respectively. Assume that the aluminum behaves elastically. *a)* What are the normal stresses in the tank wall in the directions of the gages? *b)* What are the longitudinal and circumferential stresses in the tank wall? *c)* What is the pressure in the tank?

**8-17.** A thin hoop of a material for which $E = 30 \times 10^6$ psi and $\alpha = 6.5 \times 10^{-6}$ in./in.°F has an inside diameter of 19.99 in., a thickness of $\frac{1}{4}$ in., and a width of 1 in. The hoop is to be sweated onto a rigid mandrel 20 in. in diameter. *a)* What will be the hoop stress in the hoop? *b*) What will be the bearing stress between the hoop and the mandrel?

**8-18.** The member in the illustration is composed of a section of steel and a section of aluminum. If $P$ is 3000 lb, what is the over-all elongation due to $P$ and neglecting the weight?

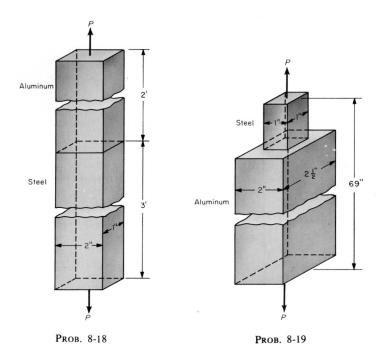

PROB. 8-18     PROB. 8-19

**8-19.** The member in the illustration is composed of a section of steel and a section of aluminum. The over-all elongation is not to exceed 0.05 in. The elongation of the aluminum should be five times that of the steel. Assuming that both materials behave elastically, determine the length of each section.

**8-20.** Derive an expression for the total elastic deformation of a homogeneous bar of constant cross-sectional area $A$ and length $L$ due to its own weight $W$ when suspended from one end.

**8-21.** A homogeneous isotropic bar is made of a material whose approximate stress strain law is $\sigma = k\epsilon^{1/2}$ where $k = 2 \times 10^5$. The original dimensions of the bar are $0.5'' \times 1.5'' \times 36''$. What centroidal longitudinal load is required to cause an elongation of the bar of a quarter of an inch?

**8-22.** A homogeneous isotropic bar is made of a material whose approximate stress-strain law is $\sigma = k\epsilon^{1/2}$, where $k = 10^5$. The original dimensions of the bar are $1'' \times 2'' \times 20''$. *a*) If the bar is subjected to a centroidal longitudinal tensile load of 20,000 lb, what is the change in length? *b*) What is the change in volume, if $\mu = \frac{1}{4}$?

**8-23.** The composite bar shown in the illustration is made up of a section of steel ($E_s = 30 \times 10^6$ psi), a section of aluminum ($E_a = 10 \times 10^6$ psi), and a section of brass ($E_b = 15 \times 10^6$ psi). The axial deformation of the brass section is twice that of the aluminum section, and the axial deformation of the aluminum section is twice that of the steel section. The over-all length is 4 ft. *a*) Determine the length of each section. *b*) If the total deformation is 0.030 in., what is the magnitude of *P*? Assume elastic behavior and neglect stress concentrations.

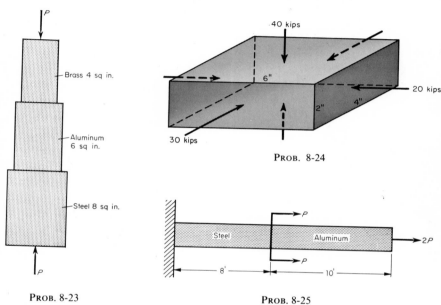

PROB. 8-24

PROB. 8-23

PROB. 8-25

**8-24.** For the material of the centroidally loaded member in the illustration, $E = 20 \times 10^6$ and $\mu = \frac{1}{4}$. *a*) Determine the total change in each dimension of the member. *b*) What is the percent change in volume? Assume elastic behavior.

**8-25.** The aluminum and steel rods in the illustration are fastened together. Each has a diameter of 4 in. Also, $E_s = 30 \times 10^6$ and $E_a = 10 \times 10^6$. The compressive stress in the steel is 10,000 psi after the loads are applied. Find the total elongation of the composite rod.

**8-26.** How much will the 165 lb load compress the plastic bar if the modulus of the plastic is 300,000 psi?

**8-27.** How much will the two tie rods in Prob. 8-26 stretch if the top one is 3/8" diameter steel and the bottom is 1/2" diameter aluminum alloy?

**8-28.** Determine the overall elongation that the member in the illustration will undergo due to its own weight. Assume elastic behavior and a modulus of elasticity of 10 million psi. The specific weight of the material is 3/4 lb/cu in.

**8-29.** The dimensions in the figure were determined with the member horizontal and fully supported. How much will the long dimension change when it is placed in the position shown? The member is made of hard yellow brass (see Table I in Appendix) and has a uniform thickness of 3 in.

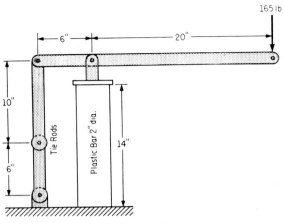

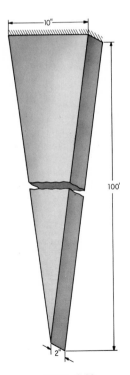

Prob. 8-26

Prob. 8-28

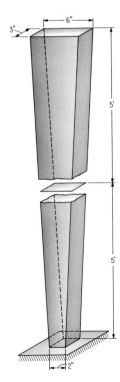

Prob. 8-29

**8-30.** Assume the same dimensions and material as in Prob. 8-29. How much will the member elongate when supported at the top as shown?

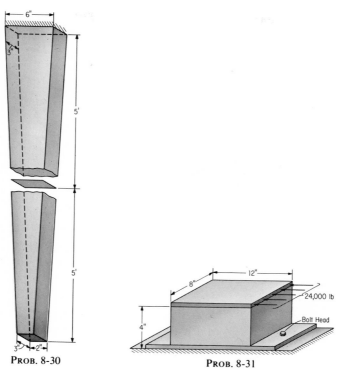

PROB. 8-30                    PROB. 8-31

**8-31.** The mounting device shown in the illustration is composed of two parallel steel plates and a block of hard rubber. A horizontal force of 24,000 lb is applied to the upper plate. Determine the horizontal displacement of the upper plate relative to the lower one: *a*) If the rubber has a linear $\tau$-$\gamma$ relationship with a shear modulus of $2 \times 10^4$ psi; *b*) if the rubber has a $\tau$-$\gamma$ relationship given by $\tau = k\gamma^{1/2}$, where $k = 10^3$.

**8-32.** When the load $P$ is applied to the composite member in the illustration, the overall contraction of the member is 0.004 in. Determine: *a*) the maximum shear stress in the steel; *b*) the longitudinal strain in the brass.

**8-33.** The steel member in the illustration is centroidally loaded with a load of 100,000 lb. If the allowable stresses are $\sigma_t = 10,000$ psi, $\sigma_c = 8000$ psi, and $\tau = 4000$ psi, and the elongation may not exceed 0.04 in., what is the minimum value the dimension $b$ may have?

**8-34.** The member in the illustration is centroidally loaded and has the following specifications: the allowable stresses are $\sigma_t = 10,000$ psi, $\sigma_c = 8000$ psi, $\tau = 4000$ psi; the elongation may not exceed 0.02 in. Assuming elastic behavior and a modulus of elasticity of $30 \times 10^6$ psi, determine the maximum allowable value of $P$.

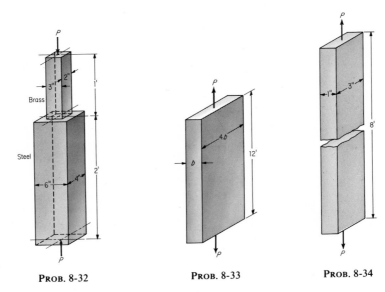

PROB. 8-32           PROB. 8-33           PROB. 8-34

**8-35.** The structure in the illustration is made of a material which has an ultimate tensile stress of 75,000 psi, an ultimate compressive stress of 90,000 psi, an ultimate shear stress of 40,000 psi, a tensile yield strength (0.2%) of 48,000 psi, and a modulus of elasticity of $30 \times 10^6$ psi. If the overall elongation of member $AB$ must not exceed 0.14 inches and a factor of safety of 3 is required, determine the minimum required cross-sectional area of member $AB$.

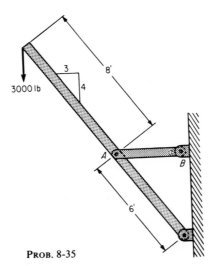

PROB. 8-35

**8-36.** In the pin-connected structure shown in the illustration, determine *a*) the maximum bearing stress on pin $B$, *b*) the shearing stress in pin $B$.

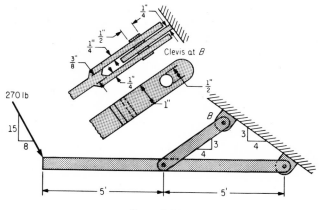

PROB. 8-36

**8-37.** A tensile member of a truss is a simple eye bar, 8 ft long, with one pin joint at each end. By mistake the bar is made 0.02 in. too short. The truss is left outside overnight, and the eye bar is heated uniformly indoors. In the morning, the truss is at $+5°$ and the eye bar is at $+85°$. The eye bar is tried again and is now too long. For how many minutes must the engineer wait before the bar will fit, if the outside air remains at $+5°F$ and the bar cools down uniformly at a rate of $1°F$ every 5 minutes? The coefficient of thermal expansion is $8 \times 10^{-6}$ in./in./°F.

**8-38.** The two members in the illustration are fastened together and to the walls. If the members are stress-free before they are loaded, what will be the stress in each after the two 60-kip loads are applied? Neglect stress concentrations at the ends, and assume elastic behavior.

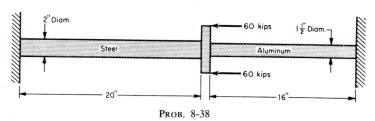

PROB. 8-38

**8-39.** A uniform homogeneous rod, 1 in. square and 14 ft long, is firmly anchored at each end in a vertical position while unstressed. An axial load of 40,000 lb is then applied downward at a position 8 ft from the bottom of the rod. Determine the reaction at each end, if $E = 10 \times 10^6$ psi, $\mu = \frac{1}{3}$, and $\sigma_{PL} = 30,000$ psi.

**8-40.** A 7075 T6 aluminum alloy bar having a cross-sectional area of 4 sq. in. is placed between two rigid immovable supports while 29.99 in. long and at a temperature of $-20°F$. The supports are 30.00 in. apart. The bar is then heated uniformly. *a*) At what temperature will the gap first close? *b*) Determine the stress in the bar when the temperature reaches 90°F. The coefficient of thermal expansion is $12 \times 10^6$ in./in./°F.

**8-41.** The round prismatic bar with fixed ends, shown in the illustration, carries an axial load of *P* of 18,000 lb. *a*) What are the reactions at the ends? *b*) How far will plane *m-n* be moved downward by the load? Assume that $E = 30 \times 10^6$.

**8-42.** The bars *A* and *C* in the illustration are made of brass ($E = 15 \times 10^6$), and bar *B* is made of steel ($E = 30 \times 10^6$). Each bar has a cross-sectional area of 4 sq in. What load *P* will cause the stress in the brass to be one-fourth of the stress in the steel?

**8-43.** The load *P* placed on the short reinforced concrete pier in the illustration causes the stress in the steel to reach 18,000 psi. *a*) Calculate *P* and the stress in the concrete at this load if the concrete conforms to a linear stress-strain law with $E = 2 \times 10^6$ psi. *b*) What is the value of *P* and the stress in the concrete if the concrete has the compressive stress-strain curve of Fig. 8-14(*a*)?

**8-44.** The vertical shaft in the illustration is supported by a collar bearing. Assume that the load *P* causes a compressive stress of 30,000 psi in the shaft, and

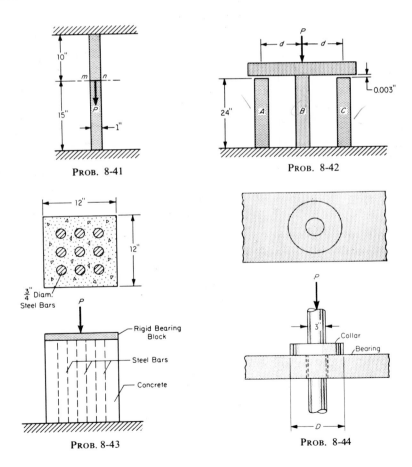

PROB. 8-41

PROB. 8-42

PROB. 8-43

PROB. 8-44

that the bearing stress between the collar and the bearing is 6000 psi. Determine the diameter of the collar.

**8-45.** The connection in the illustration transmits 5,000 lb. *a)* Find the average shear stress in piece *B*. *b)* Find the maximum average tensile stress in piece *C*. *c)* Find the average bearing stress between piece *B* and piece *C*.

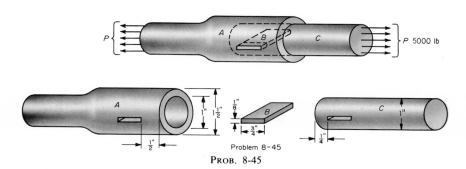

Problem 8-45

PROB. 8-45

**8-46.** For the connection in the illustration for Prob. 8-45, find *a)* the maximum average tensile stress in piece *A*; *b)* the average bearing stress between piece *A* and piece *B*.

**8-47.** For the connection in the illustration for Prob. 8-45, find *a)* the average shear stress in piece *A*; *b)* the average shear stress in piece *C*.

**8-48.** A single bolt connects two plates which are ½ in. thick and 3 in. wide. The centerlines of the bolt holes are 4 in. from the end of each plate. Design the bolt (determine the required diameter) if the joint must transmit 5 tons and the material of the bolt is such that the shear "failure" stress is 40,000 psi, the bearing "failure" stress is 80,000 psi, and the tensile "failure" stress of the plates is 60,000 psi. Use a factor of safety of 3.

**8-49.** The two plates in the illustration, which are ⅜ in. thick and 2 in. wide, are joined by four 3/16-in. diameter rivets. The joint transmits 8000 lb. *a)* Find the average shear stress in the rivets. *b)* Find the largest average tensile stress in the plates.

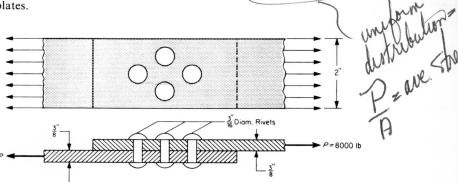

PROB. 8-49

**8-50.** How many $^3/_4$-in. diameter rivets are required to transmit 50,000 lb from the two angles in the illustration to the gusset plate, if the allowable average stresses in the rivets are 15,000 psi for shear and 40,000 psi for bearing. Neglect the possibility of tensile failure, and assume that each rivet carries the same amount of load.

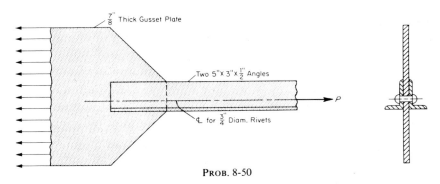

PROB. 8-50

### SUPPLEMENTARY PROBLEMS

**8-51.** The eccentricity $e$ in the illustration is 0.15 in. The strain gage axes are vertical. *a*) If the strain gage at *B* reads 933 micro-inches per inch, compressive, what should the gage at *A* read? Assume that there is linear strain variation from *A* to *B* and that the material is mild steel with a yield-point stress of 30,000 psi. *b*) Determine the load *P*.

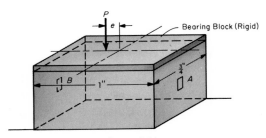

PROB. 8-51

**8-52.** The member in the illustration for Prob. 8-51 is made of mild steel with a yield-point stress of 30,000 psi. *a*) If the gage at *A* reads 120 micro-inches per inch, compressive, and *P* is 10,000 lb, what should the gage at *B* read? Assume linear strain variation. *b*) Determine the eccentricity *e*.

**8-53.** Solve Prob. 8-51 if the gage at *B* reads 1200 micro-inches per inch.

**8-54.** Solve Prob. 8-52 if *P* is 13,000 lb.

**8-55.** What will be the readings of the gages at *A* and *B* in the illustration for Prob. 8-51, if *P* is 7000 lb and *e* is 0.10 in.? The material is mild steel with a yield-point stress of 30,000 psi.

**8-56.** Solve Prob. 8-55 if *P* is 15,000 lb.

**8-57.** A thin-walled spherical pressure having a constant wall thickness *t* in. and an initial inside radius *R* in. is subjected to an internal gage pressure *p* psi. Calculate *a*) the change in radius and *b*) the change in volume of the sphere due to the pressure. Assume elastic behavior.

**8-58.** A cylindrical rocket motor case is made from an 0.035 in. thick high strength steel alloy sheet which has a uniaxial tensile yield strength of 200,000 psi and an ultimate tensile strength of 240,000 psi at the operating temperature. The case is 70 in. in diameter and 20 ft long. Eight of these motors are used in a cluster to propel a 1,000,000 lb thrust rocket. Each motor case must transmit one-eighth of this total compressive thrust. *a*) What is the allowable operating pressure in the case if it ceases to function properly when the maximum shear stress exceeds 57.7% of the uniaxial tensile yield strength? A factor of safety of 1.15 for both the pressure and thrust loads is required. Assume that failure does not occur at the welded seams. *b*) In a hydrostatic laboratory test with no thrust, how much pressure would be required in the case to make the maximum shear stress reach 57.7% of the tensile yield?

**8-59.** A thin-walled cylindrical pressure vessel has an inside radius of 12 in. and a wall thickness of $\frac{1}{4}$ in. Initially it contains a gas at atmospheric temperature and pressure (60°F and 14.7 psia). The temperature of the gas is then increased to a temperature *t*. Express the approximate maximum tensile stress in the wall as a function of the temperature *t*. Assume a constant-volume heating process.

**8-60.** The homogeneous conical-shaped member shown in the illustration has a specific weight of *w* lb/cu. in. Find an expression for the vertical deformation caused by its own weight. Assume elastic behavior.

**8-61.** The homogeneous member shown in the illustration has a specific weight of *w* lb/cu in. Find an expression for the vertical deformation caused by its own weight. Assume elastic behavior.

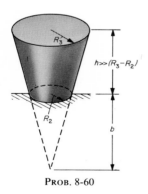

PROB. 8-60

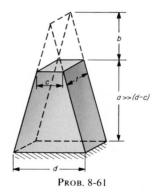

PROB. 8-61

**8-62.** In the pin-connected structure shown in the illustration, pin *A* is $\frac{3}{8}$ in. in diameter and pin *B* is $\frac{1}{2}$ in. in diameter. Pin *B* slips into a hole in the concrete base. Determine *a*) the shearing stress in pin *A*; *b*) the shearing stress in pin *B*, neglecting friction; *c*) the bearing stress between pin *B* and the structure; *d*) the bearing stress between the structure and the concrete base at *B*.

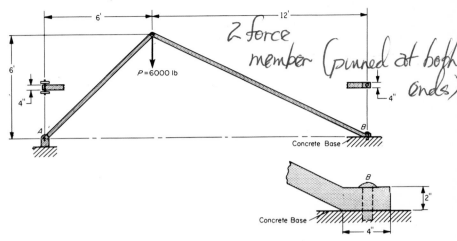

PROB. 8-62

**8-63.** The isolator shown in the illustration is composed of an outer steel ring, a disk of hard rubber, and an inner steel core. If the isolator is supported at its outer ring and loaded longitudinally at its inner core, express the shear stress in the rubber as a function of the distance *r* from the center of the core?

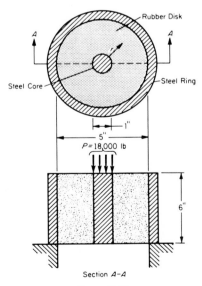

Section *A-A*

PROB. 8-63

**8-64.** For the isolator in the illustration for Prob. 8-63, determine the vertical displacement of the core *a*) if the rubber has a linear $\tau$-$\gamma$ relationship with

$G = 2 \times 10^4$ psi; *b*) if the rubber has a $\tau$-$\gamma$ relationship given by $\tau = K\gamma^{1/2}$, where $K = 10^3$.

**8-65.** The prismatic rod shown in the illustration, which has a length $L$, a constant cross-sectional area $A$, and a uniform density $\rho$, is rotating at a constant angular velocity $\omega$ radian/sec. An element of the rod with volume $A \; dl$ undergoes an inward radial acceleration of $l\omega^2$, where $l$ is the distance from the axis of rotation to the element. Find *a*) an expression for the stress at any position along the rod and *b*) the over-all change of length of the rod, due to its angular velocity. Assume elastic behavior and a modulus of elasticity of $E$ psi.

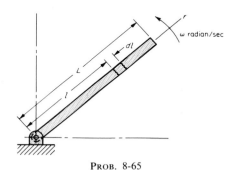

PROB. 8-65

**8-66.** The connector shown in the illustration is to carry an axial thrust of 18,000 lb. Determine the average tensile and shear stresses in the $\frac{3}{4}$-in. diameter bolts *a*) if the flange angle $\theta$ is 0°; *b*) if $\theta$ is 90°; *c*) if $\theta$ is 30°. Neglect stress concentrations in the bolt, and use a net inside thread diameter 0.620 in.

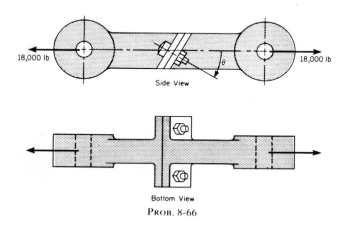

Bottom View

PROB. 8-66

**8-67.** The structure shown in the illustration is composed of a circular shell of cast iron and a circular core of brass. Initially, the shell is 10 in. long and the core is 10.005 in. long. Also, $E_b = 12 \times 10^6$, $E_{c.i.} = 15 \times 10^6$, and $\mu_{c.i.} = \mu_b = \frac{1}{4}$. If a compressive load of 50,000 lb is applied to the structure by means

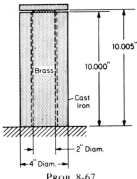

PROB. 8-67

of a flat plate, determine *a*) the maximum compressive stress in the brass and *b*) the maximum shearing stress in the cast iron. Assume elastic behavior.

**8-68.** What will the pointer in the illustration read on the scale after the temperature of both the steel bar and the aluminum bar has been increased by 100°F? The coefficients of thermal expansion are $\alpha_s = 6.5 \times 10^{-6}$ and $\alpha_a = 13 \times 10^{-6}$. The top of pointer is attached to the plate allowing the pointer to move either right or left.

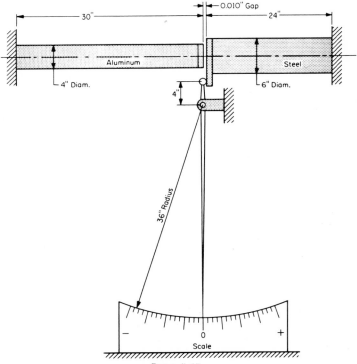

PROB. 8-68

## chapter 9

# Torsion

## 9-1.  Introduction

In Chapter 8 we considered members subjected to axial loads.  First, the basic load-stress relationship $P = \int_{\text{area}} \sigma \, da$ was established by using the method of sections.  We found that before this relationship could be evaluated, it was necessary to determine the stress distribution over the area in question by investigating the strain distribution and relating the stress to the strain.  Then the procedure for deriving load-deformation relationships for axially loaded members was illustrated.  This chapter will present a similar treatment of members subjected to torsion by loads which tend to twist the members about their longitudinal centroidal axes.

Most of this chapter deals with members in the form of concentric circular cylinders, solid and hollow, subjected to torques about their longitudinal geometric axes.  Although this may seem like a somewhat special case, a quick reflection would show that many torque-carrying engineering members are cylindrical in shape.  Examples are drive shafts, bolts, and screwdrivers.  A brief discussion of noncylindrical members is also presented.  The last section on hollow, thin-walled torsion members introduces the concept of shear flow, which will be utilized in subsequent chapters.

## 9-2.  Kinematic Response of Circular Cylindrical Members

Before we attempt to derive any load-stress or load-deformation relationships for torsion members, we shall investigate the kinematic response of a circular cylindrical member subjected to a pure torque about its longitudinal axis.  The most logical way to make this investigation would be to go into the laboratory and twist a cylindrical member.  However, since it is impractical for everyone to do this, we shall attempt to describe and simulate the experimental conditions so that you may "observe" the results.  When we refer to the surface of a cylinder or to a cylindrical surface in subsequent explanations, we will mean the long curved longitudinal surface, and not a transverse plane surface.

Let us first circumscribe two circumferential lines in parallel planes on the surface of a cylindrical bar made of a homogeneous material.  These planes are at right angles to the longitudinal axis of the cylinder and are some distance $L$ apart.  A grid pattern of small rectangles whose sides are parallel and perpendicular to the longitudinal axis is then scribed on the surface between these circumferential lines. Such a pattern of lines is shown in Fig. 9-1(*a*), where it is assumed that the dimensions of the rectangles are very small in comparison with the over-all dimensions of the bar. This assumption will minimize the effect of the curved surface on our observations. Let us now apply a torque *T* to the bar somewhat beyond the scribed lines (to eliminate localized effects due to load within the length *L*), and observe its effects on the grid pattern. The outlines of the rectangles on the twisted bar are shown in Fig. 9-1(*b*), where the amount of twist is greatly exaggerated.

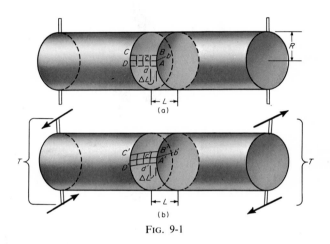

FIG. 9-1

The following important observations can be noted from this experiment:

1. The distance *L* between the outside circumferential lines does not change significantly as a result of the application of the torque. However, the rectangles become parallelograms whose sides have the same length as those of the original rectangles.

2. The circumferential lines do *not* become zigzag; that is, they remain in parallel planes.

3. The original straight parallel longitudinal lines, such as *AD* and *BC*, remain parallel to each other but do *not* remain parallel to the longitudinal axis of the member. These lines become helixes (a helix is the path of a point which moves longitudinally and circumferentially along the surface of a cylinder at a uniform rate).

Although hindsight is easier than foresight, these results might have been anticipated from the geometry and loading conditions. For example, since the member was not subjected to any axial loading, we might expect its length to remain unchanged. Also, if we draw a free-body diagram of any section of the member $\Delta L$ units long, we see that it is loaded in exactly the same manner as any other $\Delta L$ section of the member. Hence, we would expect the kinematic response of the member to be uniform with respect to the length.

If the experimental procedure were repeated on cylindrical bars of different radii and other homogeneous materials, it would be found that while the *amount* of twist would perhaps vary from test to test, the geometric characteristics of the deformations would be the same in all cases.

From the first two of the geometric observations, it is logical to draw the following conclusions:

1. An element on the surface of a cylindrical member subjected to pure torque is in a state of *pure shear*. There are no significant axial deformations in an element oriented parallel to the longitudinal axis.

2. Even though the second observation is only a *surface* observation of bars of different radii, it may be concluded that the bar does not warp; i.e., *plane cross sections* perpendicular to the longitudinal axis of the bar *remain plane* even after the bar is twisted.

Let us now investigate more thoroughly the third observation. In order to obtain some geometric relationships, we redraw the grid pattern on a "flattened out" surface, such as that shown in Fig. 9-2. The circumferential lines $A'B'$ and $D'C'$ are parallel and straight, and the "flattened" helixes, such as $D'A'$ and $C'B'$, become parallel straight lines inclined at an angle $\phi$ with the longitudinal axis of the bar. The shear deformation occurring over the length $L$ is denoted by $e_s$, and the shear deformation occurring over the length $\Delta L$ is $\Delta e_s$. By similar triangles, we have ~~shear deformation~~

$$\tan \phi = \frac{e_s}{L} = \frac{\Delta e_s}{\Delta L}$$

If we now limit $\phi$ to very small values, we can say that

$$\tan \phi = \frac{\Delta e_s}{\Delta L} \approx \sin \phi \approx \phi$$

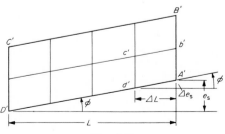

Fig. 9-2

Then, if we think of making the rectangles smaller and smaller, we obtain by definition

$$\lim_{\Delta L \to 0} \frac{\Delta e_s}{\Delta L} = \frac{de_s}{dL} = \gamma$$

From this result we arrive at a third conclusion, which follows:

3. In a cylindrical member subjected to pure torque the helix angle $\phi$ can be interpreted as the *maximum* shear strain at the surface of the member. This value is the same at all surface points within the length $L$.

These three conclusions will be utilized quite extensively in the succeeding sections. However, it is imperative for you to realize that if the geometry of the member or the manner of loading were significantly altered, any one or all of the previously discussed observations and conclusions may no longer be true. Once again, we see that the characteristics of the strain in a member are governed primarily by its geometry and the manner in which the load is applied.

## 9-3. Basic Load-Stress Relationship for Cylindrical Torsion Member

The preceding deformation analysis has led us to certain conclusions concerning the response of a circular cylindrical member subjected to a pure torque about its longitudinal geometric axis. Now, with the aid of these ideas, we will write the basic equilibrium equation for a cylindrical torsion member.

The cylindrical member in Fig. 9-3(a) is in static equilibrium. Let us pass a cutting plane *mm* perpendicular to the longitudinal axis of the member, and then separate the two portions. When the entire member is in equilibrium, each portion must also be in equilibrium. Figure 9-3(b) shows a free-body diagram of the left-hand portion, where $T$ is the resultant external torque on this portion and $T_r$ is the resultant internal resisting torque which is exerted *on* this portion *by* the right-hand portion over the area of the member cut by plane *mm*.

How is this resultant internal resisting torque produced? We can answer the question by recalling the first two conclusions in Sec. 9-2. From

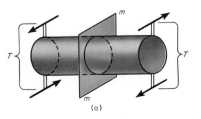
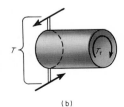

(a)     (b)

FIG. 9-3

these we know that an element oriented parallel to the longitudinal axis is in pure shear and also that any plane section perpendicular to the longitudinal axis remains plane. We combine these two ideas and use the cylindrical coordinates $x$, $\rho$, and $\theta$ to show the forces on a three-dimensional infinitesimal element, as in Fig. 9-4. This element is in a state of

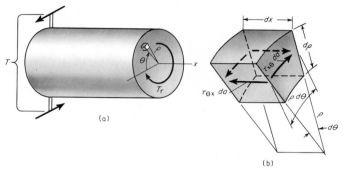

FIG. 9-4

pure shear, and we can show that $\tau_{x\theta}$ is equal to $\tau_{\theta x}$ in the following manner. Taking moments of the forces on the element about the $\rho$ axis, we obtain

$$\sum M_\rho = 0$$

$$(\tau_{\theta x}\, da)\rho\, d\theta - (\tau_{x\theta}\, da)dx = 0$$

$$(\tau_{\theta x}\, dx\, d\rho)\rho\, d\theta - (\tau_{x\theta}\, \rho\, d\theta\, d\rho)dx = 0$$

$$\tau_{\theta x} = \tau_{x\theta}$$

In the succeeding analyses, we shall refer to the shear stress simply as $\tau$. You should realize that this shear stress exists simultaneously on areas

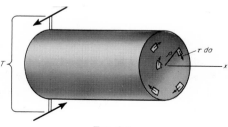

FIG. 9-5

in radial and transverse planes; i.e., it acts parallel and perpendicular to the longitudinal axis of the cylinder.

Now we can say that $T_r$ is the resultant torque produced by an infinite number of infinitesimal *shear* forces acting on the plane area perpendicular to the axis of the cylinder. Such shear forces are shown in Fig. 9-5. When moments are taken about the $x$ axis, we get

$$+ \oint\!\!\!-x$$

$$\sum M_x = 0$$

$$T - \int_{area} (\tau \, da)\rho = 0$$

$$T = \int_{area} \tau \rho \, da \qquad (9\text{-}1)$$

This is the basic load-shear stress relationship for any cylindrical torsion member. The integral is the total internal resisting torque acting on the cut cross-sectional area.

Of course, as was the case in Chapter 8 for axially loaded members, before this basic load-stress relationship can be of practical value, we must be able to perform the necessary integration. So it will be necessary to determine the *shear-stress distribution* over the cross-sectional area in question. To do this we will first determine the *strain distribution*, and will then relate the stress to the strain to get the desired stress distribution.

EXAMPLE 9-1. What torsional load will cause a stress of 70,000 psi at the outer surface of an aluminum alloy shaft with a diameter of $\frac{1}{2}$ in., if the stress distribution is assumed to be as shown in Fig. 9-6(a)?

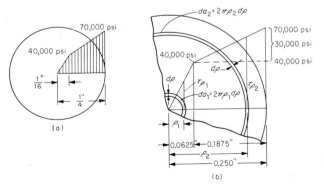

FIG. 9-6

*Solution:* The torsional load, or torque, is

$$T = \int_{area} \tau \rho \, da$$

The given stress distribution must be expressed in mathematical form in terms of $\rho$. Since the stress-distribution "curve" has a slope change at $\rho = 0.0625$ in., we must derive two expressions for $\tau$. The conditions are shown in Fig. 9-6(*b*), and we must write an expression for each interval, as follows:

For $0 \leqq \rho_1 \leqq 0.0625$,

$$\frac{\tau_{\rho_1}}{\rho_1} = \frac{40,000}{0.0625}$$

$$\tau_{\rho_1} = 640,000 \, \rho_1$$

For $0.0625 \leqq \rho_2 \leqq 0.250$,

$$\frac{\tau_{\rho_2} - 40,000}{\rho_2 - 0.0625} = \frac{30,000}{0.1875}$$

$$\tau_{\rho_2} = 160,000 \, (\rho_2 - 0.0625) + 40,000$$

Therefore,

$$T = \int_0^{0.0625} (640,000 \, \rho_1)(\rho_1)(2\pi\rho_1 \, d\rho)$$

$$+ \int_{0.0625}^{0.250} [160,000(\rho_2 - 0.0625) + 40,000]\rho_2(2\pi\rho_2 \, d\rho)$$

$$T = 1960 \text{ in-lb}$$

## 9-4. Shear-Strain Distribution in Cylindrical Torsion Member

Having established the basic load-stress relationship for a cylindrical torsion member, we will now determine how the shear strain varies throughout the member. Once again we will simulate a laboratory experiment in order to observe and interpret the results therefrom.

As was pointed out in Chapter 2, shear strain is an angular distortion, and is extremely difficult to measure experimentally. However, by utilizing two ideas which were presented in the preceding sections, we can determine the shear strain on the surface of a torsion member in this way:

1. In Sec. 9-2 it was shown that the maximum shear strain on the surface of a cylindrical member has the same value at all points on that surface. So we may make our measurements, and expect the same results, at *any* point on the surface sufficiently distant from the section where the torque is applied.

2. In Secs. 9-2 and 9-3 it was demonstrated that a rectangular element whose sides are parallel and perpendicular to the longitudinal axis of the

member is in a state of pure shear and is subjected only to shear forces (see Fig. 9-4). This means that the principal normal stresses and the principal *axial* strains occur along the axes that are oriented at an angle of 45° with the longitudinal axis of the member.[1] Mohr's circle of plane stress for the case of *pure* shear shows that the maximum shear stress, the maximum tensile stress, and the maximum compressive stress all have the same magnitude. Also, the values of the axial *strains* along the principal axes will be the coordinates of the extremities of Mohr's circle of plane *strain,* and the *diameter* of this circle will represent the magnitude of the *maximum* shear strain. Thus, if we measure the *principal axial* strains at any point on the surface of a cylinder subjected to pure torque, we will in effect be measuring the maximum *shear* strain on that surface.

For our experimental work, let us use a hollow cylindrical member having arbitrary inside and outside radii and made of a homogeneous and isotropic material. Two electrical resistance strain gages[2] are bonded to both the outer and inner surfaces of the member along lines inclined 45° to the longitudinal axis, as shown in Fig. 9-7. The cylinder is then placed in a

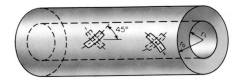

Fig. 9-7

torsion machine whereby a pure torque is applied. If the resulting axial strains measured by the gages are converted to shear strains of the cylinder, we would find that for any particular value of torque the ratio of shear strain to radius is a constant; that is,

$$\frac{\gamma_o}{r_o} = \frac{\gamma_i}{r_i} = \text{constant}$$

This relationship would be true, not just for elastic strains, but for any strain ordinarily encountered in an engineering member.

If the experimental procedure were repeated on several cylinders of different homogeneous and isotropic materials and with various inside and outside radii, similar results would be obtained. From these experiments it can be concluded that the shear *strain* in a cylindrical torsion member is directly proportional to the radius. In mathematical form this

---

[1] The third principal axis on a three-dimensional element is directed perpendicular to the free surface of the cylindrical torsion member, and the value of the third principal stress is, of course, zero.

[2] See Chapter 16 for description and theory of electrical resistance strain gages.

becomes

$$\gamma \propto \rho$$

$$\gamma_\rho = K\rho \tag{9-2}$$

where $\gamma_\rho$ is the shear strain at any arbitrary distance $\rho$ from the geometric center of the member, and $K$ is the ratio of $\gamma_o$ to $r_o$ for any particular applied torque.

At first glance, the experimental indication of a linear shear strain distribution might appear to be a rather fortuitous and unexpected result. However, with adequate hindsight, one can make rather convincing arguments based on geometrical symmetry as to why this result should occur. Such an argument will be demonstrated in the next chapter on bending but our "experiment" will suffice for this discussion of torsion.

Now that we have established both the basic load-shear stress relationship, given by Eq. 9-1, and an expression for the shear-strain distribution, given by Eq. 9-2, we can derive specific relationships between shear stress and torque for cylindrical torsion members of different sizes and materials. In studying the following cases, your attention should be concentrated on the *derivation procedure* as well as the results obtained.

**Elastic Behavior.** The most common condition encountered in engineering is elastic behavior, for which it is assumed that a linear stress-strain relationship exists. In the case of shear stress and shear strain, this linear relationship is of the following form:

$$\tau = \gamma G$$

where $G$ has been defined as the shear modulus of elasticity. To the engineer this relationship means that the shear stress-strain curve is a straight line with a slope $G$.

Since we have already shown that the shear strain varies linearly with the radius, it follows that for the special case of a linear stress-strain relationship, the shear stress must also vary linearly with the radius. The distributions for shear strain and *elastic* shear stress are shown in Fig. 9-8(a) and (b), respectively.

Applying the relationships in Fig. 9-8 to the basic load-stress relationship given by Eq. 9-1, we obtain

$$T = \int_{area} \tau_\rho \rho \, da$$

$$= \int_{area} \left(\frac{\tau_o}{r_o} \rho\right) \rho \, da$$

$$= \frac{\tau_o}{r_o} \int_{area} \rho^2 \, da$$

$$T = \frac{\tau_o}{r_o} J \quad \text{or} \quad \tau_o = \frac{Tr_o}{J} \tag{9-3}$$

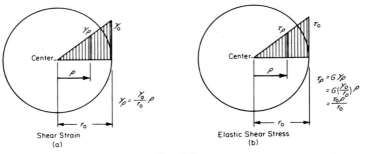

FIG. 9-8

Also,

$$\tau_\rho = \tau_o \frac{\rho}{r_o} = \frac{T\rho}{J} \tag{9-3a}$$

where $\tau_o$ is the maximum shear stress at the outer surface of the cylinder; $r_o$ is the outside radius; $J$ is the centroidal polar moment of inertia[3] of the cross section of the cylinder (either hollow or solid); and $\tau_\rho$ is the shear stress at any distance $\rho$ from the center. Equation 9-3 is popularly known as the *elastic torsion formula*. Although this is a very useful relationship, it is often used *incorrectly* in cases for which it is not valid. Be careful that you do not make this very common but serious mistake. It would be well for you to review the assumptions and limitations made in the derivation.

**Ideally Plastic Behavior.** Since a great many engineering bodies are made of mild steel, whose idealized shear stress-strain curve is shown in Fig. 9-9(a), we are justified in devoting some time to this special case.

Let us follow the progress of the stress distribution in a *solid* cylindrical torsion member as it is subjected to larger and larger shear strains. In Fig. 9-9(b), (c), and (d) are shown the three distinctive stages of the stress distribution as it passes through the elastic and elastic-plastic, or semi-plastic, stages to the idealized fully plastic stage, for which it is assumed that the entire cross section is subjected to the yield-point shear stress. The basic load-stress relationship for a *solid* cylindrical member in the idealized *fully plastic* condition becomes

$$T_{fp} = \int_{area} \tau \rho \, da$$

Since $\tau = \tau_{yp}$ at all points and the value of $da$ is as shown in Fig. 9-9(e),

$$T_{fp} = \tau_{yp} \int_0^{r_o} \rho(2\pi\rho \, d\rho)$$

$$T_{fp} = \tau_{yp} \frac{2\pi}{3} r_o^3$$

[3]For geometrical properties of various areas refer to Table II of Appendix.

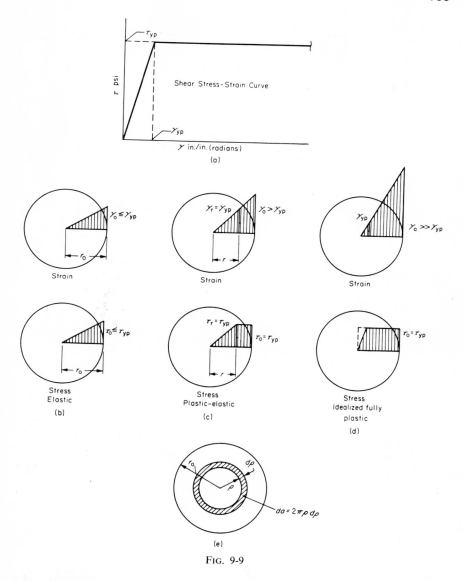

FIG. 9-9

This relationship is often given in the following equivalent form:

$$T_{fp} = \frac{4}{3} \frac{\tau_{yp}}{r_o} J \qquad\qquad (9\text{-}4)$$

A comparison of Eq. 9-3 and Eq. 9-4 shows that for a *solid* cylindrical member, the torque required for the fully plastic condition is four thirds of the maximum torque possible for elastic behavior; that is, the torque at which $\tau_o$ first reaches $\tau_{yp}$ is only three fourths of the torque required for the entire cross section to reach $\tau_{yp}$.

## 9-5. Angular Deformation or Angle of Twist

The angle of twist in a torsion member is defined as the relative rotation between two transverse plane cross sections. In Fig. 9-10(a) the left-hand end of the member is assumed to be fixed and the right end is caused to rotate by some torque $T$. The surface $ABCD$, which was originally a *longitudinal plane*, becomes the helical surface $ABCD'$. The important feature of this distortion is that the radial line $CD$ remains a straight radial line in the position $CD'$, even after twisting. This result is to be expected, since we have already shown that the shear strain varies linearly with the radius. The angle $\theta$ is the angle of twist between the left-hand and right-hand ends of the member.

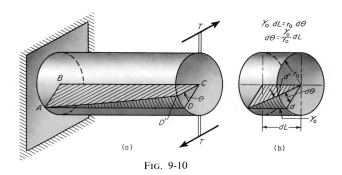

Fig. 9-10

In order to evaluate $\theta$ for any cylindrical torsion member, let us consider a portion of the member having an infinitesimal length $dL$, as shown in Fig. 9-10(b). For small values of shear strain $\gamma_o$, the arc length $dd'$ can be expressed as follows:

$$dd' = \gamma_o dL = r_o d\theta$$

Hence,

$$d\theta = \frac{\gamma_o}{r_o} dL$$

where $d\theta$ is the angle of twist occurring over the length $dL$. A finite angle of twist occurring over a finite length can be obtained by direct integration. Thus,

$$\theta = \int_{\text{over } L} \frac{\gamma_o}{r_o} dL \tag{9-5}$$

This relationship is geometrically valid for all values of $\gamma_o$ ordinarily encountered in engineering torsion members.

EXAMPLE 9-2. The member in Fig. 9-11 is subjected to the torques shown. Through how many degrees and in what direction will the left-hand end twist with respect to the fixed right-hand end? The stresses have already been checked, and elastic action is assured. Neglect any stress concentrations at the joints.

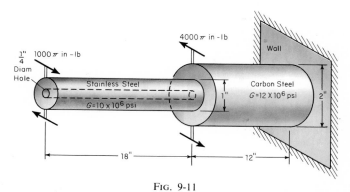

FIG. 9-11

*Solution*: The basic relationship between strain and twist is

$$\theta = \int_{\text{over } L} \frac{\gamma_o}{r_o} \, dL$$

In this elastic case, $\gamma_o = \tau_o/G$ and $\tau_o = Tr_o/J$. So

$$\theta = \int_{\text{over } L} \frac{T \, dL}{GJ}$$

In Fig. 9-11, $T$, $G$, and $J$ are not constants or continuous functions of $L$ from one end of the member to the other. In this case, we must find the twist of each part separately. Then

$$\theta = \int_0^{18} \frac{T_{ss} \, dL}{G_{ss} J_{ss}} + \int_0^{12} \frac{T_{cs} \, dL}{G_{cs} J_{cs}}$$

For the stainless-steel part, $T_{ss} = +1000\pi$ in-lb and is constant for all sections of the 18-in. length. The values of $G_{ss}$ and $J_{ss}$ are also constant throughout that length. For the carbon-steel part, $T_{cs}$ has a constant value of $-3000\pi$ in-lb (plus torque is assumed to act clockwise when you look in toward the wall). If you are not sure about $T_{cs}$, pass a plane through the carbon-steel part at right angles to its longitudinal axis and draw a free-body diagram. The values of $G_{cs}$ and $J_{cs}$ are also constant for the carbon-steel part. Therefore,

$$\theta = \frac{1000\pi}{(10 \times 10^6) \frac{\pi}{32} [1^4 - (\frac{1}{4})^4]} \int_0^{18} dL - \frac{3000\pi}{(12 \times 10^6) \frac{\pi}{32} (2^4)} \int_0^{12} dL$$

$$\theta = +0.0576 - 0.0060 = +0.0516 \text{ radian}$$

$$= \frac{180}{\pi} (+0.0516) = +2.95 \text{ degrees}$$

The plus sign means that the net twist of the left-hand end is clockwise when you look in toward the wall.

## 9-6. Summary of Cylindrical Torsion Members

When determining the mode of failure of any load-carrying member, you should keep clearly in mind the mechanical properties of the material involved and the existing state of stress in the member. Perhaps no other simple structural member illustrates the importance of the state of stress better than a cylindrical torsion member.

Consider, for example, a cylindrical torsion member made of a very ductile and homogeneous material, such as mild steel. If the member were twisted until it ruptured, the rupture would result primarily from the *shear* stresses on a plane perpendicular to the axis of the member. For a simple experiment, take a "tootsie roll" and twist it until it breaks in two, and observe the characteristics of the rupture.

As another observation, at one time or another you have probably seen someone break a wooden broom handle or mop handle by twisting it too much. The handle probably "split" along the grain, which usually runs somewhat parallel to the longitudinal axis. This splitting was caused by a combination of the *shear* stresses parallel to the longitudinal axis and the relative weakness of the wood in shear parallel to its grain.

On the other hand, consider a torsion member made of a brittle material, such as gray cast iron, which is relatively much weaker in tension than in shear or compression. If the member were twisted until it ruptured, the rupture would occur along a helical plane inclined approximately 45° to the longitudinal axis, since the maximum *tensile* stresses occur on this plane. For a simple test, twist a piece of blackboard chalk, and observe the rupture. Try to apply *pure* torque without bending the piece.

As a final consideration, hollow thin-walled cylindrical members are often used to transmit torque. If there is too much torque and the wall is too thin, the *compressive* stresses on a 45° plane may cause the wall to buckle and collapse.

So far in this chapter we have considered circular cylindrical members subjected *only* to pure torques about their longitudinal axes. In engineering structures it is very common to encounter members which are subjected to axial, torsion, and bending loads simultaneously. Even though we have not yet studied bending, our present knowledge permits us to consider members subjected to a combination of axial and torsion loads. The simplest way to analyze such a member is to utilize the principle of superposition; that is, we can superpose the effects due to torsion onto the effects due to axial loading, or vice versa. Under conditions of combined loading, the principle of superposition may be used to determine either stresses or strains. The order of superposition is unimportant, as long as the deformations are small. This will usually be the case if the stresses are less than the elastic limit. We will consider only the superposition of elastic stresses and deformations in this text.

Finally, since we have analyzed both the stresses and deformations occurring in cylindrical torsion members, we can combine this knowledge with the equations of static equilibrium to consider some statically indeterminate torsion members. The following examples illustrate the application of some of the ideas which have been discussed in this section.

EXAMPLE 9-3. The hollow closed cylinder of Fig. 9-12(a) is 10 in. in diameter and 12 ft long and has a wall thickness of $\frac{1}{8}$ in. It is subjected to an internal pressure of 60 psi, an axial tensile thrust of 8000 lb, and a torque of 1000 ft-lb. Find the maximum principal stress and the absolute maximum shearing stress and the orientation of each. Assume that the action is elastic, and neglect localized stress concentrations at the ends. Take $J$ as $2\pi t r^3$ (approximation for thin-walled circular section).

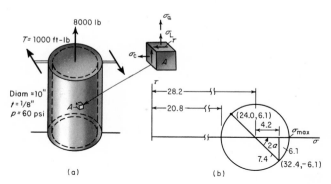

FIG. 9-12

*Solution*: Any element on the *outside* surface would have the same state of stress as element $A$ in Fig. 9-12(a). We may superpose the effects of internal pressure, torque, and axial load to get the complete state of stress. Since the longitudinal stress $\sigma_L$ due to internal pressure and the stress $\sigma_a$ due to axial load are in the same direction, we may add these two algebraically. Thus,

$$\sigma_L + \sigma_a = \frac{pD}{4t} + \frac{P}{a}$$

$$= \frac{(60)(10)}{(4)(1/8)} + \frac{8000}{\pi(10)(1/8)} = 3240 \text{ psi, tensile}$$

The circumferential stress $\sigma_c$ is

$$\sigma_c = \frac{pD}{2t} = \frac{(60)(10)}{(2)(1/8)} = 2400 \text{ psi, tensile}$$

The shear stress $\tau$ due to torque is

$$\tau = \frac{Tr_o}{J} = \frac{(1000)(12)(5)}{2\pi(1/8)(5^3)} = 610 \text{ psi}$$

From Mohr's stress circle in Fig. 9-12(b), in which one unit represents 100 psi we get

$$\sigma_{max} = 2820 + 740 = 3560 \text{ psi}$$

$$2\alpha = \tan^{-1}(6.1/4.2) = 55.4°$$

$$\alpha = 27.7°$$

The maximum principal stress acts in a direction making an angle of 27.7° counterclockwise with the longitudinal axis of the cylinder. Since the three principal stresses at any point on the outside surface are +3560 psi, +2080 psi, and 0 psi, the absolute maximum shear stress acts on a plane that bisects the angle between the planes of the +3560 psi and 0 psi stresses. This shear stress is equal to

$$\tau_{max} = \frac{1}{2}(\sigma_{max} - \sigma_{min})$$

$$= \frac{1}{2}(3560 - 0) = 1780 \text{ psi}$$

NOTE: On the *inside* surface of the cylinder, the third principal stress may be considered to be −60 psi (due to the internal pressure). However, the value of $\sigma_{max}$ would actually be smaller if the inside radius of the cylinder were used in the torsion formula. So, actually, the maximum shear stress on an element on the *inside* surface of this thin walled vessel would have nearly the same value as was found for the stress on the outside surface.

EXAMPLE 9-4. In the solid steel shaft in Fig. 9-13(a), what is the maximum shear stress? Neglect stress concentrations and assume elastic action.

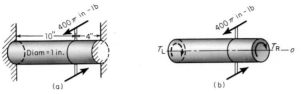

FIG. 9-13

*Solution:* Consider the free-body diagram of Fig. 9-13(b). If moments are taken about the longitudinal axis,

$$+ \text{⌐}\!-o$$

$$\sum M_o = 0$$

$$400\pi - T_L - T_R = 0$$

Since elastic action is assumed, $\tau = Tr/J$. Also, the left-hand and right-hand sections have the same values for $r$ and $J$. Therefore,

$$400\pi = \frac{\tau_L J}{r} + \frac{\tau_R J}{r} = (\tau_L + \tau_R)\frac{J}{r} \tag{a}$$

This is the only statics equation available, and it has two unknowns. So we must relate the unknowns with a deformation equation. Geometric compatibility requires that the left-hand portion of the shaft twist the same amount as the right-hand portion; that is, $\theta_L$ must be equal to $\theta_R$. For elastic action and where $T$, $G$, and $r_o$ are constant for each side, Eq. 9-5 gives

$$\theta_L = \theta_R$$

$$\frac{\tau_L L_L}{G_L r_L} = \frac{\tau_R L_R}{G_R r_R}$$

$$\tau_L = \frac{L_R}{L_L}\tau_R \tag{b}$$

Combining equations (a) and (b), we obtain

$$\frac{400\pi r}{J} = \frac{L_R}{L_L}\tau_R + \tau_R$$

The maximum shear stress is

$$\tau_R = \frac{4000\pi\,(0.5)}{\pi\!\left(\dfrac{1^4}{32}\right)\!\left(1+\dfrac{4}{10}\right)} = 4570 \text{ psi}$$

Also,

$$\tau_L = \frac{4}{10}\,(4570) = 1828 \text{ psi}$$

EXAMPLE 9-5. Suppose that a new type of plastic has the following properties: Each of its stress-strain relationships is linear to rupture, and the rupture stresses are 3000 psi in tension, 4000 psi in compression, and 5000 psi in shear. At what speed (rpm) would a solid shaft of this material rupture, if it were $\frac{1}{4}$ in. in diameter and were delivering $\frac{1}{4}$ horsepower? Assume a pure torque loading.

*Solution:* The pure torque loading will result in a pure shear state of stress at any point in the shaft. So the maximum shear, tensile, and compressive stresses are equal for any value of the torque (recall Mohr's stress circle for pure shear). Therefore the shaft will rupture in tension when all these stresses reach 3000 psi on the outer surface. Since the stress-strain relationship is linear to rupture, the elastic torsion formula is applicable and the torque required for rupture is

$$T = \frac{\tau J}{r} = \frac{(3000)\,\dfrac{\pi}{32}\left(\dfrac{1}{4}\right)^4}{\dfrac{1}{8}} = 9.22\ \text{in-lb} = 0.768\ \text{ft-lb}$$

We know that power equals torque times angular velocity. Thus,

$$\text{Power} = \text{ft-lb/min} = (\text{ft-lb})(\text{rad/min})$$

$$\left(\frac{1}{4}\ \text{hp}\right)\!\left(33{,}000\ \frac{\text{ft-lb}}{\text{min}}\Big/\text{hp}\right) = (0.768\ \text{ft-lb})(n\ \text{rpm})(2\pi\ \text{radian/rev})$$

The speed at rupture is
$$n = 1713\ \text{rpm}$$

## 9-7.  Noncircular Torsion Members

In general, the analysis of a noncircular torsion member is far more complicated than that of a circular cylindrical member. The major difficulty lies in determining the shear-strain distribution, since the discussion which we have presented in regard to the strain distribution in circular cylindrical members is not applicable to torsion members of other geometric cross sections. For example, we have concluded that in a cylindrical member plane transverse sections remain plane and the shear strain varies linearly from the geometric center. A simple experiment shows that these conclusions are not true for a torsion member having a rectangular cross section.

In Fig. 9-14(a) is shown a rectangular bar on which has been drawn a grid pattern of small rectangles. When the bar is twisted by a pure torque, the grid pattern assumes the form indicated in Fig. 9-14(b). Several observations can be made from a close examination of the pattern.

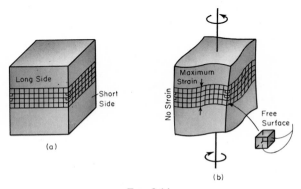

FIG. 9-14

1. The *transverse* lines do not remain straight when the torque is applied. Therefore, we can expect that transverse planes will not remain plane; that is, the transverse planes warp.

2. The maximum distortion, and, therefore, the maximum shear strain, occurs at the middle of each *longer* side. Also, the largest distortion on the shorter side occurs at its middle.

3. No distortion, and therefore no shear strain, occurs at the corners of the member. Of course, this result could be anticipated from the analysis of a free-body diagram of a corner element. Such an element has two free surfaces which are perpendicular to each other. For it to be in equilibrium, no shear force can exist on any other face of the element.

Several ingenious methods have been devised to determine the shear-strain distribution in noncircular torsion members. Perhaps the foremost of these is the *membrane analogy*.[4] However, the mathematics required to pursue this discussion to any meaningful conclusions is beyond the level of this book. The solutions of many problems for solid noncircular torsion members can be found in more advanced books.[5]

## 9-8. Hollow Thin-Walled Torsion Members

In the case of some noncircular torsion members, a careful study of the geometric and equilibrium requirements can lead to an *approximate* relationship between torque and shear stress. One such example is a hollow thin-walled member whose wall thickness is very small compared with its other dimensions; the wall thickness is not necessarily constant along the entire circumference. A portion of such a member of arbitrary shape and variable thickness is shown in Fig. 9-15(*a*). A free-body diagram of an element of the wall is shown in Fig. 9-15(*b*). For this element we know

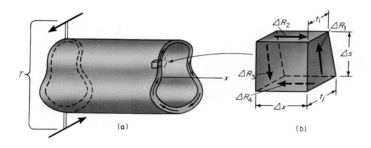

Fig. 9-15

[4]See *Advanced Mechanics of Materials,* by F. B. Seely and J. O. Smith, 2d ed., John Wiley & Sons, Inc., 1952.

[5]See *Theory of Elasticity,* by S. Timoshenko and J. N. Goodier, 2d ed., McGraw-Hill Book Company, Inc., 1951.

that no forces exist on the free inside and outside surfaces, and we make the simplifying *assumption* that no significant normal forces are present on the other four faces. This will lead to no serious error, provided that the member is not twisted too severely.

In the analysis of thin sections subjected to torsion or bending, it is often convenient to introduce a quantity called *shear flow*, denoted by $q$. The shear flow at any point in the thin section is defined as the longitudinal shear force across the thickness of the section per unit of length parallel to the longitudinal axis of the member. For example, at $i$ in Fig. 9-15, the average intensity of the longitudinal shear force per unit longitudinal length would be $\Delta R_2 / \Delta x$, and the shear flow across this thickness of section would be

$$q_i = \lim_{\Delta x \to 0} \frac{\Delta R_2}{\Delta x} = \frac{dR_2}{dx}$$

Similarly, the shear flow across the thickness at $j$ is

$$q_j = \lim_{\Delta x \to 0} \frac{\Delta R_4}{\Delta x} = \frac{dR_4}{dx}$$

Within the limits of our original assumption, equilibrium in the $x$ direction requires that $\Delta R_2 = \Delta R_4$. So it follows that $q_i = q_j$. Also, since points $i$ and $j$ were chosen arbitrarily, we can conclude that *the shear flow has the same value at any point around the circumference of the wall of a thin-walled torsion member.* Note that this statement does *not imply* that the *shear stress* is constant around the circumference of the cross section.

We shall now express the shear flow for a thin-walled torsion member in terms of the applied torque. In Fig. 9-16(*a*) and (*b*) are shown the location of an infinitesimal element of the wall and the forces acting on it.

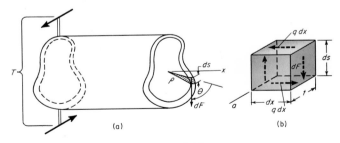

FIG. 9-16

Here the forces parallel to the $x$ axis are expressed in terms of the shear flow. If moments are taken about the $a$ axis and counterclockwise moments are considered positive, we get

$$\Sigma M_a = 0$$

$$(dF)(dx) - (q\,dx)ds = 0$$

$$dF = q\,ds$$

Now, if we take moments about the longitudinal $x$ axis of the member in Fig. 9-16(a) and substitute the value of $dF$ from Eq. 9-6, we obtain a relationship between the shear flow and the torque. Thus,

$$\Sigma M_x = 0$$

$$T - \int \rho(dF \sin \theta) = 0$$

$$T = \int_{\substack{\text{around} \\ \text{circumference}}} \rho q\,ds \sin \theta \qquad (9\text{-}7)$$

where $\theta$ is the angle between the tangential force $dF$ and the radius $\rho$.

The indicated integration operation requires some advanced theories of calculus.[6] However, we can interpret the integral in a practical manner as follows: Since $q$ is constant,

$$T = q \int_{\substack{\text{around} \\ \text{circumference}}} \rho(ds \sin \theta) \qquad (9\text{-}8)$$

But $\rho\,(ds \sin \theta)$ represents twice the area of the shaded triangle in Fig. 9-16(a). Hence,

$$\int_{\substack{\text{around} \\ \text{circumference}}} \rho(ds \sin \theta) = 2A \qquad (9\text{-}9)$$

---

[6] In vector calculus the expression inside the integral would be written as the vector product $\rho \times (ds)$; and by definition its magnitude would be equal to the area of a parallelogram with sides of length $\rho$ and $ds$. This area would be equal to twice the area of the shaded triangle in Fig. 9-16(a). In rectangular coordinates, with $\rho = i0 + jy + kz$ and $ds = i0 + j\,dy + k\,dz$,

$$\int_{\substack{\text{around} \\ \text{circumference}}} \rho \times ds = i \oint_{\substack{\text{around} \\ \text{circumference}}} z\,dy - y\,dz$$

From Green's lemma,

$$\oint_{\substack{\text{around} \\ \text{circumference}}} z\,dy - y\,dz = \iint_{\substack{\text{over} \\ \text{enclosed area}}} 2\,dy\,dz = 2A$$

where $A$ is the area enclosed by the wall of the member.

where $A$ is the area enclosed by the wall of the member i.e., the area of the hollow part of the member. It may be argued that $A$ should be the area enclosed by the "mean" wall. However, if the "mean" area were significantly different from the internal area, the member would probably not be classified as *thin-walled*, and our entire discussion would not be applicable. Also, by using the inside area the result will be on the conservative side. By combining Eqs. 9-8 and 9-9, we have

$$q = \frac{T}{2A} \tag{9-10}$$

It is significant to recall that the derivation of Eq. 9-10 was based solely on equilibrium requirements. No reference was made to the mechanical properties of the material involved or to the *actual* shear stresses and strains existing in the wall of the member.

A so-called approximate or average shear stress at a point in the wall can be evaluated by dividing the shear flow by the wall thickness at that particular point. Thus, the average shear stress at $i$ in Fig. 9-15 is

$$\tau_{i\,\text{avg}} = \frac{T}{2At_i} \tag{9-11}$$

Although this is only an approximate relationship between torque and shear stress, it is sufficiently accurate to make it applicable to many design problems. For your own benefit you should compare the results found by using Eq. 9-3 with those found by using Eq. 9-11 for a thin-walled cylindrical tube. Note that for this case Eq. 9-11 can be written as follows:

$$\tau_{\text{avg}} = \frac{T}{2At}\frac{r}{r} = \frac{Tr}{2tAr}$$

where $2tAr$ is an approximate value of $J$ which was used in Example 9-3.

### PROBLEMS

**9-1 to 9-6.** For each of the shafts, the cross sections of which are shown in the illustrations, the shear stress distribution has been determined to be as indicated. Find an expression in each case for the torque in terms of the radius $r$, $\tau_{\max}$, and other constants.

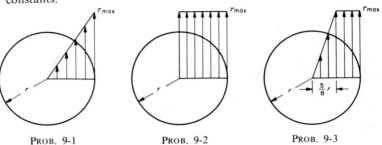

PROB. 9-1        PROB. 9-2        PROB. 9-3

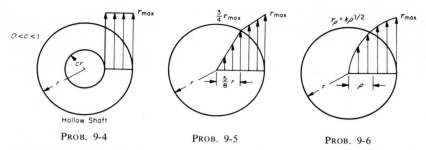

PROB. 9-4                    PROB. 9-5                    PROB. 9-6

**9-7 to 9-9.** If solid circular shafts of radius $r$ were made from materials whose shear stress-strain curves are shown in the illustrations, find the pure torque in each case that will cause a maximum shear strain $\gamma_r$ in the shaft. Assume in each case that the shear strain varies linearly from zero at the center to a maximum $\gamma_r$ at the outer surface, according to the relation $\gamma_\rho = k\rho$. Express answers in terms of the values $\tau_r$, $r$, and $n$; and also in terms of $\gamma_r$, $r$, $G$, and $n$.

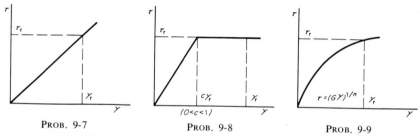

PROB. 9-7                    PROB. 9-8                    PROB. 9-9

**9-10.** A $\frac{3}{4}''$ diameter solid steel shaft is subjected to a pure torque of 180 foot pounds. What maximum shear stress is developed? What will be the maximum tensile stress?

**9-11.** In Prob. 9-10 what shear stress will occur $\frac{1}{8}''$ below the surface?

**9-12.** A 1 in. diameter maple dowel is used as a torsion member. What torque can it carry if the allowable stresses are: tension and compression 1600 psi and shear (parallel to the grain) 200 psi? The grain of the dowel is longitudinal.

**9-13.** *a*) Prove that the elastic strength of a solid shaft is reduced by only 1/16 if an axial hole having a diameter equal to one-half the outside diameter is drilled through the shaft. *b*) What percent of the cost is saved if it is assumed that the cost varies linearly with the weight? Assume elastic action only.

**9-14.** How much is the elastic strength (torque carrying capacity) of a solid steel shaft reduced by boring an axial hole through the center, if the area of the hole is one-half of the original shaft area?

**9-15.** Find the fully plastic torque in inch-pounds for a hollow shaft with outside and inside diameters of 4 in. and 2 in., respectively. The material of the shaft is ideally elastic-plastic (flat top shear stress-strain curve) and has a shear yield-point stress of 20,000 psi.

**9-16.** A solid circular shaft made of mild steel is twisted by pure torque until 40 percent of the cross-sectional area yields. What is the required torque, in terms of the yield-point of the material and the radius of the shaft?

**9-17.** A steel shaft 2 in. in diameter is subjected to a pure torque which twists the shaft through an angle of 0.12 radians in a length of 10 ft. Assuming elastic behavior, find the maximum tensile stress in the shaft.

**9-18.** A tubular aluminum alloy shaft 5 ft long has an outside diameter of 2 in. and an inside diameter of 1.5 in. It is to be used as an elastic torsion spring. What is the ratio of applied torque to angle of twist?

**9-19.** A solid circular shaft having a radius of 1.5 in. and a length of 8 in. is twisted until its maximum shear strain is 0.01 in./in. (radian). Determine the angle of twist and the required torque if the shaft is made of the material for which the stress-strain curve is shown in the illustration.

**9-20.** Solve Prob. 9-19 if the shaft is made of the material for which the stress-strain relationship is shown in the illustration for Prob. 9-20.

**9-21.** The composite shaft shown in the illustration is made up of a section of steel, a section of aluminum, and a section of brass. The total length is 6 ft. Also, $G_s = 12 \times 10^6$, $G_a = 4 \times 10^6$, and $G_b = 6 \times 10^6$. The angular deformation of each section should be the same, and the total deformation from end to end is to be 3°. Determine the length of each section and the torque $T$. Assume elastic behavior and neglect stress concentrations.

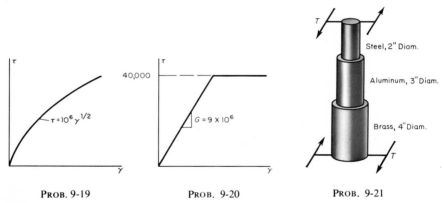

PROB. 9-19          PROB. 9-20          PROB. 9-21

**9-22.** A solid circular shaft with a uniform diameter of 2 in. is 12 ft long. At its midpoint 65 hp are delivered to the shaft by means of a belt passing over a pulley. This power is used to drive two machines, one at the left end consuming 25 hp and one at the right end consuming the remaining 40 hp. Determine *a*) the maximum shearing stress in the shaft and *b*) the relative angle of twist between the two extreme ends of the shaft. The shaft turns at 250 rpm, and the material is steel for which $G = 12 \times 10^6$.

**9-23.** A hollow shaft with an outside radius of 2 in. and an inside radius of 1 in. is rotating at 500 rpm. Using a factor of safety of 3, determine the maximum horsepower the shaft may safely transmit *a*) if the shaft is 36 in. long, it is made of a linear brittle material for which $\sigma_{ult} = 20,000$ tension, $\sigma_{ult} = 80,000$ compression, $\tau_{ult} = 30,000$, and $G = 8 \times 10^6$, and failure is due to rupture; *b*) if the shaft is 48 in. long, it is made of a ductile material for which $\tau_{yp} = 15,000$, $\sigma_{yp} = 30,000$, and $G = 12 \times 10^6$, and failure is due to excessive inelastic deformation.

**9-24.** How much will the right end of the shaft in the illustration twist relative to the fixed left end if $G_s = 12 \times 10^6$ and $G_b = 5 \times 10^6$? Assume elastic behavior and neglect stress concentrations.

**9-25.** Find the angle of twist of the left end of the member in the illustration with respect to the fixed right end if $G_s = 12 \times 10^6$ and $G_a = 3.8 \times 10^6$. Assume elastic behavior and neglect stress concentrations.

**9-26.** Find the angle of twist of the left end of the member in the illustration with respect to the fixed right end if $G_s = 12 \times 10^6$ and $G_a = 3.8 \times 10^6$. Assume elastic behavior and neglect stress concentrations.

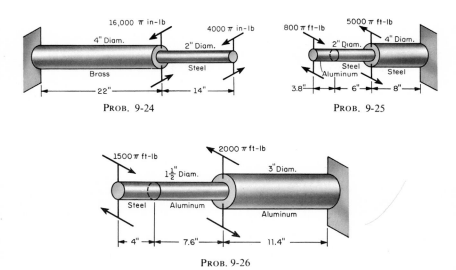

PROB. 9-24          PROB. 9-25

PROB. 9-26

**9-27.** How much will the right end of the shaft in the illustration twist relative to the fixed left end if $G_s = 12 \times 10^6$ and $G_a = 4 \times 10^6$? Assume elastic behavior and neglect stress concentrations.

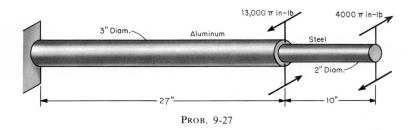

PROB. 9-27

**9-28.** For the shaft shown in the illustration, $G_s = 12 \times 10^6$ and $G_a = 4 \times 10^6$. Determine *a)* the maximum shear stress in the steel, *b)* the maximum shear stress in the aluminum, and *c)* the total angle of twist.

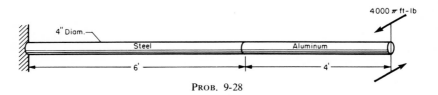

PROB. 9-28

**9-29.** What must be the length of a steel rod 0.2 in. in diameter so that it can be twisted through one complete revolution without exceeding an elastic shear stress of 8,000 psi?

**9-30.** The circular stepped shaft in the illustration is composed of a section of steel ($G_s = 12 \times 10^6$) and a section of brass ($G_b = 6 \times 10^6$). The over-all angular deformation from end to end is not to exceed 0.1 radian. The angle of twist of the brass section should be twice that of the steel section. Assuming that the elastic limits are not exceeded, determine the maximum torque that this shaft can carry between its ends. Neglect stress concentrations at the junction.

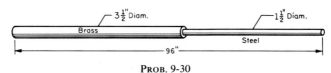

PROB. 9-30

**9-31.** A shaft 1 in. in diameter is made of a brittle material with an allowable shear stress of 15,000 psi and an allowable tensile stress of 10,000 psi. If the stress-strain relationship is linear, what is the maximum horsepower the shaft may safely transmit when it is rotating at 400 rpm?

**9-32.** A shaft 1 in. in diameter is made of a very ductile material with an ultimate shear stress of 20,000 psi and an ultimate tensile stress of 30,000 psi. What is the maximum horsepower the shaft may safely transmit when rotating at 400 rpm? Failure is probably due to excessive inelastic deformation.

**9-33.** A solid steel shaft 1 ½ in. in diameter and 20 in. long transmits power by means of a pure torque at a rate of 396,000 in-lb/sec when turning at 31,500 rpm. If $G = 12 \times 10^6$, determine *a*) the magnitude and direction of the maximum tensile stress and *b*) the total angle of twist. Assume elastic behavior.

**9-34.** A hollow circular shaft 20 in. long is rotating at 1750 rpm. Its outside diameter is 4 in. and its inside diameter is 2 in. Based on a factor of safety of 3, determine the maximum horsepower the shaft may safely transmit if *a*) the shaft is made of a linear brittle material, for which $G = 10 \times 10^6$; $\tau_{ult} = 100,000$ psi; $\sigma_{ult} = 60,000$ psi, tension; and $\sigma_{ult} = 80,000$ psi, compression; and the over-all angular deformation may not exceed 2°; *b*) the shaft is made of a ductile material for which the $\tau$-$\gamma$ curve is shown in the illustration and the angular deformation may not exceed 5°.

**9-35.** The solid circular shaft shown in the illustration is rotating at 630 rpm and is made of a linear homogeneous material which has an ultimate tensile stress of 60,000 psi, an ultimate shear stress of 40,000 psi, and a shear modulus of 12 ×

$10^6$. If the over-all angle of twist is not to exceed 0.72 radian, and a factor of safety of 2 is required, what is the maximum horsepower the shaft may safely transmit?

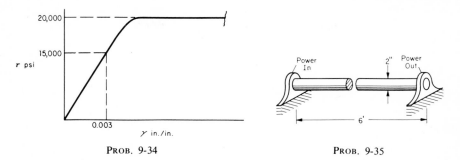

PROB. 9-34          PROB. 9-35

**9-36.** The elastic circular shaft in the illustration carries a uniformly distributed torque of $T$ in-lb/in. along its entire length. *a*) What is the maximum shearing stress at a section $L/4$ from the fixed left end? *b*) What is the total relative angle of the twist between its ends?

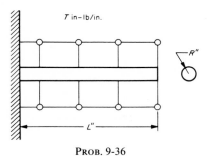

PROB. 9-36

**9-37.** Derive an expression for the maximum shear stress in the steel portion of the member shown in the illustration. Assume that there is elastic behavior and that the member is stress-free before being loaded. Neglect stress concentrations.

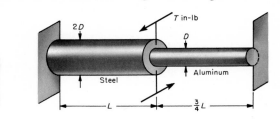

PROB. 9-37

**9-38.** The tip of the weightless pointer in the illustration moves 1 in. on the scale as a result of the application of the torque. What is the maximum shearing stress in the steel if $G_s = 12 \times 10^6$ and $G_a = 4 \times 10^6$?

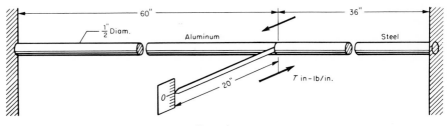

Prob. 9-38

**9-39.** Between two rigid parallel vertical plates which are 40 in. apart, two rods are to be welded with their axes on the same line. One rod is 2 in. in diameter and 30 in. long, and the other rod is 1 in. in diameter and 10 in. long. First the two rods are welded at the plates so that the other two ends come together 30 in. from one plate or 10 in. from the other plate. Before these two free ends are welded together, the free end of the 30-in rod is twisted 4°; and the weld is then made while a torque holds this twist in the larger rod. When the weld is completed, the torque is removed and the composite member is allowed to come to equilibrium. What is the resulting maximum shear stress in each rod if $G = 5.9 \times 10^6$ for the material of both rods? Assume elastic action, and neglect localized stress concentrations.

**9-40.** The brass and steel torsion members are fixed at the top and bottom and bolted at the flange with six bolts. The bolts are each 2 ½ inches from the center of the shafts. What maximum torque $T$ can be applied if the average shearing stress in the bolts must not exceed 14,000 psi?

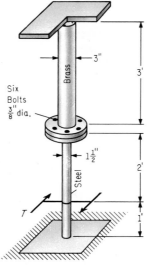

Prob. 9-40

**9-41.** Two solid steel shafts 2 in. in diameter are to be joined by means of a flanged coupling with four ¼-in. diameter steel bolts equally spaced as shown in the illustration. Determine the diameter $d$ of the circle through the bolts required to make the coupling and the shafts equally strong in resisting torsion. Assume that the shafts are elastic and that the allowable shear stress for steel is 12,000 psi.

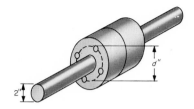

PROB. 9-41

**9-42.** A stud bolt $\frac{7}{8}$ in. in diameter is to be tightened with a 600 in-lb couple. If the axial tensile force in the unthreaded shank of the bolt at the time of the development of the maximum torque is 6000 lb, determine the maximum tensile and shear stresses in the bolt. Neglect the stresses in the threaded portion of the bolt.

**9-43.** *a*) Determine the maximum principal stress in the composite member in the illustration. Where does this stress occur, and what is its orientation? *b*) Determine the maximum shear stress in the member. Where does this shear stress occur, and what is its orientation? Neglect stress concentrations at the joint and at the fixed end and assume elastic behavior.

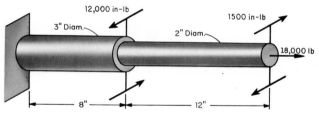

PROB. 9-43

**9-44.** For the hollow member in the illustration, find the magnitudes and directions of the principal stresses and the maximum shear stress. Assume elastic behavior, and neglect localized stresses at the ends.

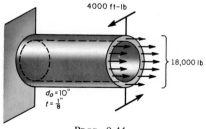

PROB. 9-44

**9-45.** The engine of a light airplane is delivering 180 hp directly to the propeller, which is turning at 2360 rpm during take-off. The forward thrust developed by the propeller is 120 lb. How big must the solid propeller shaft be, if the allowable stress in tension is 12,000 psi and that in shear is 7200 psi?

**9-46.** What would be the reading of a single axial strain gage in micro-inches per inch, if it were attached to the surface of a solid circular 2024-T3 aluminum alloy member which is 2 in. in diameter and subjected to an axial pull of 4 tons and a torque of 0.5 ft-tons. The gage axis makes an angle of 20° with the longitudinal axis of the member. The loads are in such directions that each alone would cause a tensile strain in the gage. $E = 10.6 \times 10^6$ psi and $\mu = \frac{1}{3}$.

**9-47.** A thin-walled circular brass tube with a wall thickness of 0.05 in. is to be used to transmit $500\pi$ in-lb of torque. Assuming that the tube does not buckle and that the shear stress is not to exceed 9000 psi, determine the required diameter of the tube.

**9-48.** A thin-walled elliptical tube is used to transmit 1800 in-lb of torque. The cross section of the tube is shown in the illustration. What is the shear stress in the wall of the tube?

**9-49.** An extruded aluminum tube has the cross section shown in the illustration. A piece of this tubing is to be used to transmit a torque. If the shear stress may not exceed 10,000 psi, what is the maximum permissible torque based on a factor of safety of 4?

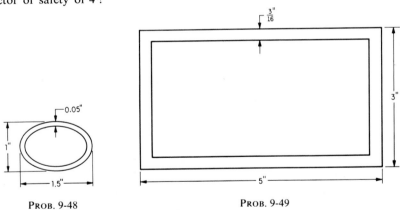

PROB. 9-48          PROB. 9-49

**9-50.** A thin-walled tube has a square cross section and a wall thickness of 0.10 in. Approximately what torque can a $1'' \times 1''$ (outside) tube transmit, if the shear stress in the wall is not to exceed 4500 psi?

**9-51.** A steel tube with closed ends has an outside diameter of $2\frac{1}{2}$ in. and a wall thickness of 0.05 in. If it is subjected to an internal gage pressure of 800 psi and a torque of 3140 in-lb, determine the magnitude and direction of the maximum principal stress.

## SUPPLEMENTARY PROBLEMS

**9-52.** Derive an expression for shear stress, as a function of pure torque and the distance from the center of the shaft, for a solid circular shaft of radius $r$ made from a material for which the shear stress-shear strain relation is $\tau = (1/K)\gamma^{2/3}$. Assume that the shear strain increases linearly from zero at the center to a maximum at the outer fiber, $\gamma_\rho = C\rho$, where $\rho$ is any distance from the center.

**9-53.** Under certain conditions the polar moment of inertia $J$ of a hollow thin-walled circular cylinder may be approximated as $J = 2\pi r^4 h$, where $h$ is the ratio of the thickness to the outside radius. a) If $h = 0.08$, what percent difference is introduced in the polar moment of inertia by using the approximate formula? b) If $\theta$, the angle of twist, is inversely proportional to $J$, what is the percent difference in the elastic angle of twist when the approximate value of $J$ is used instead of the exact value?

**9-54.** The circular shaft of constant radius shown in the illustration is acted upon by a torque which varies from zero at the right end to $T$ in-lb/in. at the wall. Assume a linear shear stress-strain relationship for the material. What is the total angle of twist of the right end of the shaft in terms of $T, r, L$, and $G$?

**9-55.** A solid circular shaft is made of material with a shear stress-strain curve shown in the illustration. The shaft is 100 in. long and has a radius of 2 $\frac{1}{2}$ in. If a torque of 500,000 $\pi$ in-lb is applied, what will be the angular twist of the shaft? Assume linear strain distribution.

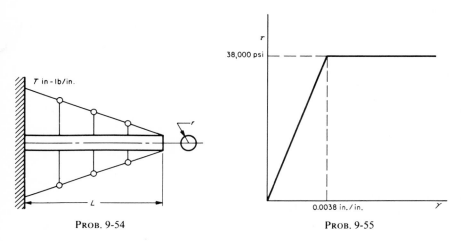

PROB. 9-54                                      PROB. 9-55

**9-56.** A solid circular shaft is made of a material having a shear stress-strain relationship of $\tau = k\gamma^{1/2}$, where $k = 400,000$. The shaft is 120 in. long and 3 in. in diameter, and is subjected to a torque of 200,000 in-lb. Assuming that the strain varies linearly from the center of the shaft, determine the angular twist due to the torque.

**9-57.** A hollow steel shaft that is 8 ft long must transmit a torque of 175,000 in-lb. The total angle of twist in this length must not exceed 2.5°, and the allowable shear stress is 12,000 psi. What is the inside and outside diameter of a shaft that meets these requirements simultaneously?

**9-58.** An electrical resistance axial strain gage is attached to the surface of a $\frac{1}{4}$-in. steel shaft so that the gage axis makes an angle of 55° with the shaft axis. When the shaft is rotating at 1720 rpm, the strain gage reads 1200 micro-inches per inch tensile (the electrical leads are taken away from the rotating shaft through slip rings). What horsepower is the shaft delivering? Assume only pure torque loading on the shaft.

**9-59.** What is the twist at the center cross section of the shaft in the illustration? Assume elastic action, and express your answer in terms of $T$, $L$, and $G$; $G$ is the same for both sides.

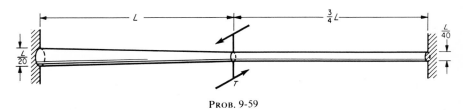

PROB. 9-59

**9-60.** Find the angle of twist of the outer shell, relative to the inner shell, of

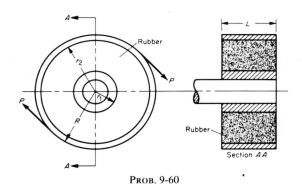

PROB. 9-60

the torsilastic spring shown in the illustration. Assume a linear shear stress-strain relationship for the rubber.

## chapter 10

# *Bending*

## 10-1.  Introduction

In Chapter 6 the word beam was used when referring to any member which supported transverse loads.  During that discussion we were primarily interested in making a static analysis of load distribution, and we did not concern ourselves with the manner in which the member responded to loads. In Chapters 8 and 9 we investigated the response of members subjected primarily to axial and torsion loads, respectively.  We will now turn our attention to relatively slender members subjected to transverse loads which produce significant bending effects.

The development of this chapter will closely follow that in the chapter on torsion.  First we shall suppose that a certain type of loading exists on a member with a somewhat special physical configuration.  Then we shall examine the geometry of the resulting deformation.  From this examination we will make certain assumptions, on the basis of which we can proceed to develop desired load-stress and load-deformation relationships.  In order to facilitate the development of the theory, we will consider the bending, or flexural, stresses and the transverse-shear stresses separately.

At this point it is well for you to refresh your memory on some of the topics discussed in Chapter 6—particularly the shear and moment diagrams and the classification of beams according to the manner in which they are loaded and supported.

## 10-2.  Pure Bending of Beams with Symmetrical Sections

**Initial Restrictions.**  Generally, when a beam having an arbitrary cross section is subjected to transverse loads, the beam will bend.  In addition, twisting and buckling may be present, and a problem that includes the combined effects of bending, twisting, and buckling can become a complicated one.  In order that we may investigate the bending effects alone, we will place certain restrictions on the geometry of the beam and on the manner of loading.

First of all we shall assume that the beam is straight, has a constant cross section, is made of a homogeneous material, and its cross section *has a longitudinal plane of symmetry.*  Second, we shall assume that the *resultant* of the *applied loads lies in this plane of symmetry.* These conditions will eliminate the possibility that the beam will twist. Later we shall see that these strict requirements can be lessened somewhat, but for the present it is convenient to assume that the above conditions exist. Also, we shall assume for the present that the geometry of the over-all member is such that bending and *not* buckling is the primary mode of failure. In Fig. 10-1 are shown several common shapes of beams, each of which has a longitudinal plane of symmetry.

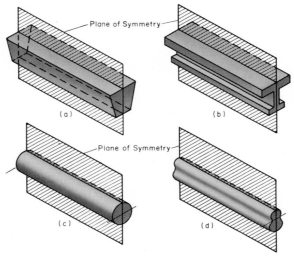

FIG. 10-1

We saw in Chapter 6 that the internal reactions on any cross section may consist of a resultant normal force, a resultant shear force, and a resultant couple. In order that we may examine the bending effects alone, we will restrict the loading to one for which the resultant normal and shear forces are zero on any section perpendicular to the longitudinal axis of the member. The zero shear force implies that the bending moment is the same at every cross section of the beam; that is, $dM/dx = 0$. We may visualize this beam, or some portion of the beam, as being loaded only by pure couples at its ends, remembering that these couples are assumed to be applied in the plane of symmetry. When a member is so loaded, it is said to be in *pure bending* and the plane of symmetry is called the *plane of bending*. The problem as now stated has sufficient symmetry to permit reasonable arguments about the deformation of the beam.

**Longitudinal Deformation.** Figure 10-2(*a*) shows one portion of a homogeneous straight beam with a longitudinal plane of symmetry.

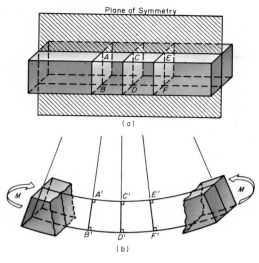

Fig. 10-2

For convenience a rectangular cross section was selected. At $AB$, $CD$, and $EF$ are represented three equally spaced plane sections perpendicular to the longitudinal axis of the beam. Figure 10-2(b) shows the shape of the longitudinal plane of symmetry of the beam after the application of the couples at its ends, with the deformation greatly exaggerated. The portion of the beam between the plane sections at $AB$ and $CD$ was originally geometrically identical in length and cross section with the portion between the sections at $CD$ and $EF$. Also, each of these portions is loaded in exactly the same manner, that is, by pure couples. Therefore, in order that each of these portions may be geometrically compatible with its adjacent portions in the deformed state, we may expect that these portions would undergo similar compatible deformations. The same argument can be extended to the entire length of the beam (except of course to a part in the immediate vicinity of the externally applied loads). We can now draw several important conclusions in regard to the deformation in Fig. 10-2(b), which is characteristic of the deformation of any beam having a symmetrical cross section and subjected to pure bending.

1. Plane sections originally perpendicular to the longitudinal axis of the beam *remain plane* and perpendicular to the longitudinal axis after bending; that is, in Fig. 10-2(b), the cross sections at $A'B'$, $C'D'$, and $E'F'$ do not become curved or warped.

2. In the deformed beam, the planes of these cross sections have a common intersection; that is, any line originally parallel to the longitudinal axis of the beam becomes an arc of a circle.

Note that the analysis so far has dealt only with the deformation in any plane parallel to *the vertical plane of symmetry*. No remarks have yet been made in regard to the deformations perpendicular to the plane of symmetry. They will be considered shortly.

Let us now examine more closely the deformation of an element of the beam having a length $\Delta x$. A small portion of the unloaded beam is shown in Fig. 10-3(*a*), and the unloaded element is shown in (*b*). Also, a part of the deformed beam and the deformed element are shown in Fig. 10-3(*c*) and (*d*), where the deformation is greatly exaggerated.

Under the influence of the couples $M$, the upper fibers of the beam are shortened while the lower fibers are elongated. However, there is one surface (containing the arc $E'F'$) in which the fibers undergo *no* elongation or contraction. This longitudinal surface is called the *neutral surface*. For a beam with a symmetrical cross section and under pure bending, the neutral surface is perpendicular to the longitudinal plane of bending. Also, the intersection of the neutral surface with the longitudinal plane of bending of a beam, as arc $E'F'$ in Fig. 10-3(*d*), is called the *neutral axis* of the beam.

Although we have now defined the neutral surface and the neutral axis for a beam, we have not as yet attempted to determine their locations in the beam. This will be done shortly, but for the present we shall assume that the neutral surface can be located.

In Fig. 10-4(*a*), *u* represents the distance between the neutral axis *EF* and some other parallel line *GH* in the plane of symmetry. We now make the assumption that the distance between these lines in the deformed beam, Fig. 10-4(*b*), will not be significantly different from the original distance in

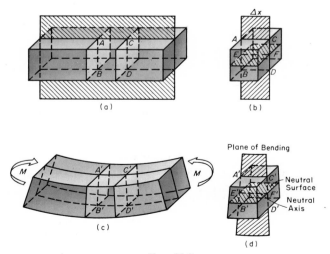

Fig. 10-3

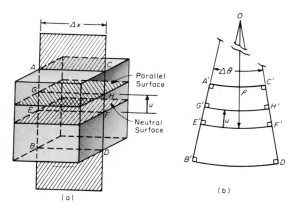

Fig. 10-4

the unloaded beam. Thus, if the radius of curvature of $E'F'$ is $\rho$, that of $G'H'$ will be $\rho - u$. From the definition of neutral surface and the geometry of Fig. 10-4($b$), we have

$$\Delta x = E'F' = \rho(\Delta\theta)$$

Also, the contraction of a fiber whose original position was $GH$ becomes

$$G'H' - GH = (\rho - u)(\Delta\theta) - \Delta x = (\rho - u)(\Delta\theta) - \rho(\Delta\theta) = -u(\Delta\theta)$$

By definition, the axial strain of the fiber $GH$ becomes

$$\epsilon = \frac{-u(\Delta\theta)}{\Delta x} = \frac{-u(\Delta\theta)}{\rho(\Delta\theta)} = -\frac{u}{\rho} \tag{10-1}$$

Since for pure bending the radius of curvature is constant for the entire length of the beam, we see that *the longitudinal axial strain is directly proportional to the distance u from the neutral surface.* The negative sign indicates compressive strain for a positive value of $u$ (fibers above the neutral surface), and tensile strain for a negative value of $u$ (fibers below the neutral surface). It is assumed, of course, that the radius of curvature is positive (the beam is concave upward).

**Transverse Deformation.** Before concluding this section, we shall consider briefly the deformations perpendicular to the longitudinal plane of symmetry. For convenience a beam with a rectangular cross section will again be used for the discussion. See the sections $ABba$ and $CDdc$ in Fig. 10-5($a$).

We have already seen that the longitudinal axial strain is directly proportional to the distance from the neutral surface, which is represented in Fig. 10-5($b$) by the shaded surface $E'F'f'e'$. We also know that an elonga-

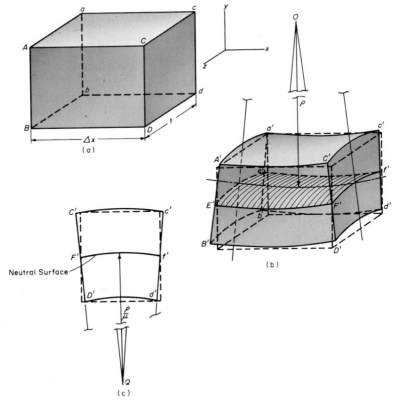

FIG. 10-5

tion in a longitudinal direction will result in a contraction in the transverse direction because of the Poisson effect. Thus the fibers above the neutral surface will be elongated in a direction perpendicular to the plane of symmetry, and those below will be shortened. If the longitudinal strain is denoted by $\epsilon_x$, the strain $\epsilon_z$ in the $z$ direction becomes

$$\epsilon_z = -\mu\,\epsilon_x = -\mu\left(-\frac{u}{\rho}\right) = \mu\left(\frac{u}{\rho}\right) = \frac{u}{\rho/\mu}$$

Hence, the original rectangular cross section $CDdc$ deforms as shown in Fig. 10-5(c), and $F'f'$ has a radius of curvature of $\rho/\mu$.

The neutral surface actually undergoes a double curvature, as indicated in Fig. 10-5(b); strictly speaking, therefore, our previous analysis of longitudinal strain is valid only for the plane of symmetry. However in applying the theory of this section in succeeding sections, we shall neglect the curvature due to Poisson's effect, and we shall assume that the longitudinal axial strain throughout the length and thickness of the beam is proportional to the distance from the neutral surface. The existence of linear strain distribution can be easily verified by a simple laboratory experiment in which several strain gages are used on a beam section.

## 10-3. Basic Load-Stress Relationship for Bending

**Equilibrium Requirements.** The entire analysis in the preceding section is based on a member subjected to pure bending. This condition exists when the only resultant internal reaction is a pure couple. For example, in Fig. 10-6(a), the portion of the beam between the two equal loads $P$ is in pure bending, because the *resultant normal* and *shear* forces on any cross section in this portion will be zero. The resultant internal reactions on a cross section are shown in Fig. 10-6(c). There is a resultant bending

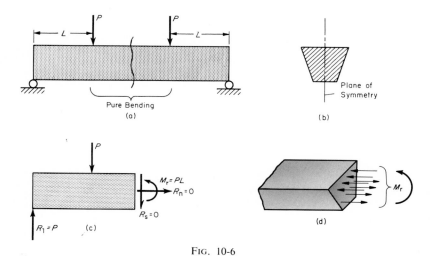

FIG. 10-6

moment, and this moment must be the net effect of the normal tensile and compressive forces exerted by the individual fibers of the beam. Thus, as shown in Fig. 10-6(d), normal stresses do exist on each cross section, but from the previous deformation analysis and the fact that there is no resultant shear force, we can assume that there are no shear stresses on the cross section. Normal stresses due to bending are commonly called *flexure stresses*.

Now, although we do not yet know the location of the neutral axis, we will arbitrarily establish a reference coordinate system in which the $x$ axis is taken as the neutral axis. As shown in Fig. 10-7, the $xy$ plane is then the plane of symmetry of the beam, and the neutral surface lies in the $xz$ plane. Since the resultant internal moment $M_r$ is actually a moment about the $z$ axis, we have

$$M_r = \sum M_z$$

$$M_r = -\int_{\substack{\text{cross} \\ \text{section}}} y\,(\sigma\,da) \qquad (10\text{-}2)$$

Also, if forces to the right are considered positive,

$$R_n = \sum F_x = 0$$

$$R_n = \int_{\substack{\text{cross} \\ \text{section}}} \sigma\,da = 0 \qquad (10\text{-}3)$$

Finally, we know that the resultant moment about the $y$ axis is zero. If counterclockwise moments about this axis are considered positive, we get

$$\sum M_y = 0$$

$$\sum M_y = \int_{\substack{\text{cross} \\ \text{section}}} z\,(\sigma\,da) = 0 \qquad (10\text{-}4)$$

Equations 10-2, 10-3, and 10-4 represent the equilibrium requirements which must be satisfied by the stress distribution over each cross section of the beam. Actually, because of the symmetry of this problem with respect to the $xy$ plane, Eq. 10-4 will automatically be satisfied whenever a beam with a symmetrical cross section is loaded in the plane of symmetry.

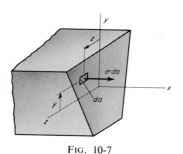

Having expressed the strain distribution by Eq. 10-1 and the equilibrium requirements by Eqs. 10-2 and 10-3, we are now ready to determine specific load-stress relationships. Of course, any such relationships will be dependent on the stress-strain relationship for the material involved. The following important special cases are intended to illustrate the procedure for deriving specific load-stress relationships.

Fig. 10-7

**Case I: Elastic Bending of Beams with Symmetrical Sections.** Elastic behavior of flexure members plays an important role in design procedure. For instance, a loaded beam which has not been stressed beyond the elastic limit of the material will exhibit no permanent "sag" when the loads are removed. Also, for most materials, elastic deformations are very small, and the deflection of an elastic beam will therefore be small. In many design problems small elastic deformations can be tolerated but large inelastic deformations would produce failure. For these reasons we are quite justified in studying elastic behavior of beams in considerable detail.

We have already shown in Sec. 10-2 that the longitudinal axial *strain* varies linearly from the neutral surface. Such a distribution can be illustrated graphically by means of a diagram such as that in Fig. 10-8 where the upper fibers at section *cc* are in compression and the lower fibers are in

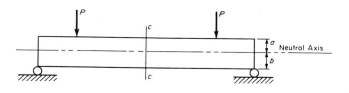

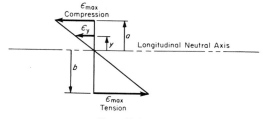

Fig. 10-8

tension. The strain at any point at a distance *y* from the neutral surface can be determined by the following simple ratio:

$$\frac{\epsilon_y}{y} = \frac{\epsilon_{max}(\text{tension})}{b} = \frac{\epsilon_{max}(\text{compression})}{a} \tag{10-5}$$

So far, no restriction on the behavior of the material has been imposed. Therefore, this strain distribution implies neither elastic nor inelastic behavior, and it is usually assumed that this distribution exists at least until the beam has been deformed beyond use.

*For a uniaxial state of stress,* elastic response of a homogeneous isotropic material usually manifests itself in a linear stress-strain relationship having the form

$$\sigma = E\epsilon \tag{10-6}$$

Since the axial strain in a beam varies linearly from the neutral surface, elastic behavior will be characterized by a *linear stress distribution,* since the stress is simply the strain multiplied by the modulus of elasticity. However, it must be remembered that in a bent beam some of the fibers are in tension and others in compression. Fortunately, the modulus of elasticity for most structural materials is the same in tension as in compression. Figure 10-9(a) shows the stress distribution for this case, while Fig. 10-9(b)

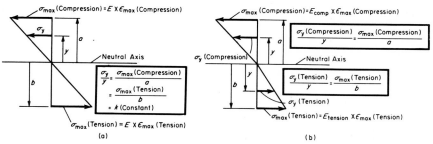

FIG. 10-9

shows the stress distribution for the case where the moduli differ. We will now continue our development for the former case.

Substituting the relationship of Fig. 10-9(a) in Eq. 10-3, we obtain

$$\int_{\substack{\text{cross} \\ \text{section}}} \sigma \, da = 0$$

$$\int ky \, da = k \int_{\substack{\text{cross} \\ \text{section}}} y \, da = 0 \qquad (10\text{-}7)$$

Since $k$ is not zero, the integral, which represents the first moment of the cross-sectional area about the neutral surface, must be zero. This requirement implies that the neutral surface of the beam must pass through the horizontal centroidal axis of the cross section. Thus, *for elastic action the strain and stress vary linearly from the transverse centroid of the cross section.*

Substitution in Eq. 10-2 yields the following result:

$$M_r = -\int_{\substack{\text{cross} \\ \text{section}}} y \, (\sigma \, da) = -\int_{\substack{\text{cross} \\ \text{section}}} y \, (ky \, da)$$

$$M_r = -k \int_{\substack{\text{cross} \\ \text{section}}} y^2 \, da \qquad (10\text{-}8)$$

Since we have just seen that $y$ is measured from the centroidal axis, the integral in Eq. 10-8 represents the second moment of the cross-sectional area about its horizontal centroidal axis. This second moment is commonly called the rectangular moment of inertia although no masses are involved, only cross sectional areas. These second moments of area should not be confused with mass moments of inertia encountered in dynamics.

If the second moment of area is denoted by $I_c$, the equation becomes

$$M_r = -kI_c \qquad (10\text{-}9)$$

The minus sign merely indicates that a positive bending moment $M_r$ produces compression in the upper fibers and tension in the lower fibers. In particular, from Fig. 10-9(a) we have

| For upper fibers, | For lower fibers, |
|---|---|

$$\frac{\sigma_{max}(\text{compression})}{a} = k \qquad \frac{\sigma_{max}(\text{tension})}{b} = k$$

$$\sigma_{max}(\text{compression}) = \frac{M_r a}{I_c} \qquad \sigma_{max}(\text{tension}) = \frac{M_r b}{I_c} \qquad (10\text{-}10)$$

In general,

$$\frac{\sigma_y}{y} = k$$

$$\sigma_y = -\frac{M_r y}{I_c} \qquad (10\text{-}10a)$$

Equations 10-10 and 10-10a are various forms of the famous and widely used (and misused) *elastic flexure formula*. It would be well for you to go back over the derivation and note the limitations which were imposed in the procedure.

**Case II: Ideally Plastic Behavior.** It was pointed out at the beginning of the preceding derivation that the strain distribution of Fig. 10-8 is independent of the behavior of the material. Since many structural members are made of mild steel, for which the idealized stress-strain curve is shown in Fig. 10-10(a), we shall devote some time to ideally plastic behavior.

Let us follow the progress of the stress distribution as a beam is subjected to larger and larger strains. In Fig. 10-10(b), (c), and (d) are shown three distinctive stages of the stress distribution as the material passes through the elastic, the elastic-plastic, and the fully plastic stages. For the *idealized* fully plastic condition, all the fibers above the neutral surface (wherever it may be) are subjected to a *uniform* compressive stress $\sigma_{ypc}$, and the lower fibers are subjected to a *uniform* tensile stress $\sigma_{ypt}$. If compressive stress is denoted by a minus sign, Eq. 10-3 becomes

$$\int_{\substack{\text{cross}\\\text{section}}} \sigma\, da = 0$$

$$\int_{\substack{\text{upper}\\\text{area}}} (-\sigma_{ypc})da + \int_{\substack{\text{lower}\\\text{area}}} \sigma_{ypt}\, da = 0$$

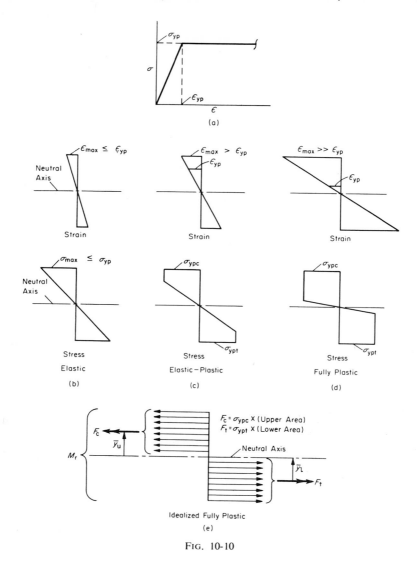

FIG. 10-10

If the yield-point stress is the same for tension and compression, we obtain

$$\sigma_{yp}(\text{upper area} - \text{lower area}) = 0$$

Hence, the area above the neutral surface must be equal to the area below the neutral surface. In other words, for fully plastic action the neutral surface splits the cross section in half, area-wise. The position of the neutral surface depends on the geometry of the cross section; the neutral surface *may or may not* pass through the centroidal axis.

The moment-stress relationship can be obtained from Fig. 10-10(e). The resultant forces $F_t$ and $F_c$ act at the centroids of the corresponding distributed forces. Thus,

$$M_r = [\sigma_{ypc} \times (\text{upper area}) \times (\bar{y}_u)] + [\sigma_{ypt} \times (\text{lower area}) \times (\bar{y}_l)]$$

where $\bar{y}_u$ and $\bar{y}_l$ are the upper and lower centroidal distances, respectively, and each is measured from the fully plastic neutral surface.

EXAMPLE 10-1. A beam is fabricated of three $6'' \times 1''$ steel plates welded together to form a symmetrical I-shaped section, as shown in Fig. 10-11(a). If the material has a yield-point stress of 40,000 psi, determine *a*) the maximum elastic resultant bending moment and *b*) the fully plastic resultant bending moment, assuming an idealized stress distribution.

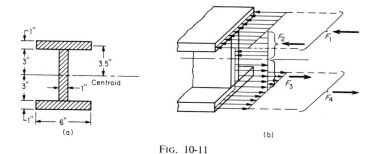

FIG. 10-11

*Solution: a*) From Fig. 10-11(a), the second moment of area (or the moment of inertia) about the centroidal axis is

$$I_c = \frac{(1)(6^3)}{12} + 2\left[\frac{(6)(1^3)}{12} + (6)(1)(3.5^2)\right]$$

$$= 18 + 148 = 166 \text{ in.}^4$$

For elastic behavior the flexure formula is applicable. Since the maximum stress (tensile or compressive) occurs at the outside fibers, the use of Eq. 10-10 gives

$$M_{\text{elastic}} = \frac{\sigma_{\text{max}} I_c}{y_{\text{max}}} = \frac{(40,000)(166)}{4} = 1,660,000 \text{ in-lb}$$

*b*) For the fully plastic case, the idealized stress distribution is shown in Fig. 10-11(b). This distribution can be thought of as four uniformly distributed forces, each of which is equal to the yield-point stress multiplied by the corresponding cross-sectional area. Hence, assuming that the lower fibers are in tension, we obtain

$$M_{\text{plastic}} = F_1(3.5) + F_2(1.5) + F_3(1.5) + F_4(3.5)$$
$$= 2[F_1(3.5) + F_2(1.5)] = 2(40,000)[6(3.5) + 3(1.5)]$$
$$= 2,040,000 \text{ in-lb}$$

NOTE: In this particular example, both the neutral surface for elastic action and that for plastic action pass through the geometric center. In general, this condition will *not* be true.

## 10-4. Combined Shear and Bending

Although it has been convenient to confine our analysis of beams thus far to pure bending, this type of loading condition is seldom encountered in an actual engineering problem. It is much more common for the resultant internal reaction to consist of a bending moment and a shear force. The presence of the shear force indicates a variable bending moment in the beam. The relationship between the shear force and the change in bending moment is then as follows (see Chapter 6):

$$V = \frac{dM}{dx} \tag{6-2}$$

Strictly speaking, the presence of the shear force and the resulting shear stresses and shear deformation would invalidate some of our statements in Sec. 10-2 in regard to the geometry of the deformation and the resulting axial strain distribution represented by Eq. 10-1. Plane sections would no longer remain plane after bending, and the geometry of the actual deformation would become considerably more involved. Fortunately, for a beam whose length is large in comparison with the dimensions of the cross section, the deformation effect of the shear force is relatively small; and it is *assumed* that the longitudinal axial strains are still distributed in the same manner as for pure bending. When this assumption is made, the load stress relationships developed in Sec. 10-3 are still considered valid. Experience and experimental work indicate that this assumption is sufficiently accurate for most practical purposes.

The following question immediately arises: When are the shearing effects so large that they cannot be ignored as a design consideration? It is difficult to answer this question. Probably the best way to begin is to try to approximate the shear stresses on the cross section of the beam. This can be done by examining the necessary equilibrium requirements.

Figure 10-12(a) shows a simply supported beam with a single transverse load P. Once again, for convenience, a rectangular cross section has been chosen, as shown in Fig. 10-12(b). Figure 10-12(c) shows the free-body diagram of that portion of the beam between sections ① and ②

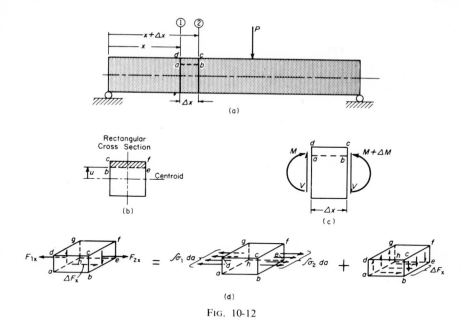

Fig. 10-12

in Fig. 10-12 (*a*). The vertical shear forces on these sections produce a change in bending moment denoted by $\Delta M$ in (*c*). Figure 10-12 (*d*) shows a free-body diagram of a slice of the beam extending between sections ① and ② and having its lower surface *abeh* at a distance *u* above the neutral surface of the beam.

Since there is a change in bending moment between sections ① and ②, it is reasonable to expect that the normal stresses $\sigma_1$ and $\sigma_2$ on the areas *ahgd* and *befc* may be different. Therefore, the resultant horizontal force $F_{1x}$ on the area *ahgd* would be different in magnitude from the resultant force $F_{2x}$ on the area *befc*. Equilibrium would require the existence of a horizontal shear force $\Delta F_x$ on the surface *abeh*. This horizontal shear force can be expressed in terms of the normal stresses as follows:

$$\overset{+}{\rightarrow} \sum F_x = 0$$

$$F_{2x} + \Delta F_x - F_{1x} = 0$$

$$\Delta F_x = F_{1x} - F_{2x} = \int_{ahgd} \sigma_1 \, da - \int_{befc} \sigma_2 \, da$$

$$\Delta F_x = \int_{befc} (\sigma_1 - \sigma_2) \, da \tag{10-11}$$

The existence of such longitudinal shear forces can easily be illustrated by means of a simple experiment. Suppose that several planks or slabs are stacked one on top of the other without fastening them together, as shown in Fig. 10-13(a). When a bending load is applied, the stack will deform as indicated in Fig. 10-13(b). Since the slabs were free to slide on one another, the ends do not remain even but become staggered. Each of the slabs behaves as an independent beam, and the total resistance to bending of *n*

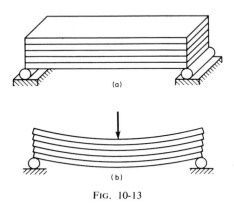

FIG. 10-13

slabs is approximately *n* times the resistance of one slab alone. If the experiment is repeated after the slabs have been fastened together so as to prevent their sliding on one another, the entire assembly would behave as a single beam having a thickness equal to *n* times the thickness of one slab. In the case of elastic action, the bending resistance of the assembly would be approximately $n^3$ times the bending resistance of one slab. From this simple experiment you can see the importance of a beam being able to resist longitudinal shear forces so that this "slipping" will not occur.

In order to evaluate the longitudinal shear force as given by Eq. 10-11, it is usually *assumed* that the normal stresses $\sigma_1$ and $\sigma_2$ on their respective areas are given by the *elastic* flexure formula (Eq. 10-10). For this assumption, expressions for $\sigma_1$ and $\sigma_2$ are

$$\sigma_1 = -\frac{My}{I} \qquad \sigma_2 = -\frac{(M + \Delta M)y}{I} \qquad (10\text{-}12)$$

where *y* is measured from the neutral surface, and *I* is the second moment of the *entire* cross-sectional area about its horizontal centroidal axis.

Substituting these expressions for $\sigma_1$ and $\sigma_2$ in Eq. 10-11, we obtain

$$\Delta F_x = -\int_{befc} \left[ \frac{My}{I} - \frac{(M + \Delta M)y}{I} \right] da$$

$$\Delta F_x = \frac{\Delta M}{I} \int_{befc} y \, da \qquad (10\text{-}13)$$

The integral, which will be denoted by $Q$, represents the first moment of area *befc* about the centroidal axis of the entire cross section. This area *befc* is merely that *part* of the total cross section which lies above the surface on which the longitudinal shear force acts.

We shall now reintroduce the concept of shear flow, which was first discussed in Sec. 9-8 in connection with thin-walled torsion members. Shear flow, denoted by $q$, is defined as the longitudinal shear force across the thickness of the cross section per unit length of the beam. According to this definition, the shear flow across the thickness of the beam at any cross section is

$$q = \lim_{\Delta x \to 0} \frac{\Delta F_x}{\Delta x} \tag{10-14}$$

If the value of $\Delta F_x$ given by Eq. 10-13 is substituted, we have

$$q = \lim_{\Delta x \to 0} \frac{\Delta M}{\Delta x} \frac{Q}{I} = \frac{dM}{dx} \frac{Q}{I}$$

$$q = \frac{VQ}{I} \tag{10-15}$$

where $V$ is the *resultant vertical* shear force at the cross section under consideration.

The expression in Eq. 10-15 represents the shear force per unit length of the beam across the thickness of the beam at the cross section under consideration. For example, suppose that it is desired to determine the shear flow across the geometric middle of the portion of a beam shown in Fig. 10-14(*a*). The procedure would be as follows:

First, determine the vertical shear force $V$ from the free-body diagram in Fig. 10-14(*b*). Thus, $V = R_1 - P_1$.

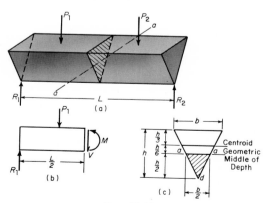

Fig. 10-14

Second, locate the centroid of the entire cross section. In this case the distance from the top to the centroid is $h/3$.

Third, the distance from the top to the geometric middle of the depth is $h/2$.

Fourth, the moment of inertia, or the second moment, of the entire triangular cross section about the *centroid* is $bh^3/36$.

Fifth, determine $Q$. The area *above* the specified longitudinal surface *aa* at the middle of the depth of the beam is the *unshaded* area in Fig. 10-14(c). However, since the first moment of the entire area about the centroid is zero, the first moment of the unshaded area about the centroid is the same (except for sign) as that of the shaded area below line *aa*. For the shaded area,

$$Q = \frac{1}{2}\left(\frac{b}{2}\right)\left(\frac{h}{2}\right) \times \left(\frac{h}{6} + \frac{1}{3}\frac{h}{2}\right)$$

$$= \left(\frac{1}{8}\,bh\right)\left(\frac{1}{3}\,h\right) = \frac{1}{24}\,bh^2$$

Finally, the shear flow across line *aa* is

$$q = \frac{VQ}{I} = \frac{(R_1 - P_1)\left(\frac{1}{24}\,bh^2\right)}{\dfrac{bh^3}{36}} = \frac{3}{2}\frac{(R_1 - P_1)}{h}\ \text{lb/in.}$$

## 10-5. Average Longitudinal and Transverse Shear Stresses

You should note that the development of the shear-flow equation (Eq. 10-15) was based solely *on equilibrium requirements and the elastic flexure formula*. No consideration was given to the manner in which the shear strain and shear stress are distributed throughout the beam.

From the free-body diagram in Fig. 10-15, most of which is the same as Fig. 10-12(d), we see that shear forces exist on both the longitudinal surface *abeh* and the transverse surfaces *ahgd* and *befc*. These shear forces naturally produce shear stresses on their respective surfaces. Also, any infinitesimal element having an edge in common with line *be* has equal

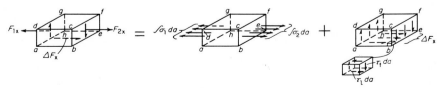

FIG. 10-15

shear stresses on two mutually perpendicular surfaces; that is $\tau_l = \tau_t$. Thus, we see that simultaneous *and equal* longitudinal and transverse shear stresses exist in a beam acted on by transverse forces. However, it is difficult to determine the distribution of these stresses throughout the length, depth, and thickness, or width, of the beam.

It is relatively easy to show experimentally by means of a grid pattern on the side of a *rectangular* beam, as in Fig. 10-16, that the maximum distortion—and, therefore, the maximum shear strain—occurs at the neutral axis of bending, while no distortion occurs at the free edges of the beam.

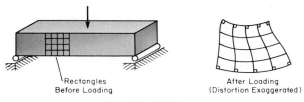

Rectangles
Before Loading

After Loading
(Distortion Exaggerated)

FIG. 10-16

However, this observation serves only as a *qualitative* analysis of the distribution of the shear strain. We still do not know the exact variation of the shear stresses throughout the beam, but it is usually assumed that the shear stresses are constant across the thickness of the beam. For example, in Fig. 10-15 we assume that the shear stresses at $b$ have the same value as those at $e$ and at all points on the line between $b$ and $e$. We are *not* saying that the stresses at $b$ are equal to those at $c$, since our previous experiment has indicated that the shear stresses are not constant throughout the vertical depth of the beam.

We have already argued that the longitudinal shear stress $\tau_l$ at any point is equal to the transverse shear stress $\tau_t$ at the point. If the shear force per unit length of the beam at any particular point is given by the shear flow (Eq. 10-15), and the thickness of the beam is denoted by $t$, then the average longitudinal or transverse shear stress across the thickness of the beam is

$$\tau_{l\text{(avg)}} = \tau_{t\text{(avg)}} = \frac{q}{t} = \frac{VQ}{It} \qquad (10\text{-}16)$$

This expression for the shear stress is consistent with the previously noted observation of Fig. 10-16 in that for a *rectangular* cross section the shear stress becomes maximum at the centroidal axis (neutral surface) and is zero at the outer edges, since $Q$ behaves in this manner (see Example 10-2 which follows).

Although the shear flow found by Eq. 10-15 *always* is maximum at the centroid and zero at the edges, it is important to note that the shear stress found by Eq. 10-16 may or may not be maximum at the centroid. Where the thickness of the member varies throughout the depth, the position of the maximum transverse shear stress depends on the shape of the section. For instance, it is left as an exercise for the student (Prob. 10-19) to show that the maximum shear stress for a triangular cross section, like that in Fig. 10-14, occurs at the geometric middle of the depth, and *not* at the centroid.

EXAMPLE 10-2. Determine an expression for the shear-stress distribution on a rectangular cross section of height $h$ and thickness $t$.

*Solution:* Refer to Fig. 10-17. If it is assumed that the shear stresses are uniform across the thickness, the shear stress at a surface at a distance $y_0$ from the centroidal axis is, by Eq. 10-16,

$$\tau = \frac{VQ}{It} = \frac{V}{It} \int_{\substack{\text{area} \\ abef}} y \, da$$

$$= \frac{V}{(th^3/12)t} \int_{y_0}^{h/2} y(t \, dy) = \left(\frac{12V}{t^2 h^3}\right) t \left.\frac{y^2}{2}\right]_{y_0}^{h/2}$$

$$\tau = \frac{6V}{th^3}\left[\frac{h^2}{4} - y_0^2\right]$$

From this last expression we get the following results: at $y_0 = h/2$, $\tau = 0$; at $y_0 = 0$, or at the centroidal axis, $\tau$ has its maximum value, which is $3V/2A$. Also, since $\tau$ is a function of $y_0^2$, the distribution is parabolic, as shown in Fig. 10-17($b$).

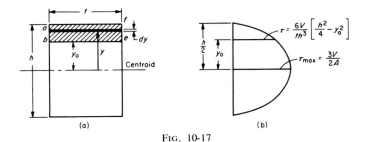

(a)    (b)

FIG. 10-17

## 10-6. Elastic Bending of Unsymmetrical Sections

So far we have dealt only with beams which have a longitudinal plane of symmetry and whose bending loads are applied in this plane of symmetry. We shall now see that for elastic behavior these strict requirements of symmetry can be somewhat relaxed.

Consider the beam of arbitrary cross section shown in Fig. 10-18(*a*), where the *x* axis is the longitudinal centroidal axis of the beam and the *y* and *z* axes are the *principal* centroidal axes of the cross section. For this choice of axes, we have

$$\int_{\substack{\text{cross} \\ \text{section}}} y\, da = 0 \qquad \int_{\substack{\text{cross} \\ \text{section}}} z\, da = 0 \qquad \int_{\substack{\text{cross} \\ \text{section}}} yz\, da = 0 \quad (10\text{-}17)$$

Let us assume that a *pure couple* about *either principal* axis is applied to the beam, as in Fig. 10-18(*b*), where for convenience a couple about the *z* axis was chosen. Under this loading condition, the arguments in Sec. 10-2 are still reasonable and plane sections originally perpendicular to the *x* axis will remain plane after loading. Hence, the longitudinal axial strains will vary linearly from some neutral surface.

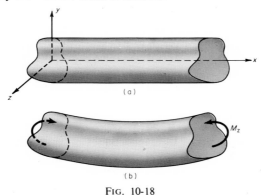

(a)

(b)

Fig. 10-18

Now it will be assumed that the beam is *elastic*, homogeneous, and isotropic. The axial stress is then given by the expression in Fig. 10-9(*a*), namely,

$$\sigma = ky \qquad (10\text{-}18)$$

Substituting this expression in Eqs. 10-2, 10-3, and 10-4, we have

$$\sum M_z = M_z$$

$$M_z = -\int_{\text{area}} y\,(\sigma\, da) = -\int_{\text{area}} y\,(ky\, da) = -k\int_{\text{area}} y^2\, da = -kI_z$$

$$k = -\frac{M_z}{I_z}$$

$$\sigma = -\frac{M_z y}{I_z} \qquad (10\text{-}19)$$

Also,

$$\overset{+}{\underset{}{\longrightarrow}} \sum F_x = R_n = 0$$

$$= \int_{\text{area}} \sigma \, da = \int_{\text{area}} (ky) da = k \int_{\text{area}} y \, da = 0$$

by Eq. 10-17. Finally,

$$\sum M_y = 0$$

$$= \int_{\text{area}} z(\sigma \, da) = \int_{\text{area}} z(ky \, da) = k \int_{\text{area}} yz \, da = 0$$

by Eq. 10-17.

Thus, if the bending couple is applied about the $z$ axis, we see that the elastic axial stress will be given by Eq. 10-19 and that the neutral surface will lie in the $xz$ plane. In a similar manner, if a couple is applied about the $y$ axis, the stress will be

$$\sigma = - \frac{M_y z}{I_y} \tag{10-20}$$

By using Eqs. 10-19 and 10-20, we are now able to handle the more general problem in which the couple producing the bending moment acts about some axis which is not parallel to either principal axis. Such a condition is illustrated in Fig. 10-19, where once again the $y$ and $z$ axes are *principal centroidal* axes and the couple acts about the $l$ axis. From elementary mechanics we know that $M_r$ can be resolved into components about the $y$ and $z$ axes. These components are

$$M_z = M_r \cos \alpha$$

$$M_y = M_r \sin \alpha$$

The stresses produced by these components can be calculated by applying Eqs. 10-19 and 10-20. The resultant stress at any point on the cross section is then obtained by superposition.

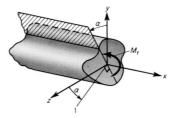

FIG. 10-19

You should note that throughout the discussion of elastic bending of beams with unsymmetrical sections, it was assumed that the bending moment was produced by a *pure couple.* In general, if there is a resultant shear force on any cross section resulting from transverse loading of the beam, the bending will be accompanied by twisting of the beam. The problem then becomes one of combined bending and torsion of an unsymmetrical section. Such problems are discussed in Chapter 12.

## 10-7. Summary of Beam Stresses

The preceding sections have dealt with methods of evaluating the flexure stresses and the average *elastic* shear stresses in beams subjected to various loading conditions. Also, it has been shown that at any particular point in a beam the stress on an element may be composed of longitudinal normal stresses (flexure stresses) and transverse shear stresses. Under this condition, the state of stress is biaxial, and the maximum normal and shear stresses will act in directions that are neither longitudinal nor transverse. Hence, it is often necessary to employ Mohr's circle in order to completely determine the principal stresses at a particular point.

Recall that the flexure stresses are maximum at the outermost fibers and are zero at the neutral surface. On the other hand, the transverse shear stresses are zero at the outside surface and have a maximum value somewhere in the interior. Also, both the flexure and shear stresses vary with the loading on the beam. Hence, the problem of determining the point or points at which the absolute maximum principal normal and shear stresses occur can be a very difficult one which requires careful investigation. In determining maximum stresses in beams, don't be satisfied with your results until you have exhausted all possible combinations of flexure and shear stresses which could give the maximum principal stresses. Often, the construction of shear and moment diagrams and a comparison of the orders of magnitude of the flexure stresses and transverse shear stresses will greatly simplify this problem.

Since members subjected to axial and torsion loads were investigated in previous chapters, we can now handle problems in which axial, torsion, and bending loads are combined. As before, we shall only consider superposition of *elastic* stresses.

EXAMPLE 10-3. A cantilever beam is shown in Fig. 10-20(*a*). Determine the principal normal and shear stresses at point *A* just below the flange, and show them on a sketch. Assume elastic action and neglect any stress concentration at the wall.

*Solution*: The bending moment at the left-hand end of the beam is

$$M = (80,000) (2) (12) \text{ in-lb}$$

and the vertical shear force is $V = 80,000$ lb. Hence, the flexure stress at point *A* is

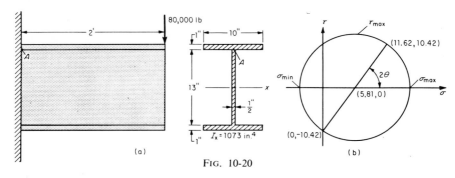

FIG. 10-20

$$\sigma = \frac{My}{I} = \frac{(80,000)(2)(12)(6.5)}{1073} = 11,620 \text{ psi (tension)}$$

and the transverse shear stress at $A$ is

$$\tau = \frac{VQ}{It} = \frac{(80,000)[(10)(1)(7)]}{(1073)(1/2)} = 10,420 \text{ psi}$$

The stresses acting on an element at $A$ oriented parallel and perpendicular to the longitudinal axis of the beam are shown in Fig. 10-21(a). Also, Mohr's circle for this state of stress is shown in Fig. 10-20 (b). From the geometry of the circle, we have the following results:

$$\sigma_{max} = 17,750 \text{ psi (tensile)} \qquad \sigma_{min} = 6,130 \text{ psi (compression)}$$

$$\tau_{max} = 11,940 \text{ psi} \qquad \theta = 30.5°$$

Thus, the desired orientations are as shown in Fig. 10-21(b) and (c).

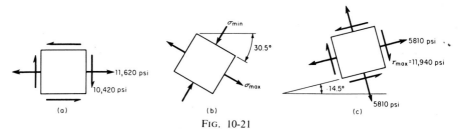

FIG. 10-21

NOTE: In such a short member as this with its thin-web cross section, the validity of the flexure formula is questionable. Notice, for example, that the shear and normal stresses are of the same order of magnitude.

EXAMPLE 10-4. The dimensions of a T-shaped member and the loads on it are shown in Fig. 10-22(a) and (b). The horizontal 5000-lb tensile force acts through the centroid of the cross-sectional area. Determine the maximum principal normal stress and the maximum shear stress in the beam, assuming elastic behavior. *occur at outermost fiber*

*Solution:* Sketches of the shear and moment diagrams are shown in Fig. 10-22(c) and (d), respectively. Hence, both the maximum *flexure* stress and the maximum *transverse* shear stress occur at the extreme right-hand end of the beam. Also, the axial stress produced by the tensile load will be uniform throughout the entire beam.

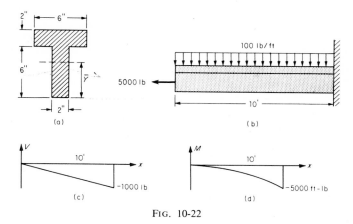

FIG. 10-22

The distance from the bottom of the beam to the centroid of the cross section is found to be

$$\overline{Y} = \frac{(12)(3) + (12)(7)}{12 + 12} = 5 \text{ in.}$$

Also, the moment of inertia or second moment of the area about the centroid is

$$I = \left[\frac{(6)(2^3)}{12} + (12)(2^2)\right] + \left[\frac{(2)(6^3)}{12} + (12)(2^2)\right]$$

$$= 52 + 84 = 136 \text{ in.}^4$$

The longitudinal normal stress on the uppermost fibers is

$$\sigma_{up} = \frac{P}{A} + \frac{My_{up}}{I} = \frac{5000}{24} + \frac{(5000)(12)(3)}{136}$$

$$= 208 + 1323 = 1531 \text{ psi (tension)}$$

while the normal stress on the lowest fibers is

$$\sigma_{low} = \frac{P}{A} - \frac{My_{low}}{I} = \frac{5000}{24} - \frac{(5000)(12)(5)}{136}$$

$$= 208 - 2205 = -1997 = 1997 \text{ psi (compression)}$$

The maximum longitudinal or transverse shear stress occurs at the centroid.

This stress is

$$\tau = \frac{VQ}{It} = \frac{(1000)[(2)(5)(2.5)]}{(136)(2)}$$

$$= 91.9 \text{ psi}$$

At any point between either outer surface and the centroid, there will be both a normal stress and a shear stress. However, a comparison of their orders of magnitude shows that the normal stresses at the outside far exceed the transverse shear stresses at the centroid. Hence, we can conclude that the critical fibers will be those at the *lower* outside surface, where the maximum principal normal stress will be

$$\sigma_{max} = 1997 \text{ psi (compression)}$$

and, recalling Mohr's circle for uniaxial stress, the maximum shear stress will be

$$\tau_{max} = \frac{\sigma_{max} - \sigma_{min}}{2} = \frac{1997 - 0}{2} = 999 \text{ psi}$$

This shear stress will act at 45° to the longitudinal axis of the beam.

## 10-8. Basic Geometric Relationships for Deflection

The word "deflection" generally refers to the deformed shape and position of a member subjected to bending loads. More specifically, however, we will use deflection in reference to the deformed shape and position of the longitudinal neutral axis of a beam. In the deformed condition the neutral axis, which is initially a straight longitudinal line, assumes some particular shape which is called the *deflection curve*. The deviation of this curve from its initial position at any point is called the *deflection* at that point. In developing the theory for determining the deflection of a beam, we will assume that the shear strains do *not* significantly influence the deformation.

Consider the cantilever beam of Fig. 10-23(*a*). For convenience, the origin of the coordinate system was chosen at the left-hand end, and the *x* axis is assumed to be the undeflected neutral axis of the beam. From the definition of neutral axis, we know that the small increment of length $\Delta s$ between points *A* and *B* will remain unchanged in length after deformation. Thus, the distance *A'B'* on the deflection curve in Fig. 10-23(*b*) will also be $\Delta s$. The geometry of an element of the deflected beam is shown in Fig. 10-23(*c*).

From elementary calculus, the relationships in Fig. 10-23(*d*) are valid for any smooth curve, such as the curve formed by the deflected neutral axis (deflection curve). If it is assumed that the slope at any arbitrary point is very small compared with unity, then the relationships in Fig.

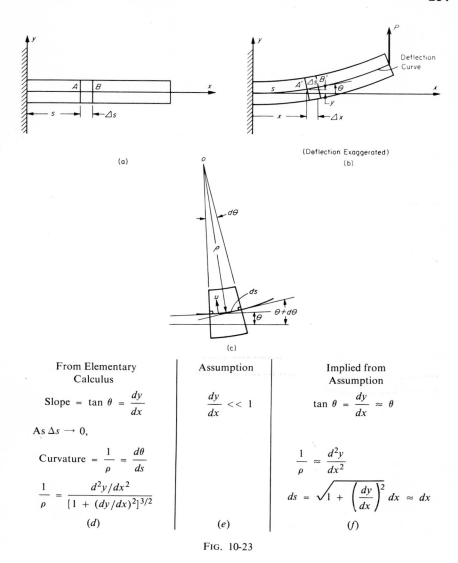

(a)

(Deflection Exaggerated)

(b)

(c)

| From Elementary Calculus | Assumption | Implied from Assumption |
|---|---|---|
| Slope $= \tan\theta = \dfrac{dy}{dx}$ | $\dfrac{dy}{dx} \ll 1$ | $\tan\theta = \dfrac{dy}{dx} \approx \theta$ |
| As $\Delta s \to 0$, | | |
| Curvature $= \dfrac{1}{\rho} = \dfrac{d\theta}{ds}$ | | $\dfrac{1}{\rho} \approx \dfrac{d^2 y}{dx^2}$ |
| $\dfrac{1}{\rho} = \dfrac{d^2 y/dx^2}{[1+(dy/dx)^2]^{3/2}}$ | | $ds = \sqrt{1+\left(\dfrac{dy}{dx}\right)^2}\,dx \approx dx$ |
| (d) | (e) | (f) |

Fig. 10-23

10-23(*f*) are valid approximations. Thus,

$$\theta = \frac{dy}{dx} \qquad \frac{1}{\rho} = \frac{d^2 y}{dx^2} = \frac{d}{dx}\left(\frac{dy}{dx}\right) = \frac{d\theta}{dx} \qquad (10\text{-}21)$$

This assumption implies that the deflection of any point is very small in comparison with the length of the beam, and that the horizontal projected length of the deflection curve is the same as its undeformed length. Such an assumption is quite reasonable when it is realized that relatively few engineering members can tolerate large deflections without their failing. There are, of course, exceptions. Even in most of these cases, however, large deflections would result in inelastic behavior of the material, and failure would probably be based on some criterion other than deflection.

Since we have already restricted the analysis to small deflections, it is reasonable to assume further that the material behaves elastically; that is, the flexure stress on any fiber is given by the relationship

$$\sigma = -\frac{Mu}{I}$$

where $u$ is the distance from the neutral surface to the fiber. (Recall the sign convention used in the derivation of Eq. 10-10a.) When this assumption is made and Hooke's law is used, Eq. 10-1 yields the following result:

$$\epsilon = -\frac{u}{\rho} = \frac{\sigma}{E} = -\frac{Mu}{EI}$$

$$\frac{1}{\rho} = \frac{M}{EI} \tag{10-22}$$

This result relates the curvature at any point to the bending moment at that point. Note that a positive moment produces a positive curvature (concave upward), and a negative moment produces a negative curvature (concave downward).

Now by combining Eq. 10-21 with Eq. 10-22, we obtain the basic relationships for determining the characteristics of the deflection curve, such as its slope or the deflection at any point, as well as its general shape. These relationships are as follows:

$$\frac{d^2y}{dx^2} = \frac{d\theta}{dx} = \frac{1}{\rho} = \frac{M}{EI} \tag{10-23}$$

where $\theta$ is the slope of the deflection curve at any point at distance $x$ from the origin;

$y$ is the deflection at any point at distance $x$ from the origin;

$M$ is the bending moment, usually expressed as a function of the distance $x$ and the applied external loads;

$E$ is the modulus of elasticity of the material;

$I$ is the moment of inertia of the entire cross-sectional area about the neutral surface.

The product $EI$ is often called the *flexural rigidity* or the *bending modulus* of the beam. It is a measure of the stiffness of the beam, since it involves both the material and cross section of the beam.

Equation 10-23 is a linear differential equation in the independent variable $x$. It is of first order in the dependent variable $\theta$ and of second order in the dependent variable $y$. The solution of this equation will give both the proper slope and the deflection of the beam, *provided that the solution satisfies the conditions of restraint imposed by the supports.* In succeeding sections, various techniques for solving Eq. 10-23 for $\theta$ and $y$ will be presented.

## 10-9. Deflection by Direct Integration

Since Eq. 10-23 does not contain a first-order term $dy/dx$, successive ordinary integrations with the appropriate constants of integration will give the slope $\theta$ and the deflection $y$ as functions of $x$. The step-by-step procedure will be outlined and then illustrated by several examples.

1. Set up a reference coordinate system consistent with Eq. 10-23. The origin of a right-hand system at the left-hand end of the beam is one such coordinate system.

2. Derive from equilibrium requirements an expression (or expressions) for the moment as a function of $x$. In cases where there are discontinuities in the loading, several expressions may be necessary (see Example 10-6, which follows).

3. Determine the bending modulus $EI$. If the cross section varies with the length, the moment of inertia must be expressed as a function of $x$. If there are abrupt discontinuities in the cross section or in the material, then several expressions may be necessary.

4. Integrate the equation once for the slope $\theta$, and twice for the deflection $y$, being careful to *include the constant of integration in each case.* These constants are then evaluated from the conditions imposed on the deflection curve by the supports. A free-hand sketch of the deflection curve is often helpful in recognizing the conditions of restraint.

EXAMPLE 10-5. Assuming elastic action for the beam shown in Fig. 10-24(a), sketch the deflection curve. Then determine the slope at point $B$ and the deflection at point $A$.

*Solution:* A free-body diagram of any portion of the beam between points $A$ and $B$ will show that there is no bending moment in that portion. Hence, that

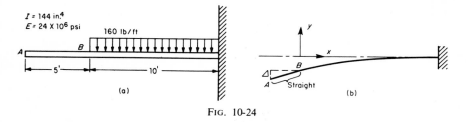

FIG. 10-24

portion will *not* bend. It will remain straight, and its slope will be equal to the slope at *B*. The other part of the beam between *B* and the wall does have an internal bending moment, and this part will deflect into some shape. Therefore the deflection curve for the entire beam will be as shown in Fig. 10-24(*b*).

For convenience, we will choose an *x-y* coordinate system with its origin at the undeformed position of *B*. Consider a free-body diagram of a portion of the beam between the origin and a point at a distance *x* from the origin. If counterclockwise moments are positive, and *x* is expressed in inches,

$$\overset{\frown}{\underset{+}{\sum}} M = 0$$

$$\left( \frac{160 \text{ lb/ft}}{12 \text{ in./ft}} \right) (x \text{ in.}) \left( \frac{x}{2} \text{ in.} \right) + M = 0$$

$$M = - \frac{160}{24} x^2 \text{ in-lb}$$

From Eq. 10-23 we have

$$\frac{d^2y}{dx^2} = \frac{d\theta}{dx} = \frac{M}{EI} = - \frac{160}{24} x^2 \frac{1}{EI} \tag{a}$$

The restraining wall requires that both the slope and deflection be zero at the wall, where $x = 120$ in. Thus,

$$y_{x = 120} = 0 \qquad \theta_{x = 120} = 0 \tag{b}$$

Integrating equation (a), we obtain

$$EI \, \theta = - \frac{160}{24} \frac{x^3}{3} + C_1 \tag{c}$$

$$EI \, y = - \frac{160}{24} \frac{x^4}{12} + C_1 x + C_2 \tag{d}$$

Using the boundary conditions (b) above, we find that

$$C_1 = \frac{160}{24} \frac{(120)^3}{3} \qquad C_2 = - \frac{160}{24} \frac{(120)^4}{4}$$

Hence, the equations for the slope and the deflection are as follows:

$$EI \, \theta = \frac{160}{24} \left[ - \frac{x^3}{3} + \frac{(120)^3}{3} \right] \tag{e}$$

$$EI \, y = \frac{160}{24} \left[ - \frac{x^4}{12} + \frac{(120)^3}{3} x - \frac{(120)^4}{4} \right] \tag{f}$$

Now, from equation (e), the slope at point $B$ is

$$\theta_B = \frac{160}{(24)(24)(10^6)(144)} \times \left[ -0 + \frac{(120)^3}{3} \right]$$

$$= \frac{160}{(12^2)(10^3)} = 1.11 \times 10^{-3} \text{ rad.}$$

Finally, the total deflection at $A$ will be

$$y_A = y_B + \Delta$$

From equation (f)

$$y_B = \frac{160}{(24)(24)(10^6)(144)} \left[ -0 + 0 - \frac{(120)^4}{4} \right]$$

$$= \frac{160}{(16)(10^2)} = -0.1 \text{ in. (downward)}$$

From Fig. 10-24(*b*),

$$\Delta = (5)(12)(\theta_B) = 60 \times \frac{160}{(12^2)(10^3)} = 0.0667 \text{ in.}$$

Hence,

$$y_A = 0.1 + 0.0667 = 0.1667 \text{ in.}$$

EXAMPLE 10-6. Find an expression for maximum deflection of the straight, homogeneous, elastic beam in Fig. 10-25(*a*). Sketch the deflection curve.

*Solution*: A free-body diagram of the entire beam is shown in Fig. 10-25(*b*). Also, the left-hand end of the beam is chosen as the origin of the coordinate system. Since the loading is discontinuous at $x = L/3$, it will be necessary to integrate and solve two sets of equations simultaneously. Thus, for $0 \le x \le L/3$, we have

$$M = \frac{2}{3} Px$$

$$EI\theta = \frac{Px^2}{3} + C_1 \tag{a}$$

$$EIy = \frac{Px^3}{9} + C_1 x + C_2 \tag{b}$$

Similarly, for $L/3 \le x \le L$, we have

$$M = \frac{2}{3} Px - P\left(x - \frac{L}{3}\right) = -\frac{P}{3}(x - L)$$

$$EI\theta = -\frac{P}{3} \frac{(x - L)^2}{2} + C_3 \tag{c}$$

$$EIy = -\frac{P}{3} \frac{(x - L)^3}{6} + C_3 x + C_4 \tag{d}$$

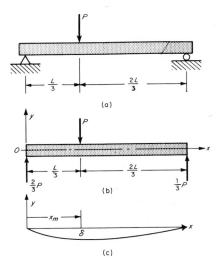

Fig. 10-25

Now, to evaluate the four constants of integration, four independent boundary conditions will be needed. Since the deflection of each support must be zero, two boundary conditions are

$$y_{x=0} = 0 \qquad y_{x=L} = 0$$

These conditions yield, from equations (b) and (d), respectively,

$$C_2 = 0 \qquad C_4 = -C_3 L \tag{e}$$

The two additional boundary conditions are not quite so obvious as were the first two. Since the beam itself is continuous, equation (b) and equation (d) must give the same value for $y$ at their only common point, at which $x = L/3$. This condition yields

$$C_1 = \frac{PL^2}{3^3} - 2C_3 \tag{f}$$

Also, since the deflection curve is smooth, equations (a) and (c) must give the same slope value at $x = L/3$. Thus,

$$C_1 = -\frac{PL^2}{3^2} + C_3 \tag{g}$$

Equations (e), (f), and (g) yield

$$C_1 = -\frac{5}{3^4}PL^2 \qquad C_2 = 0 \qquad C_3 = \frac{4}{3^4}PL^2 \qquad C_4 = -\frac{4}{3^4}PL^3$$

Finally, the deflection will be a maximum when $\theta$ is zero. Hence, from equation (c),

$$-\frac{P}{3}\frac{(x-L)^2}{2} + \frac{4PL^2}{3^4} = 0$$

$$x_m = L\left(1 \pm \frac{2}{3}\sqrt{\frac{2}{3}}\right)$$

Only the second root is applicable, and substitution of this root in equation (d) yields

$$\delta = y_{max} = -\frac{PL^3}{EI}\frac{2^4}{3^6}\sqrt{\frac{2}{3}}$$

The deflection curve is shown in Fig. 10-25(c).

## 10-10. Deflection by Use of Singularity Functions

The singularity functions were introduced in Chapter 6 in conjunction with shear and moment equations for beams. It would be well for you to review Sec. 6-5 for the definitions and properties of the Dirac delta, unit step, and doublet functions.

The singularity functions can be utilized in determining the slope and deflection of beams. One way to do this is to start with the loading function for the entire beam; then by utilizing the relations

$$w(x) = -\frac{dV}{dx} \qquad [6\text{-}1]$$

$$V = \frac{dM}{dx} \qquad [6\text{-}2]$$

$$\frac{M}{EI} = \frac{d\theta}{dx} \qquad [10\text{-}23]$$

$$\theta = \frac{dy}{dx} \qquad [10\text{-}21]$$

one can successively integrate each of these equations and, by determining each of the constants of integration from appropriate boundary conditions, eventually obtain a single deflection equation for the entire beam. However, in practice it is often easier to draw a free body diagram and write the moment equation for the entire beam in terms of singularity functions; and then integrate once to get the slope $\theta$ and once again for the deflection $y$. Usually this procedure will involve only the integration of the unit step function which was performed in Eq. 6-7 and 6-8. The following examples serve to illustrate these ideas.

EXAMPLE 10-7. Re-work Example 10-6 using singularity functions.

*Solution:* Refer to Fig. 10-25. The moment equation for the entire beam can be written in terms of singularity functions as follows:

$$M = \frac{2}{3}Px - P\left(x - \frac{L}{3}\right)u\left(x - \frac{L}{3}\right) + \frac{1}{3}P(x - L)u(x - L) \qquad (a)$$

Then, from Eq. 10-23 we have

$$EI \frac{d^2y}{dx^2} = EI \frac{d\theta}{dx} = M \qquad (b)$$

with boundary conditions

$$y_{x=0} = 0 \quad y_{x=L} = 0 \qquad (c)$$

Integrating (a) we have (recall Eqs. 6-7 and 6-8)

$$EI\theta = \frac{Px^2}{3} - \frac{P\left(x - \frac{L}{3}\right)^2}{2} u\left(x - \frac{L}{3}\right) + \frac{1}{6} P(x - L)^2 u(x - L) + C_1 \qquad (d)$$

$$EIy = \frac{Px^3}{9} - \frac{P}{6}\left(x - \frac{L}{3}\right)^3 u\left(x - \frac{L}{3}\right) + \frac{1}{18} P(x - L)^3 u(x - L)$$

$$+ C_1 x + C_2 \qquad (e)$$

The boundary conditions yield

$$y_{x=0} = 0 = 0 - 0 + 0 + 0 + C_2 \qquad (f)$$

$$C_2 = 0$$

$$y_{x=L} = 0 = \frac{PL^3}{9} - \frac{P}{6}\left(\frac{2}{3} L\right)^3 + 0 + C_1 L + 0$$

$$C_1 = - \frac{5}{3^4} PL^2$$

Let us check our result by evaluating the deflection at $x = \dfrac{L}{3}$.

$$EIy_{x=L/3} = \frac{P\left(\frac{L}{3}\right)^3}{9} - 0 + 0 - \frac{5}{3^4} PL^2 L$$

$$= - \frac{14}{3^5} PL^3$$

which is the same value as obtained from equation (b) of Example 10-6.

EXAMPLE 10-8. Determine the maximum deflection of the beam shown in Fig. 10-26(a). $E = 10 \times 10^6$ psi    $I = 14.4$ in.[4]

*Solution:*    For convenience we will work the problem using the dimensions in feet. The wall reactions are shown in Fig. 10-26(b). The tricky part of this problem is to express the moment of the distributed load in terms of the step function. Referring to Fig. 10-26(c) we see that the moment at (with respect to

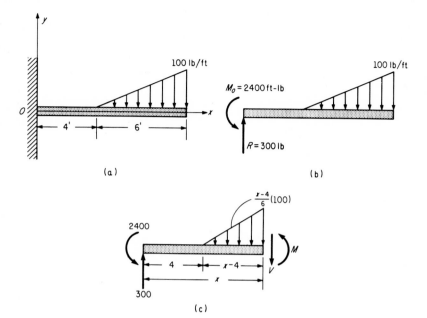

FIG. 10-26

the right end) of the partial distributed load of the figure is

$$\frac{1}{2}\left(\frac{x-4}{6}\ 100\right)(x-4)\left(\frac{x-4}{3}\right)u(x-4)\ =\ \frac{100}{36}\ (x-4)^3\ u(x-4)$$

With this, the moment equation for the entire beam is

$$M\ =\ -2400\ +\ 300x\ -\ \frac{100}{36}\ (x-4)^3\ u(x-4)$$

Thus

$$EI\theta\ =\ -2400x\ +\ 150x^2\ -\ \frac{25}{36}\ (x-4)^4\ u(x-4)\ +\ C_1$$

$$EIy\ =\ -1200x^2\ +\ 50x^3\ -\ \frac{5}{36}\ (x-4)^5\ u(x-4)\ +\ C_1x\ +\ C_2$$

The wall restrains the beam so that

$$\theta\ =\ 0,\quad y\ =\ 0\quad\text{at}\quad x\ =\ 0$$

Hence $C_1\ =\ C_2\ =\ 0$. Therefore the maximum deflection is

$$y_{\max}\ =\ y_{x=10}\ =\ \frac{1}{EI}\left[-1200\ (10)^2\ +\ 50\ (10)^3\ -\ \frac{5}{36}\ (6)^5\right]$$

$$=\ -\ \frac{12^4}{10^7\ 12^2\ 14.4}\ 7.1\ \times\ 10^4\ =\ -0.071\ \text{ft}$$

## 10-11. Moment-Area Method for Deflection

In many cases the slope and the deflection of a beam can be determined with relative ease by a semigraphical integration technique, usually called the moment-area method. The method is based on a geometrical interpretation of definite integrals.

Recall Eq. 10-23, which is

$$\frac{d\theta}{dx} = \frac{1}{\rho} = \frac{M}{EI} = \text{curvature}$$

Rewrite this equation in the form

$$d\theta = \frac{1}{\rho}\,dl = \frac{M}{EI}\,dl \tag{10-24}$$

where $d\theta$ is the infinitesimal change in the slope of the deflection curve occurring over the infinitesimal length $dl$. Notice that $(M/EI)\,dl$ can be interpreted as an infinitesimal area on the $M/EI$ (curvature) diagram. Such a curvature diagram can be obtained from the moment diagram simply by dividing the moment by the bending modulus $EI$. If the beam is homogeneous and has a constant cross section, the moment and curvature diagrams will have the same general shape. For example, see Fig. 10-27 (*a*) and (*b*).

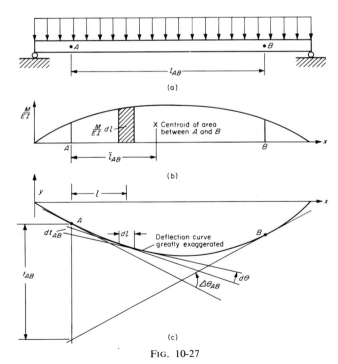

Fig. 10-27

The finite *change* in slope between any two distinct points on the beam can be obtained by integrating Eq. 10-24 between the corresponding limits. Thus,

$$\Delta\theta_{AB} = \int_A^B d\theta = \int_A^B \frac{M}{EI} \, dl \qquad (10\text{-}25)$$

From the previous statements, we see that the definite integral on the right may be interpreted as the area under the curvature curve between points $A$ and $B$. Positive areas, or those above the $x$ axis, indicate increases in slope; while negative areas, or those below the $x$ axis, indicate decreases in slope. Notice, however, that the integral, and hence the corresponding area, represents the *change* in slope and *not* the slope itself.

The deflection of the beam can be obtained indirectly by considering the *tangential deviation*. The tangential deviation $t_{AB}$ is defined as the distance, measured perpendicular to the undeformed neutral axis, between the point $A$ on the deflection curve and a tangent line to the deflection curve drawn through the point $B$. For example, see Fig. 10-27(c). From this figure, in which the deflections and slopes are greatly exaggerated, we see that

$$dt_{AB} = l \, d\theta$$

Thus,

$$t_{AB} = \int_A^B l \, d\theta = \int_A^B l \frac{M}{EI} \, dl \qquad (10\text{-}26)$$

$$t_{AB} = \bar{l}_{AB} \, a$$

where $\bar{l}_{AB}$ is the horizontal distance from point $A$ to the centroid of the area $a$ between points $A$ and $B$ on the curvature diagram. Thus, the moment of the area on the curvature diagram is always taken *about the point for which the tangential deviation is being determined.* It is from Eq. 10-26 that the method gets its name moment-area, since the equation represents the first moment of an area.

Since $\bar{l}_{AB}$ is necessarily a positive number, $t_{AB}$ will be positive when the corresponding curvature area $a$ is positive, and negative when the area is negative. A positive value for $t_{AB}$ indicates that point $A$ lies *above* the extended tangent line through point $B$; and $A$ lies *below* this tangent line if $t_{AB}$ is negative.

It is important for you to realize that $t_{AB}$ is not the deflection of either point $A$ or point $B$. However, the following examples illustrate how the tangential deviation may be used to determine the deflection. Also, you should be able to show for yourself that $t_{AB}$ will generally *not* be equal to $t_{BA}$. Construct distance $t_{BA}$ on the curve in Fig. 10-27(c).

EXAMPLE 10-9. Solve Example 10-5 by means of the moment-area method.

*Solution:*  The beam is redrawn in Fig. 10-28(*a*) and the shear, moment, curvature, slope, and deflection diagrams are shown in Fig. 10-28 (*b*), (*c*), (*d*), (*e*), and (*f*), respectively.  The beam has zero slope at the right-hand end, and thus the slope at point *B* will be equal to the *change* in slope between points *B* and *C*. Hence,

$$\theta_C - \theta_B = \Delta\theta_{BC} = \text{area of curvature diagram}$$

$$0 - \theta_B = -\frac{1}{3}(120)\left(\frac{1}{36,000}\right) = -\frac{1}{900}\text{ radian}$$

$$\theta_B = \frac{1}{900}\text{ radian}$$

The tangent at point *C* is horizontal and thus, in this case, $t_{BC}$ will represent the deflection of point *B*.  Hence,

$$t_{BC} = \frac{3}{4}(120)\left(-\frac{1}{900}\right) = -\frac{1}{10}\text{ in. }\downarrow$$

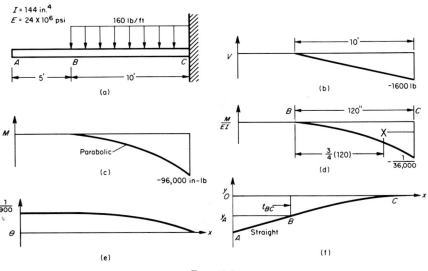

FIG. 10-28

Then, as in Example 10-5, the deflection of the left-hand end is

$$y_A = t_{BC} - 60\,\theta_B$$

$$= -\frac{1}{10} - (60)\left(\frac{1}{900}\right)$$

$$= -\frac{1}{10} - 0.0667 = -0.1667\text{ in. }\downarrow$$

EXAMPLE. 10-10. A solid circular reduced section rod shown in Fig. 10-29 (*a*) is simply supported and loaded at its midspan with a load of 600 lb. Assuming that the material is elastic and homogeneous with $E = 10 \times 10^6$, and neglecting stress concentrations at the fillet, find the maximum deflection of the rod.

*Solution:* The moment and curvature diagrams are shown in Fig. 10-29(*b*) and (*c*), respectively. From the geometry and loading of the rod, it is logical to expect that the maximum deflection will occur at some point in the reduced portion. If we consider some arbitrary point $C$ in this portion, from Fig. 10-29(*d*) we see that

$$y_C + t_{CA} = \frac{x}{72} t_{BA}$$

$$y_C = \frac{x}{72} t_{BA} - t_{CA} \tag{a}$$

Now, from Fig. 10-29(*c*),

$$t_{CA} = \text{moment of area } ACC' \text{ about } C$$

$$= \frac{x}{36} \frac{(12)(36)}{\pi(10^5)} \frac{x}{2} \frac{x}{3}$$

$$= \frac{2x^3}{\pi(10^5)} \tag{b}$$

Similarly,

$$t_{BA} = \left(\frac{36}{2}\right) \frac{(4)(64)}{3\pi(10^5)} (24) + \left(\frac{36}{2}\right) \frac{(12)(36)}{\pi(10^5)} (48)$$

$$= \frac{(2^{12})(3^2) + (2^9)(3^6)}{\pi(10^5)} \tag{c}$$

Combining equations (a), (b), and (c), we have

$$y_C = \frac{1}{\pi(10^5)} [(2^9 + 2^6 \times 3^4) x - 2x^3] \tag{d}$$

We know that the maximum deflection occurs where $dy/dx = 0$. Hence,

$$\frac{dy}{dx} = \frac{1}{\pi(10^5)} [(2^9 + 2^6 \times 3^4) - 6x^2] = 0$$

from which

$$x = (2^3)\sqrt{\frac{89}{6}} = 31 \text{ in.}$$

Substitution of this value in equation (d) gives the desired result. Thus,

$$y_{max} = \frac{1}{\pi(10^5)} [(2^9 + 2^6 \times 3^4) 31 - 2(31^3)]$$

$$= 0.372 \text{ in.}$$

These examples show that the moment-area method is quite advantageous as long as the curvature-diagram geometry is relatively simple.

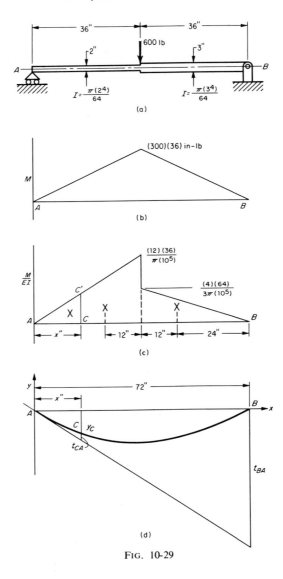

FIG. 10-29

Also it is a convenient way of determining the deflection of a particular point rather than the deflection equation of the entire beam.

## 10-12. Deflection by Superposition

When a beam is subjected to several loads placed at various positions along the beam, the problem of determining the slope and the deflection usually becomes quite involved and tedious. This is true regardless of the method used. However, most complex loading conditions are merely combinations of relatively simple loading conditions. If it is assumed that the beam behaves elastically for the combined loading, as well as for the individual loads, the resulting final deflection of the loaded beam is simply the sum of the deflections caused by each of the individual loads. This sum may be simply an algebraic one or it might be a vector sum, the type depending on whether or not the individual deflections lie in the same plane. The superposition method is illustrated in the following example.

EXAMPLE 10-11. For the beam in Fig. 10-30(a) determine the flexure stress at point $A$ and the deflection of the left-hand end.

*Solution*: The stress at point $A$ is a combination of a compressive flexure stress due to the concentrated load and a tensile flexure stress due to the distributed load. Hence,

$$\sigma_A = \frac{M_z(2)}{I_z} - \frac{M_y(3)}{I_y}$$

$$= \frac{\dfrac{(5)(6)(80^2)}{2}(2)}{\dfrac{(6)(4^3)}{12}} - \frac{(600)(80)(3)}{\dfrac{(4)(6^3)}{12}}$$

$$= 6000 - 2000 = 4000 \text{ psi (tension)}$$

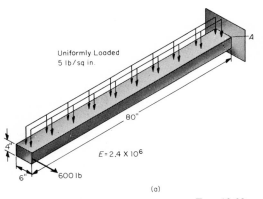

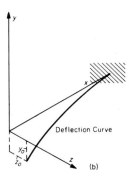

FIG. 10-30

The deflection of a cantilever beam due to a uniformly distributed load $w$ is

$$y = \frac{1}{EI_z}\left(-\frac{wx^4}{24} + \frac{wL^3}{6}x - \frac{wL^4}{8}\right)$$

where $x$ is the distance from the free end in inches and $w$ is in pounds per inch. Also, the deflection due to a concentrated load $P$ at its free end is

$$z = \frac{1}{EI_y}\left(\frac{Px^3}{6} - \frac{PL^2}{2}x + \frac{PL^3}{3}\right)$$

Thus, at the free end of the beam in Fig. 10-30(*b*), we have

$$y_0 = -\frac{(30)(80^4)}{\dfrac{(6)(4^3)}{12}(8)(2.4)(10^6)} = -2 \text{ in.}$$

$$z_0 = \frac{(600)(80^3)}{\dfrac{(4)(6^3)}{12}(3)(2.4)(10^6)} = \frac{16}{27} \text{ in.}$$

The deflection at the free end is the vector sum

$$y_0 \mapsto z_0 = 2.1 \text{ in.}$$

## 10-13. Statically Indeterminate Beams

All the flexure members which we have considered thus far in this chapter have been statically determinate; that is, the equilibrium equations were sufficient to determine all the external reactions. However, many engineering flexure members are supported by more than the minimum number of reactions which are necessary to maintain equilibrium. The presence of such additional or redundant reactions makes the member statically indeterminate, and all the reactions cannot be determined from the equations of equilibrium alone. Just as in the cases of torsion members and axial members, the equilibrium equations must be supplemented with some additional relationships based on the characteristics of the deformation of the member. For beams, the additional relationships usually arise from the restraints which the supports place on the shape of the deflection curve.

Finding the deflection curve for statically indeterminate beams requires no new theories or techniques. The unknown external reactions may be treated simply as ordinary individual external loads. The deflection caused by these external loads can be found by any of the methods previously discussed: direct integration, singularity functions, moment-area, and superposition. Since the presence of the external reactions places geometrical restrictions on the deflection curve, there will always be a sufficient number of boundary conditions to completely determine the unknown reactions. The following examples illustrate the general procedure.

EXAMPLE 10-12. Determine all the external reactions for the elastic beam shown in Fig. 10-31(a). What is the maximum deflection of the beam?

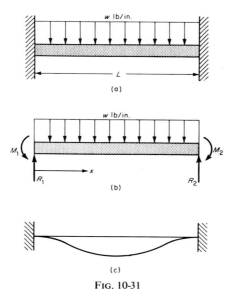

w lb/in.

$L$

(a)

w lb/in.

$M_1$    $M_2$

$R_1$    $x$    $R_2$

(b)

(c)

FIG. 10-31

*Solution:* From the free-body diagram of Fig. 10-31(b), we have

$$+\uparrow$$
$$\sum F_y = 0$$
$$R_1 + R_2 - wL = 0 \tag{a}$$

$$\overset{\curvearrowright}{+}$$
$$\sum M = 0$$
$$R_2 L - M_2 - w\frac{L^2}{2} + M_1 = 0 \tag{b}$$

The couples $M_1$ and $M_2$ are present because the beam must have a horizontal tangent at each wall. Thus, the boundary conditions are:

$$y = 0 \text{ at } x = 0 \quad \text{and} \quad x = L$$
$$\theta = 0 \text{ at } x = 0 \quad \text{and} \quad x = L$$

The deflection equation can be obtained by direct integration.

$$M = R_1 x - M_1 - \frac{wx^2}{2}$$

$$EI\theta = R_1 \frac{x^2}{2} - M_1 x - \frac{wx^3}{6} + C_1 \tag{c}$$

$$EIy = R_1 \frac{x^3}{6} - M_1 \frac{x^2}{2} - \frac{wx^4}{24} + C_1 x + C_2 \tag{d}$$

Substitution of the boundary conditions in equations (c) and (d) gives

$$C_1 = 0 \qquad C_2 = 0$$

$$R_1 \frac{L^2}{2} - M_1 L - \frac{wL^3}{6} = 0$$

$$R_1 \frac{L^3}{6} - M_1 \frac{L^2}{2} - \frac{wL^4}{24} = 0$$

From these relationships,

$$R_1 = \frac{wL}{2} \qquad M_1 = \frac{1}{12} wL^2$$

From equations (a) and (b)

$$R_2 = R_1 \qquad M_2 = M_1$$

Of course, from the symmetry of the problem, this last result should have been anticipated. The deflection equation becomes

$$EIy = \frac{wLx^3}{12} - \frac{wL^2 x^2}{24} - \frac{wx^4}{24}$$

The maximum deflection occurs at $x = L/2$ and has the value

$$y = \frac{1}{EI} \left[ \frac{wL^4}{96} - \frac{wL^4}{96} - \frac{wL^4}{(16)(24)} \right]$$

$$= -\frac{wL^4}{384\,EI} \text{ in.}$$

EXAMPLE 10-13. Determine all the external reactions on the elastic beam shown in Fig. 10-32($a$). Sketch the deflection curve.

*Solution:* The free-body diagram of Fig. 10-32($b$) shows that there are three unknowns and only two independent equilibrium equations. However, if $R_2$ is treated as an ordinary external load, this problem can be conveniently handled by the superposition method.

The deflection equation for the simply supported, uniformly loaded beam shown in Fig. 10-32($c$) is

$$EIy_1 = \frac{wLx^3}{12} - \frac{wx^4}{24} - \frac{wL^3 x}{24}.$$

Also, from Example 10-6 we know that for a simply supported beam with a downward load $P$ at a distance $L/3$ from the left-hand support, the deflection equation for the portion for which $0 \le x \le L/3$ is

$$EIy_2 = \frac{Px^3}{9} - \frac{5}{81} PL^2 x$$

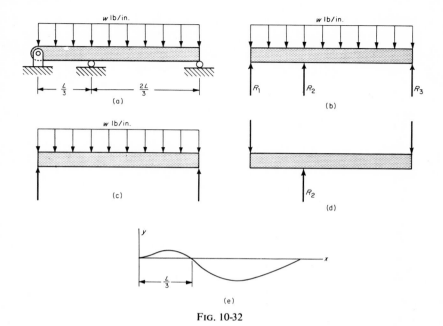

Fɪɢ. 10-32

Hence, for $0 \leq x \leq L/3$ the deflection of our original beam is

$$Ely = Ely_1 + Ely_2$$

$$= \frac{wLx^3}{12} - \frac{wx^4}{24} - \frac{wL^3x}{24} - R_2\frac{x^3}{9} + \frac{5}{81}R_2L_2x \qquad \text{(a)}$$

For the original beam, the deflection at $x = L/3$ must be zero. Thus,

$$\frac{wL^4}{(12)(27)} - \frac{wL^4}{(24)(81)} - \frac{wL^4}{(24)(3)} - \frac{R_2L^3}{(9)(27)} + \frac{5}{81}\frac{R_2L^3}{3} = 0$$

$$- \frac{22}{(24)(81)}wL^4 = - \frac{R_2L^3(4)}{(3)(81)}$$

$$R_2 = \frac{11}{16}wL$$

With this value for $R_2$ we can now determine $R_1$ and $R_3$ from the equilibrium requirements of Fig. 10-32(*b*). Thus,

$$R_1 = \frac{wL}{24} \qquad R_2 = \frac{11wL}{16} \qquad R_3 = \frac{13wL}{48}$$

Substitution of the value of $R_2$ in equation (a) gives, for $0 \leq x \leq L/3$,

$$Ely = \frac{wLx^3}{144} - \frac{wx^4}{24} + \frac{wL^3x}{(9)(144)} \qquad \text{(b)}$$

Also,

$$EI\theta = \frac{wLx^2}{48} - \frac{wx^3}{6} + \frac{wL^3}{(9)(144)} \tag{c}$$

At $x = L/3$ and $x = 0$,

$$EI\theta_{x = L/3} = - \frac{wL^3}{(81)(4)}$$

$$EI\theta_{x = 0} = \frac{wL^3}{(9)(144)}$$

With these values we can now sketch a qualitative deflection curve (exaggerated) as shown in Fig. 10-32(e). A moment diagram or curvature diagram is usually a great aid in visualizing the shape of the deflection curve, since a positive curvature indicates concave upward, etc.

It is interesting to observe that although the deflection of the portion of the beam for which $0 \leq X \leq L/3$ is upward, the left-hand reaction $R_1$ is also upward. Thus, this reaction is pushing the beam upward, and not pulling the left-hand end of the beam downward as might be anticipated.

## PROBLEMS

**10-1.** Which of the two cross-sections in the illustration would be better for use as a beam?

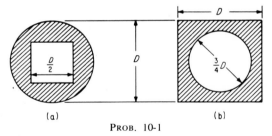

(a)                    (b)

PROB. 10-1

**10-2.** The cast-iron machine part, a section through which is shown in the illustration, acts as a beam resisting a *positive* bending moment. If the allowable

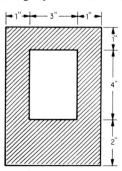

PROB. 10-2

tensile stress is 6000 psi and the allowable compressive stress is 24,000 psi, what maximum moment may be applied to the member? Assume elastic behavior.

**10-3.** An elastic straight beam has the constant cross section shown in the illustration. If a positive bending moment of 40,000 ft-lb acts about its horizontal centroidal axis, what are the maximum tensile and compressive flexure stresses?

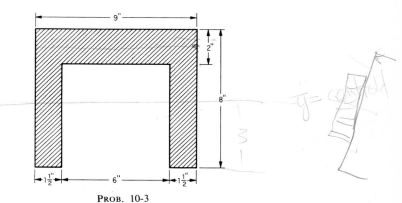

PROB. 10-3

**10-4.** What maximum tensile flexure stress will occur in a simply supported beam 14 ft long with a rectangular cross section 8″ wide by 10″ deep? It is uniformly loaded with 500 lbs per ft. Exactly where does this stress occur?

**10-5.** How long can a simply supported beam of square cross section 4″ × 4″ be if it must support a 10,000 lb load at the center and the bending stress cannot exceed 26,000 psi?

**10-6.** The elastic beam shown in the illustration is a rolled steel 10 WF 21. Neglecting the weight of the beam itself, determine the magnitude and location of the maximum flexure stress.

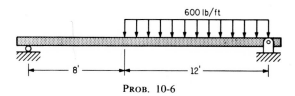

PROB. 10-6

**10-7.** The elastic beam shown in the illustration is a rolled steel 10 WF 112. What is the maximum value L can have, if the flexure stress is not to exceed 5000 psi anywhere in the beam? Do not neglect the weight of the member.

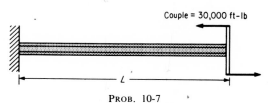

PROB. 10-7

**10-8.** The elastic beam in the illustration has an $I$ of 64 in.$^4$ about its centroidal axis. Under the load $P$ at point $A$, 2 in. from the top of the beam, there is a longitudinal strain of $-0.0005$ in./in.; and at point $B$, 2 in. from the bottom of the beam, there is a strain of $+0.00075$ in./in. The beam is made of a material with modulus $E = 15 \times 10^6$. What is the amount of the load $P$? Neglect the localized effect of the load.

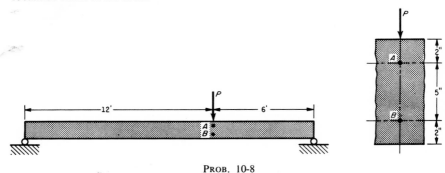

PROB. 10-8

**10-9.** What is the ratio of the ideally fully plastic bending moment $M_{fp}$ to the maximum elastic bending moment $M_{el}$ for a beam having a rectangular cross section?

**10-10.** The illustration shows the cross section of a beam subjected to pure bending about the axis $xx$. Determine the ratio $M_{fp}/M_{el}$.

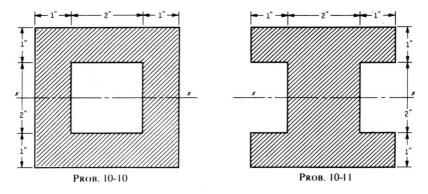

PROB. 10-10      PROB. 10-11

**10-11.** The illustration shows the cross section of a beam subjected to pure bending about the axis $xx$. Determine the ratio $M_{fp}/M_{el}$.

**10-12.** What is the ratio of the ideally fully plastic bending moment to the maximum elastic bending moment for a beam of circular cross section subjected to pure bending?

**10-13.** The illustration shows the cross section of a beam made of an ideally plastic material which has a yield-point stress of 35,000 psi. What is the fully plastic bending moment for this section?

**10-14.** Determine the fully plastic bending moment for a beam having the cross section shown in the illustration, if the yield-point stress for the material is 30,000 psi.

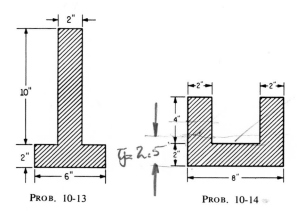

PROB. 10-13                                PROB. 10-14

**10-15.** A beam having a length $L$ and a rectangular cross section $2c$ in. in height and $b$ in. in width is subjected to a pure bending moment $M$. The tensile and compressive stress-strain law for the material is approximated by $\sigma = \lambda \epsilon^{1/4}$. In an attempt to calculate the maximum flexure stress, a designer uses the relationship $\sigma = K(Mc/I)$. What should be the value of $K$ for this application?

**10-16.** The beam shown in part (*a*) of the illustration is deformed by couples $M$ into the shape shown in part (*b*) so that there is a strain of 0.006 in./in. at point $A$. The beam is made of a material whose stress-strain curve is shown in part (*c*). Assuming that the strains vary linearly from the neutral axis, determine the value of $M$.

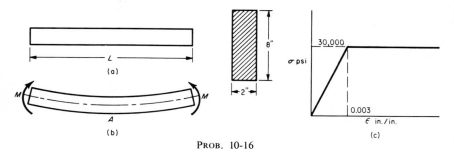

PROB. 10-16

**10-17.** A cantilever beam of negligible weight is made of a material which has a yield-point stress and has the cross section shown in the illustration. The beam is 5 ft long and is loaded with a single concentrated load at the free end. When this load reaches 20,000 lb, the beam yields and ideally fully plastic behavior occurs. What is the yield-point stress of the material?

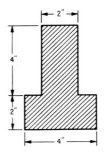

PROB. 10-17

**10-18.** A cast-iron beam has the T-shaped section shown in the illustration. If this beam transmits a vertical shear of 136 kips, find the average transverse shear stress at each of the levels indicated. Sketch a plot showing how the shear stress varies throughout the height of the beam.

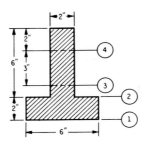

PROB. 10-18

**10-19.** An elastic homogeneous straight beam which carries a constant shear force *V*, has the constant triangular cross section shown in the illustration. Express

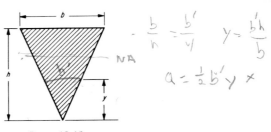

PROB. 10-19

the average transverse shear stress as a function of the distance *y* from the vertex of the triangle. Also, show that the maximum shear stress occurs at $y = h/2$ and is given by $\tau_{\max} = 3V/2A$, where *A* is the cross-sectional area.

**10-20 and 10-21.** In the beams whose cross sections are shown in the illustration, where would the maximum horizontal shear stress occur in the cross sections? Prove your answer.

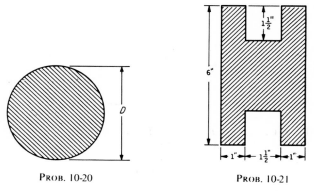

PROB. 10-20    PROB. 10-21

**10-22.** A beam is made of six planks nailed together to form the cross section shown in the illustration. Each nail is capable of resisting 120 lb of shear force. If the section is carrying a shear force of 4000 lb, determine the required longitudinal spacing of the nails connecting $A$ with $B$ or $C$.

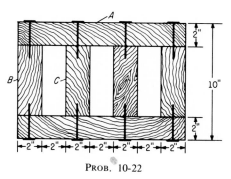

PROB. 10-22

**10-23.** A simply supported beam of 8 ft span carries uniformly distributed load of 1000 lb/ft. The cross section is $2'' \times 8''$. What is the maximum horizontal shear stress that is developed and exactly where does it occur?

**10-24.** A beam 12 ft long has a support 3 ft from each end. A uniformly distributed load of 960 lb/ft acts on the overhangs only. The cross section of the beam is a hollow symmetrical rectangle, 6 in. wide by 8 in. deep outside and 4 in. wide by 4 in. deep inside. Assuming that the material is linearly elastic, *a)* find the maximum tensile stress and its location. *b)* Find the maximum horizontal shear stress and its location.

**10-25.** A steel beam 30 ft long has simple supports at the right-hand end and 10 ft. from the left-hand end. It has a uniform load of 400 lb/ft over the right 15 ft of the beam and a load of 600 lb/ft on the overhang. In addition, a 4500-lb concentrated load acts midway between the supports. The cross section is an upright "Tee" with an 8 by 1 in. flange and a 6 by 2 in. stem. *a)* Assuming elastic behavior, determine the maximum tensile and compressive flexure stresses in the beam, and show where they occur. *b)* Find the horizontal shear stress 3 ft from the right-hand end and 4 in. from the top surface of the beam.

**10-26.** An elastic cantilever beam has the loading and dimensions shown in the illustration. Sketch the shear and bending-moment diagrams, and find *a)* the location and value of the maximum tensile flexure stress; *b)* the location and value of the maximum compressive flexure stress; *c)* the position and value of the maximum transverse shear stress.

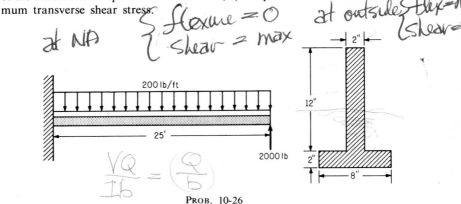

PROB. 10-26

**10-27.** A simply supported beam 5 ft long carries a uniformly distributed load of 12,000 lb/ft. The beam has a rectangular cross section 2 in. wide and 6 in. deep. Assuming elastic behavior, determine the maximum normal and shear stresses developed at a point 2 in. below the top surface and 8 in. from the right-hand support. Sketch the planes on which these stresses act.

**10-28.** For the elastic beam shown in the illustration, determine the magnitude and location of *a)* the maximum tensile flexure stress; *b)* the maximum compressive flexure stress; *c)* the maximum horizontal shear stress.

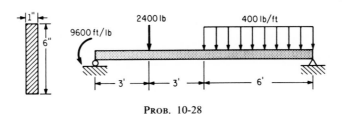

PROB. 10-28

**10-29.** For the elastic beam *AB* in the illustration, determine the magnitude and location of *a)* the maximum tensile flexure stress; *b)* the maximum compressive flexure stress; *c)* the maximum transverse shear stress.

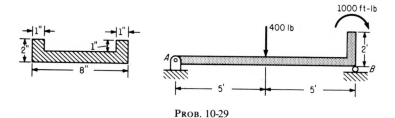

PROB. 10-29

**10-30.** A simply supported, uniformly loaded timber beam 11 ft long has a rectangular cross section 8 in. wide and 12 in. deep. If the maximum allowable flexure stress is 1200 psi and the maximum allowable horizontal shear stress is 100 psi, what is the maximum uniformly distributed load $w$ lb/ft that the beam can carry?

**10-31.** A rectangular wood beam is to be used to support a uniform load of 150 lb/ft over a 10-ft span. Determine an acceptable nominal size of timber if the flexure stress is not to exceed 1200 psi and the horizontal shear stress is not to exceed 75 psi. See Appendix.

**10-32.** The beam in part (*a*) of the illustration has the cross section shown in part (*b*). Determine the maximum allowable value of $P$ for a) failure due to the elastic flexure stresses exceeding 50,000 psi in tension or 60,000 psi in compression, b) failure due to the horizontal shear stress exceeding 20,000 psi, and c) failure due to excessive deformation resulting from the formation of a fully plastic hinge with $\sigma_{yp} = 30,000$ psi.

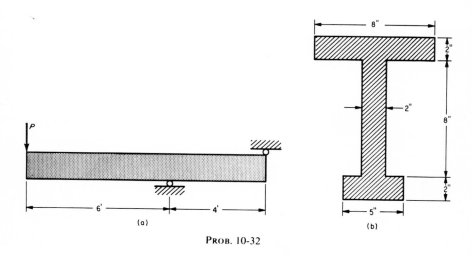

PROB. 10-32

**10-33.** If the beam in (*a*) of the illustration is made of a material which has the stress-strain curve shown in part (*b*), what load $W$ will cause complete failure of the beam? Describe the failure.

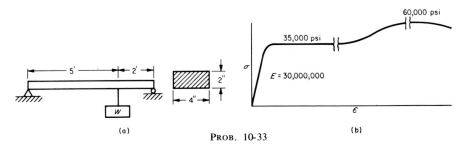

(a)

PROB. 10-33

(b)

**10-34.** The beam shown in the illustration has the dimensions and loading indicated. Sketch the shear and bending-moment diagrams for the horizontal part in terms of *P*. Determine *P* if *a*) the beam is made of a material for which the allowable elastic stresses are 40,000 psi in tension and 30,000 psi in compression, and *b*) failure is due to shear when the maximum elastic horizontal shear stress is 5000 psi. Neglect stress concentration at the wall.

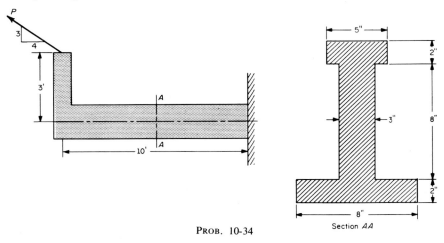

PROB. 10-34

Section *AA*

**10-35.** Assuming elastic behavior, determine the maximum normal stress in the short block in the illustration. Express your answer in terms of *P* and *d*, and indicate where this stress occurs.

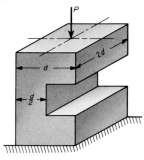

PROB. 10-35

**10-36.** What should be the thickness $t$ of the cross section in the illustration if the normal stress at point $A$ cannot exceed 1000 psi?

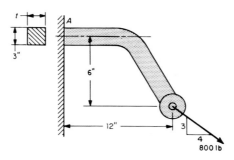

PROB. 10-36

**10-37.** A cast frame has the cross section shown in the illustration. What maximum force $P$ may be applied if the allowable tensile stress at section $AA$ is 4000 psi and the allowable compressive stress is 5000 psi?

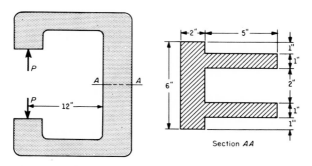

PROB. 10-37

**10-38.** A cylindrical pressure tank has an inside diameter of 20 in. and an outside diameter of 21 in. and an over-all length of 300 in. The tank is simply supported at each end, as shown in the illustration. The tank is filled with a gas under a pressure of 250 psig. The total weight of the tank and gas combined is 10,000 lb, and it can be considered uniformly distributed over the entire length of the tank. Determine the maximum normal stress in the wall of the tank.

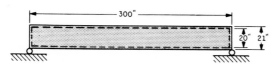

PROB. 10-38

**10-39.** Determine the maximum normal stress in the member shown in the illustration. Assume elastic behavior, and neglect stress concentrations at the wall.

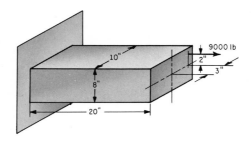

PROB. 10-39

**10-40.** A propeller shaft 3 in. in diameter is subjected to a combination of a twisting torque of 16,000 in-lb, a bending moment of 12,000 in-lb, and an axial tensile thrust of 15,000 lb. What are the maximum tensile, compressive, and shear stresses in the shaft?

**10-41.** A circular rod 2 in. in diameter is subjected to an axial tensile load $P$ and a bending moment $M$. What should be the ratio of $P$ to $M$ if the maximum tensile stress in the rod is to be two times the magnitude of the maximum compressive stress? Assume elastic behavior.

**10-42.** As shown in the illustration, a circular cantilver beam of radius $r$ is subjected to a load $P$ which acts perpendicular to the longitudinal axis of the beam. If the material is linearly elastic and the maximum allowable normal stress is $\sigma_{max}$, what is the maximum allowable load $P$ in terms of $\sigma_{max}$, $L$, and $r$?

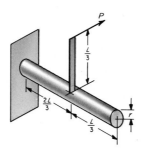

PROB. 10-42

**10-43.** A rigid bar 4 in. long is welded to a solid homogeneous shaft having a length of 8 in. and a diameter of 4 in. A load of 30,000 lb is applied as shown in the illustration. Determine the principal stresses and the maximum shear stress at point $A$ on top at the extreme left-hand end of the shaft. Assume elastic behavior, and neglect stress concentrations at the wall.

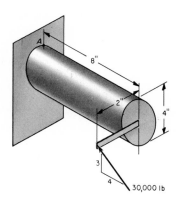

PROB. 10-43

**10-44.** A simply supported elastic beam of length $L$ is loaded with a single concentrated load $P$ at the middle of the span. Derive an expression for the maximum deflection of the beam in terms of $P$, $L$, $E$, and $I$. What is an appropriate set of units for each of the constants $E$, $L$, and $I$?

**10-45.** An elastic cantilever beam of length $L$ is fixed at the right-hand end and supports a uniformly distributed load of $w$ lb per unit length. Derive an expression for its deflection at the left-hand end in terms of $w$, $L$, $E$, and $I$. What is an appropriate set of units for each of the constants, $w$, $L$, $E$, and $I$?

**10-46.** The length of a simply supported elastic beam is $L$, its modulus is $E$ psi, and the moment of inertia is $I$ in.$^4$ Derive an expression for the deflection at the middle of the span if $a$) the length is given in feet and the beam is uniformly loaded with $w$ lb/ft, and $b$) the length is given in inches and the uniform loading is $w$ lb/in.

**10-47.** A hickory diving board 3 by 12 in. in cross section and 9 ft long is built with one end fixed. If a person standing at the middle of the board causes a deflection of 0.350 in. at the free end, what is the weight of the person? Assume that $E = 2 \times 10^6$.

**10-48.** In the construction of private dwellings contractors frequently support flooring by arranging the joists in one of the following three ways: $a$) $2'' \times 8''$ spaced 16 in. on centers, $b$) $2'' \times 10''$ spaced 16 in. on centers, $c$) $2'' \times 10''$ spaced 18 in. on centers. If deformation is the major criterion in deciding which arrangement should be used, determine the stiffness ratio of ($b$) to that of ($a$), ($c$) to ($a$), and that of ($b$) to ($c$). Which is better, ($a$) or ($c$)?

**10-49.** What is the deflection of point $A$ due to the load of 6000 lb on the beam in the illustration? Assume that the behavior is elastic and that $E = 24 \times 10^6$ psi and $I = 120$ in.$^4$

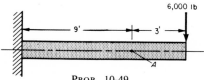

PROB. 10-49

**10-50.** Determine the deflection of point $A$ in the beam shown in the illustration. Assume that the action is elastic and that $E = 30 \times 10^6$ and $I = 80$ in.$^4$

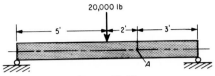

20,000 lb

PROB. 10-50

**10-51.** For the cantilever beam in the illustration, $I$ is 140 in.$^4$ and $E$ is $30 \times 10^6$ psi. If the load $P$ is 5,000 lb, what is the vertical deflection of point $A$ at the extreme right-hand end of the beam. Assume elastic behavior.

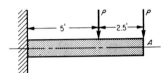

PROB. 10-51

**10-52.** The beam shown in the illustration is made of an elastic homogeneous material. Draw the complete shear and bending moment diagrams. Also derive the slope and deflection equations, sketch the slope and deflection curves.

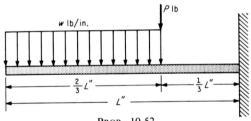

$P$ lb

$w$ lb/in.

$\frac{2}{3}L''$     $\frac{1}{3}L''$

$L''$

PROB. 10-52

**10-53.** For the beam shown in the illustration, determine the horizontal and vertical displacement of the point $A$ due to the load $P$. Assume elastic behavior and small deformations.

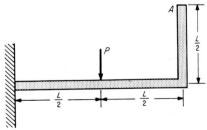

$A$

$P$

$\frac{L}{2}$

$\frac{L}{2}$    $\frac{L}{2}$

PROB. 10-53

**10-54.** The beam shown in the illustration is loaded at mid-span by the couple *PkL*. If the beam has constant stiffness *EI* and if $k = (24EI/PL^2) \times 10^{-3}$, find the slope and deflection at mid-span. Also, sketch the elastic deflection curve and label the pertinent points.

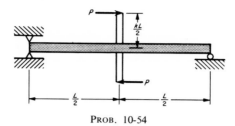

PROB. 10-54

**10-55.** The beam shown in the illustration has a constant cross section and is made of a homogeneous, linearly elastic material. Derive the shear, bending-moment, slope, and deflection equations for the beam. Sketch the shear, bending-moment, slope, and deflection diagrams for the beam.

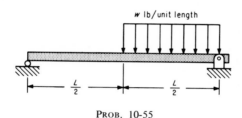

PROB. 10-55

**10-56.** The free end of the prismatic, linearly elastic cantilever beam shown in the illustration is supported at its top by a linear elastic spring with a spring constant of *k* lb/in. If the spring carries no load when the beam is vertical, how far will the spring be compressed when the beam is loaded with a uniformly distributed load of *w* lb/in.?

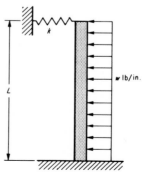

PROB. 10-56

**10-57.** Referring to the illustration, and assuming that $P$ = 2700 lb, $L$ = 5 ft, $E$ = 4 × 10$^6$ psi, and $I$ = 2.25 in.$^4$, determine the magnitude of $M_0$ required to keep the longitudinal axis of the pointer horizontal. What is the horizontal displacement of the tip of the pointer?

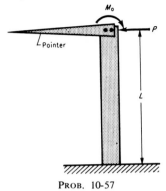

PROB. 10-57

**10-58.** The rectangular cantilever beam in the illustration carries a concentrated load at its left-hand end. The beam is 2 in. wide and has a varying height of 2,3, and 4 in. Find the deflection of the left end if $E$ = 30 × 10$^6$ psi.

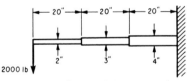

PROB. 10-58

**10-59.** A circular cantilever beam carries a concentrated load as shown in the illustration. The radius of the beam varies linearly with the length from 2 in. at the right-hand end to 4 in. at the left-hand end. What is the deflection of the right-hand end if $E$ = 30 × 10$^6$ psi?

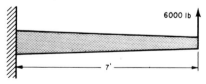

PROB. 10-59

**10-60.** The reduced rectangular-section beam shown in the illustration has a width of 2 in. Neglecting any stress concentrations and assuming that $E$ = 10 × 10$^6$ psi, find the maximum flexure stress and the maximum deflection.

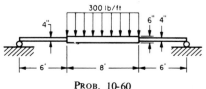

PROB. 10-60

**10-61.** Derive the elastic deflection equation for the beam shown in the illustration. (NOTE: The magnitude of $P$ is not necessarily the same as that of $Q$.)

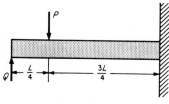

PROB. 10-61

**10-62.** An elastic cantilever beam is 8 ft long and made of a rolled steel equal-leg angle $5 \times 5 \times \frac{5}{8}$ (see Appendix). The beam is loaded at its free end as indicated in the illustration. Find the deflection of the free end, if $E = 30 \times 10^6$ psi.

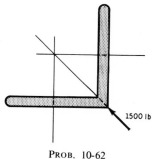

PROB. 10-62

**10-63.** An elastic beam with a rectangular cross section carries a uniformly distributed vertical load of 5 lb/ft. The beam is 12 ft long and is simply supported at its ends at an angle with the vertical axis as shown in the illustration. If $E = 3 \times 10^6$ psi, determine the magnitude and direction of its deflection at the mid-span.

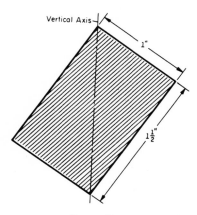

PROB. 10-63

**10-64.** An elastic beam with a length of $L$ inches and a square cross section carries a uniformly distributed vertical load of $w$ lb/in. If the beam is simply supported at its ends at an angle $\theta$ with the horizontal, determine the magnitude and direction of the deflection at the mid-span if *a*) $\theta = 60°$; *b*) $\theta = 45°$; *c*) $\theta$ is an arbitrary angle such that $0 \le \theta \le 90°$.

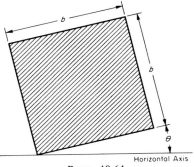

PROB. 10-64

**10-65.** Determine all the external reactions for the elastic beam shown in the illustration. Also, what is the maximum deflection of the beam?

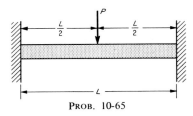

PROB. 10-65

**10-66.** Determine all the external reactions for the elastic beam shown in the illustration.

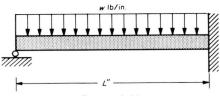

PROB. 10-66

**10-67.** Determine all the external reactions for the elastic beam in the illustration.

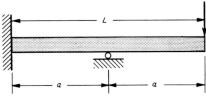

PROB. 10-67

**10-68.** Determine all the external reactions for the elastic beam in the illustration.

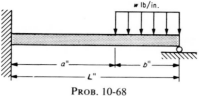

PROB. 10-68

**10-69.** Determine all the external reactions for the elastic beam shown in the illustration.

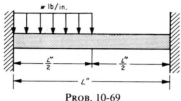

PROB. 10-69

**10-70.** Find the external reactions for the continuous elastic beam in the illustration. Sketch the deflection curve.

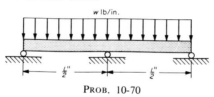

PROB. 10-70

**10-71.** The circular beam shown in the illustration has fixed ends and is subjected to a load $P$ which acts perpendicular to the longitudinal axis of the beam. The radius is $r$, and the material is linearly elastic. If the maximum normal stress is $2000\,L/\pi r^3$, what is the value of $P$?

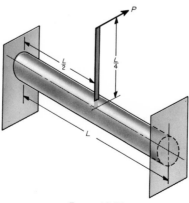

PROB. 10-71

## SUPPLEMENTARY PROBLEMS

**10-72.** What is the ratio of the ideally fully plastic bending moment to the maximum elastic bending moment for a beam subjected to pure bending and having a cross section in the shape of an upright isosceles triangle?

**10-73.** What is the ratio of the ideally fully plastic bending moment to the maximum elastic bending moment for a beam subjected to pure bending and having an eliptical cross section $c$ in. deep (major axis) and $b$ in. wide (minor axis)?

**10-74.** A rectangular beam 8 in. wide and 12 in. deep has a nonlinear stress distribution given by $|\sigma| = k|y|^{1/2}$ for both the upper and lower fibers of the beam, the upper fibers being in compression and the lower in tension. What is the magnitude of the pure bending moment producing this stress if $k = 4000$?

**10-75.** A rectangular timber 4 in. wide by 8 in. deep is to be used as a cantilever beam to carry a uniform load of 40 lb/ft which includes the beam weight. If the allowable flexure stress is 1400 psi and the allowable horizontal shear stress is 60 psi, how long can the beam be?

**10-76.** For the elastic beam shown in the illustration, determine the magnitude and location of  *a*) the maximum tensile flexure stress,  *b*) the maximum compressive flexure stress, and *c*) the maximum transverse shear stress.

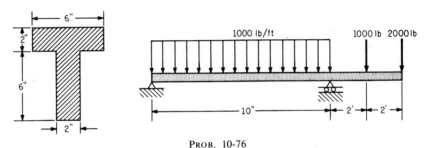

PROB. 10-76

**10-77.** A loaded beam and its original dimensions are shown in the illustration. The 10,000 lb axial load acts through the centroid of the cross section. Determine the value and location of *a*) the maximum normal stress, *b*) the maximum longitudinal shear stress, and *c*) the maximum shear stress.

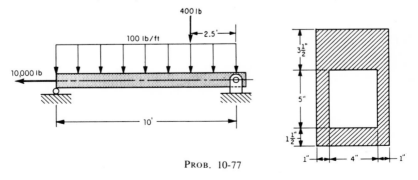

PROB. 10-77

**10-78.** The member shown in the illustration has a solid circular cross section of radius *r*. It is subjected to a load of 7,000 lb applied as shown. Find the smallest permissible radius if the maximum normal stress in the horizontal portion may not exceed 25,000 psi. Neglect stress concentrations.

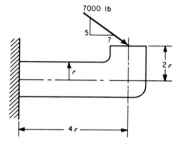

PROB. 10-78

**10-79.** For the straight elastic beam shown in the illustration derive the shear, bending-moment, slope, and deflection equations. Also, sketch the shear, bending-moment, slope, and deflection diagrams.

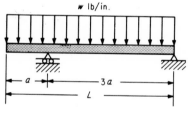

PROB. 10-79

**10-80.** The free end of the prismatic, linearly elastic cantilever beam shown in the illustration is supported at its left-hand end by a linear elastic spring with a spring constant of *k* lb/in. If the spring carries no load when the beam is horizontal, how far will the spring be compressed when the beam is loaded.?

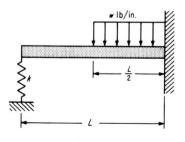

PROB. 10-80

**10-81.** For the beam shown in the illustration draw the complete shear and bending-moment diagrams. Also, assuming elastic behavior, derive the slope and deflection equations and sketch the deflection curve.

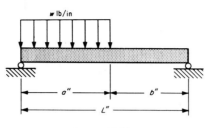

PROB. 10-81

**10-82.** For the cantilever beam shown in the illustration, find the deflection of the free end in terms of $P$, $L$, $E$, and $I$. Neglect the weight of the beam.

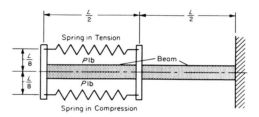

PROB. 10-82

**10-83.** The beam shown below is to be used as a weighing device with the load at $W$ and the indicator at $P$. We wish to calibrate the device so that the weight can be read directly on the indicator scale. Derive an expression between the load $W$ and the movement (in inches) of the pointer assuming $E$ and $I$ to be known. Make a sketch of the shape of the deflected beam.

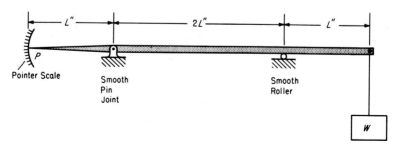

PROB. 10-83

**10-84.** If the weighing device described in Prob. 10-83 were made of mild steel and the dimension *L* were 5 ft and the cross-section was that shown below, what would you as a design engineer specify as the weighing capacity of this device?

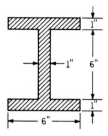

PROB. 10-84

**10-85.** For the elastic beam shown in the illustration, $L = 18$ ft. The pointer *AB* is rigidly attached to the beam directly above the right-hand support. With no load on the beam, the tip *A* of the pointer is at the neutral surface. Determine the moment $M_0$ necessary to keep the tip of the pointer at the neutral axis of the cross section of the beam when a load of *P* of 4000 lb is applied.

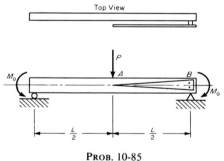

PROB. 10-85

**10-86.** A reduced circular-section beam has the dimensions and loading shown in the illustration. Neglecting any stress concentrations, find the maximum flexure stress and the deflection of the mid-point. Does this latter value represent the maximum deflection? Assume that there is elastic behavior and that $E = 10 \times 10^6$ psi.

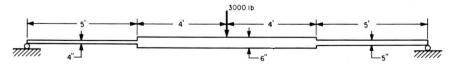

PROB. 10-86

**10-87.** A simply supported beam 24 ft long and having the cross section shown in the illustration carries a concentrated load of 10,000 lb at its mid-span. The load acts in the direction indicated. What is the maximum deflection if $E = 20 \times 10^6$ psi?

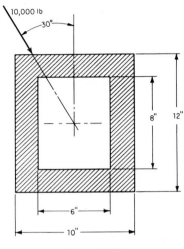

PROB. 10-87

**10-88.** Find the external reactions for the continuous elastic beam in the illustration. Sketch the deflection curve.

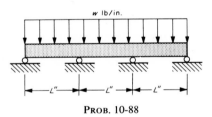

PROB. 10-88

**10-89.** Find the external reactions for the continuous elastic beam in the illustration. Sketch the deflection curve.

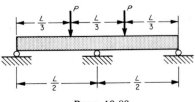

PROB. 10-89

## chapter 11

# Buckling

## 11-1. Nature of Buckling

Buckling is a mode of failure characterized generally by an unstable deflection due to compressive action on the structural member or element involved. Some common engineering structures that are likely to buckle if overloaded are metal building columns, compression members in bridge or roof trusses, the hull of a submarine, the metal skin on aircraft fuselage or wings with excessive torsional and/or compressive loading, any thin-walled torque tube, a thin web of a beam with excessive shear load, and a thin flange of a beam subjected to excessive compressive loads. Notice from these few examples that buckling is the result of compressive action. Over-all torsion or shear, as we learned in previous chapters, may cause a localized compressive action which could lead to buckling.

The distinctive feature of buckling is the catastrophic and often spectacular nature of the failure. The collapse of a column supporting stands in a stadium or the roof of a building usually draws large headlines and cries of engineering negligence. On a less grandiose scale, you can witness and get a better understanding of buckling by trying a few of the tests shown in Fig. 11-1. In Fig. 11-1(a), (b), (c), and (d) are examples of temporary or elastic buckling; while (e), (f), (g), and (h) are examples of permanent or plastic buckling.

Although plates, shells, tubes, and various kinds of structural members have a tendency to buckle under load, this initial study will consider only straight members that are axially loaded and have constant cross section. Other types are usually studied in more advanced courses. The constant cross section will be of such proportions that its principal second moments of area are of the same order of magnitude. Such compression members that are slender enough to make buckling effects important are called *columns*.

In stating that this chapter deals only with straight members that are axially loaded and have constant cross section, we do not mean to imply that it is possible in reality to have a column which is dimensionally perfect, of a perfectly homogeneous material, and exactly axially loaded. Actually, the fact that every column has some degree of imperfection in loading, dimensions, and/or material is the *cause* of the initiation of buck-

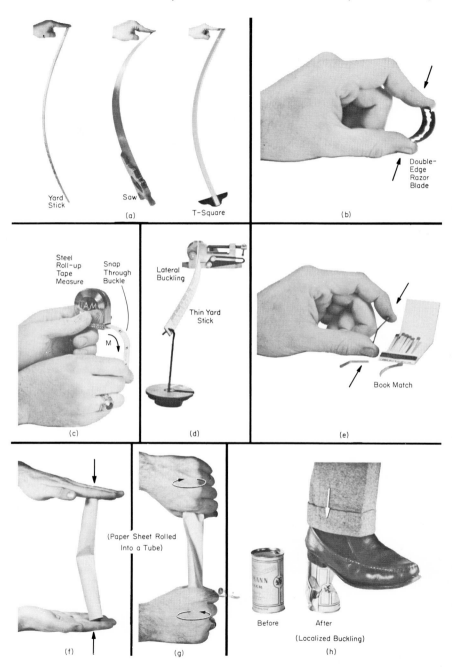

FIG. 11-1

ling if it is overloaded. If an ideal slender member did exist, and if it were ideally loaded, it would not buckle. Instead, it would fail (yield, fracture, or "flatten out") as the result of excessive compressive stress. It is sometimes convenient to discuss and analyze an ideal column and to apply a controlled "imperfection" in the analysis, as is done in the next section.

We do not mean to imply that buckling must occur in every real compression member if the load is increased until something damaging happens. A member will buckle or collapse only if it is slender enough. If the member is stocky, rather than slender, it will crush (yield or fracture) because of excessive compressive stress. How slender is slender enough to cause buckling? How do other parameters influence this unique phenomenon? The next sections answer these questions.

## 11-2. Uniqueness of Failure

In Chapters 8, 9, and 10, we related load to stress and load to deformation. For these nonbuckling cases of axial, torsional, bending, and combined loading, the stress or deformation was the significant quantity in failure. Buckling of a member is uniquely different in that the quantity significant in failure is the buckling load itself. *The failure (buckling) load bears no unique relationship to the stress and deformation at failure!* Our usual approach of deriving a load-stress relationship and a load-deformation relationship cannot be used here. The approach here will be to find an expression for the buckling load $P_f$ in terms of whatever parameters influence it.

Buckling is unique from our other structural-element considerations in that it results from a state of unstable equilibrium. All our previous studies were concerned with elements in stable equilibrium. The following discussion will help you understand the concept of instability.

To visualize the three classes of behavior and the concept of critical buckling load, let us experiment again. Follow Fig. 11-2 carefully. The examples in the upper part with the pipe and ball are simple, but important. The corresponding experiments in the lower part are performed on an "ideal" slender, rounded-end column. We know that an ideal column can't exist, but we will comment on how these experiments relate to the real thing. The term "rounded ends" means that there is no resistance to rotation of the ends. This condition may also be called pivoted ends, hinged ends, or pinned ends. It is assumed that this ideal column is slender enough to buckle *elastically*; that is, for any of the loads in Fig. 11-2 the resulting *axial* stress is below the elastic limit.

In Fig. 11-2(a) some load $P$ less than the buckling load is applied to the column. The column is then given a small (relative to its length) deflection by applying the small lateral force $F$. However, when the force $F$ is removed, the column will spring back to its original straight position of

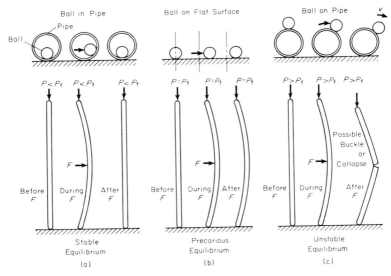

FIG. 11-2

equilibrium, just as a ball returns to the bottom of a pipe. This action indicates that the straight position is one of *stable* equilibrium.

This same procedure can be repeated for increased values of the load $P$ until some critical value $P_f$ is reached, as indicated in Fig. 11-2(*b*). When the column carries this load, and a lateral force $F$ is applied and removed, the column will remain in the slightly deflected position. The conditions are like those that exist when a ball is placed on a flat surface. The amount of deflection will depend on the magnitude of the lateral force $F$. Hence, the column can be in equilibrium in an infinite number of slightly bent positions. By definition, we shall say that for a column carrying an axial load $P_f$, the straight position is one of *neutral* or precarious equilibrium.

If the column is subjected to an axial load $P$ which exceeds $P_f$, as indicated in Fig. 11-2(*c*), and a lateral force $F$ is applied and removed, the column will "snap" to a considerably bent position. Depending on the value of $P$, the column either will remain in the bent position or will completely collapse and fracture, just as a ball will roll off a pipe. This type of behavior indicates that for axial loads greater than $P_f$, the straight position of a column is one of *unstable* equilibrium.

If you try to duplicate these experiments with a thin yard stick, T-square, or similar slender member, you probably won't need to apply any lateral force $F$, because your "real" column will have material, dimensional, and/or loading imperfections which will serve to initiate the buckling. Remember that a perfect column with perfect loading would never buckle. You might say that the initial deflection of the ideal column caused by the lateral force *F* in Fig. 11-2 has the same effect as do the imperfections which are always present in real columns.

When "hand" loading a yard stick there is a tendency to reduce the load when unstable equilibrium is reached and noticeable buckling occurs. Buckling tests are more realistic (and dramatic) when performed on a constant load or dead load type machine. Failures are then truly catastrophic.

## 11-3. Elastic Buckling

Figure 11-3 shows the relation between axial load and unit lateral deflection for an ideal column and a real column for *small elastic* unit deflections $\Delta/L$. For the *ideal* column, when the load *P* exceeds a certain value

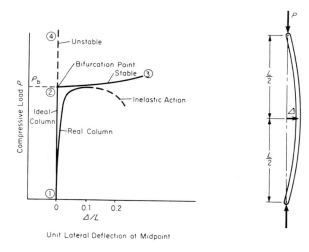

Fig. 11-3

$P_b$, two possible equilibrium positions exist. Either the column can remain straight, in which case $\Delta = 0$; or the column can become bent, in which case $\Delta > 0$. The straight position (②) to ④ on the curve) is *unstable*, while the bent portion (② to ③ on the curve) is *stable*. Point ②, at which the two branches of equilibrium join, is called a *bifurcation point*.

The shape of the curve for a real column would depend on the degree of the imperfections present. As the imperfections diminish, the curve for a real column approaches the stable branch ①-②-③ of the curve for the ideal column. That is, an *elastic real* column will attain a bent equilibrium position for all axial loads up to and somewhat beyond $P_b$. However, the theoretical load $P_b$ at the bifurcation point is usually taken as the actual buckling load. Thus, for an *elastic* column, failure is based on the assumption that

$$P_f \text{ (buckling)} = P_b \text{ (bifurcation)}$$

Hence, our problem now is to determine the lowest load at which more than one possible equilibrium position can exist. This load, $P_b$, is usually called the *Euler elastic buckling load.*

The derivation that follows will be concerned with an ideal column having pivoted ends. The influence of other types of end conditions and restraints will be considered afterward.

Figure 11-4(*a*) is a sketch of an ideal column with an axial load $P$ which is sufficiently large to hold the column in a *slightly* deflected position after a lateral load has been applied and removed. The column is represented diagrammatically in Fig. 11-4(*b*). The deflection is assumed to be

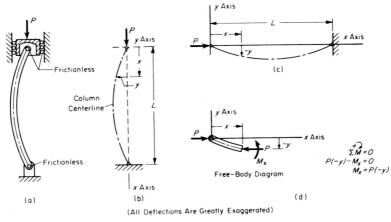

(All Deflections Are Greatly Exaggerated)

FIG. 11-4

very small, for two reasons. First, this assumption implies that the load $P$ is very close to the bifurcation load $P_b$; second, the small deflection permits us to use the elastic deflection equations derived in Chapter 10. In effect, we are assuming that the member is in a state of neutral equilibrium. Figure 11-4(*c*) is really the same as (*b*) except that the column is turned on its side so that the chosen $x$ and $y$ axes will look more familiar to you. This rotation will also make it easier for you to see the application of the now familiar general equation of the elastic curve for bending (Eq. 10-23).

From the free-body diagram of Fig. 11-4(*d*), the moment $M_x$ at any point at a distance $x$ from the end is $-Py$. The negative sign is in agreement with our usual sign convention for beams, since $y$ is negative (downward). Therefore,

$$EI \frac{d^2y}{dx^2} = M_x = -Py$$

or

$$EI\, y'' + Py = 0 \qquad (11\text{-}1)$$

This is a linear second-order differential equation with constant *positive* coefficients. There are several formal methods for obtaining the general solution of such an equation, but careful consideration shows that the solution for $y$ must be of such a form that its second derivative differs from $y$ by only a *negative* constant. The most convenient functions that meet this requirement are the elementary trigonometric functions. Hence, a general solution is

$$y = C_1 \sin kx + C_2 \cos kx \qquad (11\text{-}2)$$

This solution not only must satisfy the differential equation, but must also satisfy the physical boundary conditions. Specifically, $y$ must be zero when $x$ is equal to both zero and $L$, since the ends have no deflection. The first boundary condition gives

$$y_0 = 0 = C_1 \sin k0 + C_2 \cos k0$$

Hence,

$$C_2 = 0 \qquad \text{and} \qquad y = C_1 \sin kx \qquad (11\text{-}3)$$

Substituting for $y$ in Eq. 11-1, we get

$$EI\,(-C_1 k^2 \sin kx) + PC_1 \sin kx = 0$$

from which we obtain

$$C_1 \sin kx \,(-EIk^2 + P) = 0 \qquad (11\text{-}4)$$

This equation will be satisfied if $C_1 = 0$ or if $k = 0$. In either case, $y \equiv 0$. Hence, $y \equiv 0$ is a possible solution of Eq. 11-1. This solution represents the *unstable* straight position of the column.

Equation 11-4 will also be satisfied if $(-EIk^2 + P) = 0$. Then

$$k^2 = \frac{P}{EI} \qquad k = \pm \sqrt{\frac{P}{EI}}$$

and from Eq. 11-3 a possible second solution of Eq. 11-1 is

$$y = C_1 \sin \sqrt{\frac{P}{EI}}\, x \qquad (11\text{-}3a)$$

where $C_1$ cannot be zero. Now, in order that this solution will satisfy our second boundary condition, for which $y = 0$ and $x = L$, we must have

$$y_L = 0 = C_1 \sin \sqrt{\frac{P}{EI}}\, L$$

But since $C_1 \neq 0$, then

$$\sqrt{\frac{P}{EI}} \, L = n\pi \qquad (n = 1, 2, \ldots) \tag{11-5}$$

Hence, the *smallest* load $P$ for which more than one possible equilibrium position can exist is the one for which $n = 1$. The load corresponding to $n = 1$ is, by definition, the Euler elastic buckling load. Thus,

$$P_b = P_f = \frac{(\pi 1)^2 EI}{L^2} = \frac{\pi^2 EI}{L^2} \tag{11-6}$$

The loads corresponding to the higher values of $n$ also are bifurcation loads, indicating that there are an infinite number of bifurcation points. By definition, *the Euler load is the load corresponding to the lowest bifurcation point.* Equation 11-6 is commonly called the *Euler buckling equation.*[1]

It is significant here that we cannot evaluate $C_1$. We only make use of the fact that it is not zero. Not being able to determine $C_1$ means, physically, that the deflection of the column in its state of neutral equilibrium cannot be determined from our equation of equilibrium (Eq. 11-1) and the associated boundary conditions. The reason for this is that certain geometric approximations were made in deriving Eq. 10-23 (refer to Sec. 10-8).

## 11-4. Discussion of Elastic Buckling

Equation 11-6 is often expressed in either of the following more convenient forms:

$$P_f = \frac{\pi^2 E(ar^2)}{L^2} = \frac{\pi^2 Ea}{(L/r)^2}$$

or

$$\frac{P_f}{a} = \frac{\pi^2 E}{(L/r)^2} \tag{11-7}$$

where $a$ is the cross-sectional area of the column, and $r$ is the radius of gyration about the bending or buckling axis.

It is very important to notice that the derivation in Sec. 11-3 was based on the general *elastic* deflection equation $EI\,y'' = M_x$. Therefore, Eq. 11-7 is applicable only when $E$ is a constant, that is, when the buckling load $P_f$ causes an axial stress below the *proportional limit* at the initiation of buckling. From Eq. 11-7 we see that the buckling load $P_f$ (or the unit buckling load $P_f/a$), is dependent on the stiffness $E$ of the material, and

[1]Leonard Euler, a famous Swiss mathematician, first derived this equation in a paper published in 1759.

the dimensions $L$ and $r$. The ratio $L/r$ is defined as the *slenderness ratio* of the column.

You must develop a physical "feel" for the slenderness ratio. As you gain experience with engineering columns, you will acquire this "feel." You will then realize that a slenderness ratio greater than 200 is for members that are so slender that they are of little use as load-carrying engineering members. Members with slenderness ratios less than about 30 are so stocky that they have little or no tendency to buckle and are not considered to be columns. Therefore, most engineering columns, if they are considered as columns, are of intermediate slenderness. The limiting values of about 30 to about 200 are for real materials whose stiffness $E$ is between approximately 1 million and 30 million pounds per square inch, as is the case for most common engineering materials.

Figure 11-5 is an example of Eq. 11-7 plotted for a common engineering material—structural steel. When the slenderness ratio is over 100, the

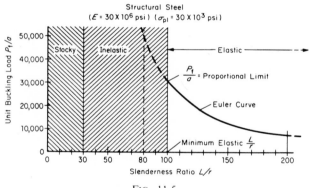

FIG. 11-5

equation is valid because the unit buckling load is below the proportional limit of the material and $E$ is a constant. For values of $L/r$ greater than 200, the curve is dashed indicating that it is unlikely that an engineering member would be this slender. The fact that the curve is dashed for values of $L/r$ below 100, or for unit loads above the proportional limit, is more important. A column for which $L/r = 80$ would buckle under an axial load *less* than the "Euler load" (load predicted from the Euler equation). It would buckle *before* the Euler load was reached, because its stiffness would be reduced when the unit buckling load exceeded the proportional limit and the material began to yield.

Obviously, the range for the slenderness ratio from 30 to 100 is very important from a practical engineering standpoint. A rational approach to the determination of the buckling loads for columns with slenderness ratios in this range, which buckle inelastically, is presented later in this chapter.

There are also many empirical approaches to the design of columns for this range of slenderness ratios. These are discussed in Sec. 11-8.

Before using Eq. 11-6 or 11-7, one question still to be answered is: What value of the moment of inertia $I$ or the radius of gyration $r$ should be used in evaluating the buckling load? The value of $I$ in Eq. 11-6 is the one carried directly through the derivation from Eq. 11-1, and therefore from Eq. 10-23; this equation in turn is based on the elastic flexure formula. The value of $I$ is therefore the second moment of the cross sectional area with respect to the axis about which the beam bends; or in the case of a column, the axis about which the column buckles. From the Euler equation (Eq. 11-6 or Eq. 11-7), we see that the buckling load $P_f$ will be least when $I$ is least. Thus, assuming no resistance or equal resistance to bending in all directions at the ends, a column will bend about its axis of least moment of inertia. Since $r = \sqrt{I/a}$, this axis will also be the axis of least radius of gyration. Of course, if the column is restrained in such a manner that buckling occurs about a particular axis, then the value of $I$ with respect to that axis must be used.

EXAMPLE 11-1. A structural steel strut with rounded ends has the rectangular cross section shown in Fig. 11-6, and is 10 in. long. At what compressive axial load will the strut fail? The modulus of the steel is 30 million psi.

*Solution:* First compute the slenderness ratio to determine if the Euler formula is applicable.

$$I_x = \frac{(1.20)(0.25)^3}{12}$$

$$I_y = \frac{(0.25)(1.20)^3}{12}$$

Since the $x$ and $y$ axes are axes of symmetry and therefore principal axes of the cross section, $I_x$ is the least $I$ of the section. Since $r = \sqrt{I/a} = 0.0722$ in., $L/r =$

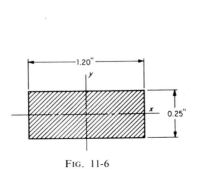

FIG. 11-6

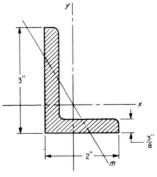

FIG. 11-7

$10/0.0722 = 138$. From Fig. 11-5, for steel, this value of $L/r$ is well into the Euler range. This slender compression member will therefore fail by elastic buckling.

By Eq. 11-7,

$$\frac{P_f}{a} = \frac{\pi^2 E}{(L/r)^2} = \frac{\pi^2(30,000,000)}{138^2} = 15,550 \text{ psi}$$

Hence,

$$P_f = (15,500)(1.20)(0.25) = 4665 \text{ lb}$$

Thus, $P_f/a = 15,550$ which is well below the proportional limit. The member is therefore behaving elastically.

EXAMPLE 11-2. If a $3'' \times 2'' \times 3/8''$ structural steel angle, for which the cross section is shown in Fig. 11-7, must support 16,000 lb in axial compression, what is the maximum length that it can have? Assume hinged ends.

*Solution:* From a steel handbook, the least radius of gyration for the angle cross section is $r_m = 0.43$ in., and the area is 1.73 sq in. The $m$ axis in this case is a principal axis and lies in the plane of the cross section. If we assume that the Euler equation is applicable, we obtain

$$\frac{P_f}{a} = \frac{\pi^2 E}{(L/r)^2}$$

$$\frac{16,000}{1.73} = \frac{\pi^2(30,000,000)}{(L/0.43)^2} = 9250$$

$$L = 77.0 \text{ in.}$$

Check the slenderness ratio to see if our assumption is really okay. Since $L/r = 77.0/0.43 = 180$, it is well into the Euler range. Also, $P_f/a = 9250$, which is well below the proportional limit for steel.

## 11-5. End Conditions

The Euler equation (Eq. 11-7) was arbitrarily derived for the case of rounded ends. This is sometimes referred to as the fundamental case. An engineering compression member may have end conditions that approximate those in the fundamental case of rounded ends, but more often the ends would have a different degree of fixity. The effect of the end fixity on the buckling load can be determined if the end conditions approximate any of those shown in Fig. 11-8(a) to (e) or if an intelligent guess can be made for $k$ in Fig. 11-8(f).

The length $L$ assumed in the derivation of the Euler equation is the distance covered by one loop of the sine curve of Eq. 11-3a when $n = 1$ in Eq. 11-5. Physically speaking, the length used in the Euler equation is the distance between points of zero moment (inflection points). Only in the fundamental case in Fig. 11-8(a) will the total length $L$ of the member be

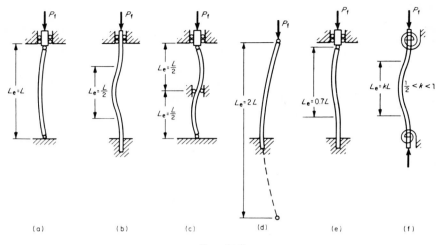

FIG. 11-8

equal to the distance $L_e$ between inflection points, sometimes called the effective length. Substituting the distance $L_e$ expressed as a function of $L$ in the Euler equation (Eq. 11-6) for each of the cases in Fig. 11-8, we get the following results:

For $L_e = L$, as in the fundamental case or as in Fig. 11-8(a),

$$P_f = \frac{\pi^2 EI}{L_e^2} = \frac{\pi^2 EI}{L^2}$$

For $L_e = L/2$, as in Fig. 11-8(b) or (c),

$$P_f = \frac{\pi^2 EI}{(L/2)^2} = \frac{4\pi^2 EI}{L^2}$$

For $L_e = 2L$, as in Fig. 11-8(d),

$$P_f = \frac{\pi^2 EI}{(2L)^2} = \frac{1}{4} \frac{\pi^2 EI}{L^2}$$

For $L_e = 0.7L$, as in Fig. 11-8(e),

$$P_f = \frac{\pi^2 EI}{(0.7L)^2} = \frac{2\pi^2 EI}{L^2}$$

For $L_e = kL$, as in Fig. 11-8(f),

$$P_f = \frac{\pi^2 EI}{(kL)^2} = \frac{1}{k^2} \frac{\pi^2 EI}{L^2}$$

Thus, it takes four times as much load to buckle a fixed-ended column as to buckle a free-ended one of the same over-all length; and only one-fourth as much load to buckle a column with one fixed end and one free end.

EXAMPLE 11-3. What is the least thickness a rectangular wood plank 4 in. wide can have, if it is used for a 20-ft column with one end fixed and one end pivoted, as in Fig. 11-8(*e*), and must support an axial load of 1000 lb? Use a factor of safety of 5. The modulus of the wood is 1.5 million psi.

*Solution:* For these end conditions $L_e = 0.7L$. Also, $I = bt^3/12$, where $b$ is the 4-in. breadth and $t$ is the desired least lateral dimension. Since $P_f$ is the allowable load times the safety factor,

$$P_f = \frac{\pi^2 EI}{(0.7L)^2} = 5,000 = \frac{\pi^2(1.5)(10^6)(4)t^3/12}{[(0.7)(12)(20)]^2}$$

Hence, $t = 3.06$ in.

EXAMPLE 11-4. Find the buckling load for the rectangular steel connecting rod in Fig. 11-9. Assume that the material behaves elastically for all values of stress encountered.

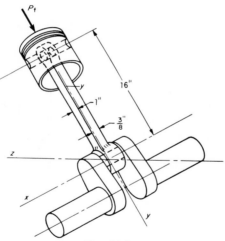

FIG. 11-9

*Solution:* If it is assumed that the bearings are frictionless, there is no resistance to buckling about the *x* axis. When buckling about the *z* axis is considered, it may be assumed that there is no side play in the bearings, and that the ends are fixed. Therefore, the load required to buckle the rod about each axis will be determined.

For the *x* axis, the ends are pivoted, as in Fig. 11-8(*a*), and $L_e = L$. Also, $I$ is a maximum. Hence,

$$P_x = \frac{\pi^2 EI_x}{L^2} = \frac{\pi^2(30,000,000)(3/8)(1^3)/12}{16^2} = 36,800 \text{ lb}$$

For the $z$ axis, the ends are fixed, as in Fig. 11-8($b$), and $L_e = L/2$. Also, $I$ is a minimum. Hence,

$$P_z = \frac{\pi^2 E I_z}{(L/2)^2} = \frac{\pi^2(30,000,000)(1)(3/8)^3/12}{(16/2)^2} = 20,600 \text{ lb}$$

The rod will buckle first about the $z$ axis at a load of 20,600 lb.

## 11-6. Inelastic Buckling

Our previous discussions have been concerned with members which buckle at a stress level below the proportional limit. These columns have relatively large slenderness ratios. As mentioned earlier, a great variety of engineering compression members have intermediate slenderness ratios, and many of these do not buckle until *after* the proportional limit has been exceeded. Thus, from an engineering point of view, there is an important class of members between those having very low slenderness ratios (stocky), which fail in direct compression with no buckling effects, and those having high slenderness ratios, which buckle *elastically*. For example, a structural steel member for which $L/r$ is 80 would buckle before the Euler load could be reached, as shown in Fig. 11-5, because the material stiffness would decrease when the axial stress exceeded the proportional limit.

To determine a rational inelastic buckling load we must take into account the reduced stiffness, or the reduced slope of the stress-strain curve. Consider an ideal column, Fig. 11-10($a$), having an intermediate slenderness ratio and pivoted ends that is to be loaded in small increments up to its critical buckling load $P_f$, which occurs after the proportional-limit stress has been exceeded. In Fig. 11-10($b$) and ($c$) are shown *uniform* stress distributions resulting from the increasing axial load.

Assume that as the last increment of axial load $\Delta P$ is applied, a *small* lateral force $F$ is also applied. This simultaneous application of both an axial force and a lateral force is necessary to make sure that the longitudinal compressive strains in the member continue to increase. With this type of loading program, the stress at any point in the column will *not decrease* at any time. On the other hand, since the stresses are in the inelastic range, if the axial load were first increased to cause the critical stress and then the lateral load were applied, the fibers on the convex side would stretch slightly and thus undergo a *decrease* in stress ("unloading") and there would be an accompanying increase in the modulus back to approximately the elastic value. This second loading program leads to a "double modulus" theory. We shall develop the theory based on the first loading assumption of monotonic increasing stress.

With our assumption of no decrease in stress, the distribution will be as shown in Fig. 11-10($d$), where the stress $\Delta\sigma$ is that due to the bending effects. The bending component of the strain distribution is relatively very

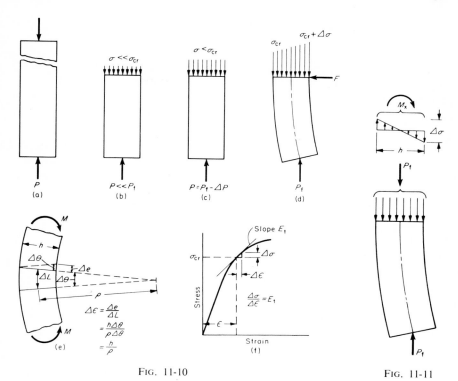

FIG. 11-10                                                FIG. 11-11

small, and it is assumed that plane cross sections remain plane. Also, since $\Delta\sigma$ is small, it is assumed that this increase in stress is directly proportional to the increase in strain. That is, $\Delta\sigma = (\Delta\epsilon)E_t$, where $E_t$ is the tangent modulus for $\sigma = \sigma_{cr}$ in Fig. 11-10($f$). From Fig. 11-10($e$) we see that the deformation due to bending is $\Delta e$, and the corresponding strain is

$$\Delta\epsilon \text{ (bending)} = \frac{\Delta e}{\Delta L} = \frac{h\,\Delta\theta}{\rho\,\Delta\theta} = \frac{h}{\rho}$$

Hence,

$$\Delta\sigma = (\Delta\epsilon)\,E_t = \frac{h}{\rho}\,E_t \tag{11-8}$$

A free-body diagram of the deflected column will be the same as that of Fig. 11-4($d$). Thus,

$$M_x = -P_f y \tag{11-9}$$

In order to express the bending moment $M_x$ in terms of the stress, we can think of the net stress distribution as being composed of two separate ones, as shown in Fig. 11-11. There is a *uniform* stress distribution due to the axial load $P_f$, and there is a *linearly varying* one due to the bending or the eccentricity of $P_f$. Then, as in the elastic flexure formula,

$$\frac{\Delta\sigma}{2} = \frac{M \times \frac{h}{2}}{I}$$

or

$$\Delta\sigma = \frac{M_x h}{I}$$

$$M_x = \frac{(\Delta\sigma)I}{h} \tag{11-10}$$

Then, from Eqs. 11-8 and 11-9,

$$M_x = \frac{h}{\rho} E_t \frac{I}{h} = \frac{E_t I}{\rho} = -P_f y \tag{11-11}$$

Finally, using the same approximation for $\frac{1}{\rho}$ as in Sec. 10-8, we have

$$\frac{d^2y}{dx^2} \approx \frac{1}{\rho} = -\frac{P_f y}{E_t I}$$

$$E_t I y'' + P_f y = 0 \tag{11-12}$$

This is the same as Eq. 11-1 for elastic buckling, but here we have the tangent modulus $E_t$ instead of the elastic modulus $E$ (review Chapter 3 for a discussion of $E_t$). Solving Eq. 11-12 just as we did Eq. 11-1, we find that the lowest inelastic buckling (bifurcation) load is

$$P_f = \frac{\pi^2 E_t I}{L^2} \tag{11-13}$$

or

$$\frac{P_f}{a} = \frac{\pi^2 E_t}{(L/r)^2} \tag{11-14}$$

This relationship is commonly called the tangent modulus equation or the Engesser equation, after the man who first proposed it.

The result based on the double modulus theory is a much more complicated expression, which gives a somewhat higher value for the buckling load than does the tangent modulus formula. Which assumption most closely approximates what actually happens in a column? This question was not fully answered until 1946, when F. R. Shanley[2] explained that the load based on the double modulus theory was an upper limit, and the

2"The Column Paradox," *Journal of Aeronautical Sciences*, Dec., 1946, p. 678.

tangent modulus theory more nearly predicted the buckling load of an actual column.

## 11-7. Discussion of Engesser Equation

Although Eq. 11-14 is very similar to the Euler equation (Eq. 11-7), its application is considerably more difficult. The main problem is to find the proper value for $E_t$. Since $E_t$ is the slope of the stress-strain curve for the unit axial stress $\sigma_{cr} = P_t/a$, it may be necessary in some problems to solve Eq. 11-14 by trial and error. In such a problem it is convenient to have a plot of $\sigma$ vs. $E_t$ for the material of the particular column. Some sample $\sigma$ vs. $E_t$ curves are shown in Fig. 11-12.

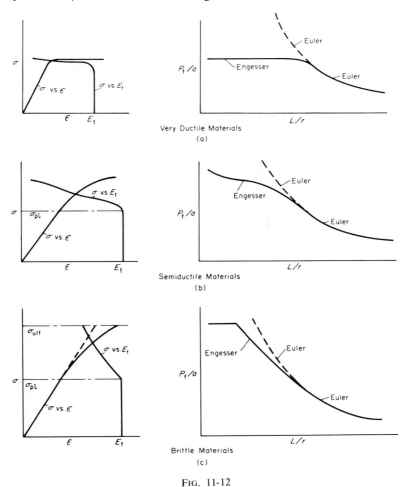

Very Ductile Materials
(a)

Semiductile Materials
(b)

Brittle Materials
(c)

Fig. 11-12

Of course, if an analytical expression for the stress-strain curve is known, then $E_t$ can be obtained as a function of the stress by the following relationship:

$$E_t = \frac{d\sigma}{d\epsilon} \qquad \text{(slope)}$$

or

$$\frac{1}{E_t} = \frac{d\epsilon}{d\sigma} \qquad \text{(evaluated at } \sigma_{cr}\text{)}$$

Then $E_t$ in Eq. 11-14 can be expressed explicitly in terms of $\sigma_{cr}$.

In conclusion, if we use the Euler equation to predict the buckling load for unit axial stresses $(P_f/a)$ below the proportional limit and we use the Engesser equation for unit stresses above the proportional limit, we can then predict the buckling load for all intermediate slenderness ratios (between the stocky members and the extremely slender members). Various plots of unit stress vs. slenderness ratio are shown in Fig. 11-12. The dashed portion of the Euler curve is included merely to illustrate the significant effect that inelastic behavior can have on the buckling load as compared with elastic behavior.

The tangent modulus equation resulted from the solution of a second-order linear differential equation similar to that used for elastic behavior. Therefore, the discussion of end conditions in Sec. 11-5 is applicable to the Engesser equation, as well as to the Euler equation.

EXAMPLE 11-5. A round tube, with 1 in. O.D. and 7/8 in. I.D., is 8 in. long and is made of a material whose stress-strain diagram is approximated in Fig. 11-13. If one end is fixed. and one end is free, what is the axial compressive buckling load?

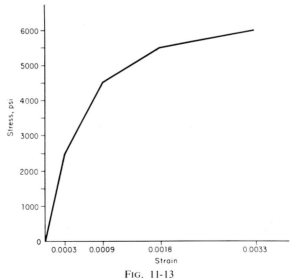

FIG. 11-13

*Solution*: The area is 0.184 sq in. and $I = \dfrac{\pi}{64}(d_o{}^4 - d_i{}^4)$. The radius of gyration is

$$r = \sqrt{I/a} = 0.331 \text{ in.}$$

and the slenderness ratio is

$$L/r = 8/0.331 = 24.2$$

For the given end conditions, the tangent modulus equation becomes

$$\frac{P_f}{a} = \sigma_{cr} = \frac{\pi^2 E_t}{(L_e/r)^2} = \frac{\pi^2 E_t}{4(L/r)^2}$$

It is not necessary to worry about whether we should use the Engesser equation or the Euler equation. The Engesser equation will automatically become the elastic Euler equation if the slenderness ratio dictates use of the elastic modulus $E$. For the computed value of $L/r$,

$$\sigma_{cr} = \frac{\pi^2}{4(L/r)^2} E_t = 0.00422\, E_t \tag{a}$$

Since we do not have an analytical expression for $E_t$ as a function of $\sigma_{cr}$ for the material, we must work from the given experimental stress-strain curve. The curve has four distinct slopes, each of which is the value of $E$ or $E_t$ for some range of values of $\sigma_{cr}$. A trial-and-error solution is employed here.

For the initial (elastic) range, the slope is

$$E = \frac{2500}{0.0003} = 8.33 \times 10^6$$

Substituting in equation (a), we get

$$\sigma_{cr} = (0.00422)(8.33 \times 10^6) = 35,200 \text{ psi}$$

As seen from Fig. 11-13, this value of $E$ is not valid above 2500 psi and certainly not anywhere near 35,200 psi.

For the second range of stress, the slope is

$$E_{t1} = \frac{4500 - 2500}{0.0009 - 0.0003} = 3.33 \times 10^6$$

and

$$\sigma_{cr} = (0.00422)(3.33 \times 10^6) = 14,050 \text{ psi}$$

As seen from Fig. 11-13, $E_{t1}$ is valid only between 2500 and 4500 psi and not near 14,050 psi.

For the third range, the slope is

$$E_{t2} = \frac{5500 - 4500}{0.0018 - 0.0009} = 1.11 \times 10^6$$

and
$$\sigma_{cr} = (0.00422)(1.11 \times 10^6) = 4680 \text{ psi}$$

As seen from Fig. 11-13, $E_{t2}$ is valid between 4500 and 5500 psi, and this range includes the value 4680 psi. Therefore, the critical stress at buckling is 4680 psi. Since $\sigma_{cr} = P_f/a$, the buckling load is

$$P_f = \sigma_{cr}a = (4680)(0.184) = 862 \text{ lb}$$

## 11-8. Empirical Column Formulas

The column theory for inelastic behavior has not been too well understood by engineers until recent years. It has therefore been common engineering practice for many years to use empirical relationships for designing columns with intermediate slenderness ratios. One reason for their popularity is that it is usually simpler to use them than the rigorously derived theoretical formulas.

An empirical relationship is generally based on data obtained by measuring the buckling loads in a series of tests on experimental columns with various slenderness ratios. The unit buckling loads corresponding to the several slenderness ratios are computed, and a convenient and mathematically simple expression fitting the results is developed. The three types of expressions most commonly used are as follows:

$$\frac{P}{a} = \sigma_l - S\frac{L}{r} \tag{11-15}$$

$$\frac{P}{a} = \sigma_l - S\left(\frac{L}{r}\right)^2 \tag{11-15a}$$

$$\frac{P}{a} = \frac{\sigma_l}{1 + \phi(L/r)^2} \tag{11-15b}$$

A plot of the first equation is a *straight line* with a negative slope $S$ and a $P/a$ intercept of $\sigma_l$. This is the simplest type of equation. The second equation represents a *parabola* with a $P/a$ intercept of $\sigma_l$ and a decreasing slope (it opens downward). The last equation is referred to as the *Gordon-Rankine* formula. The three types of curves are shown in Fig. 11-14.

Relationships of these forms, with appropriate values of the constants $\sigma_l$, $S$, and $\phi$, are commonly used in various codes and specifications relating to the design of structures. In Fig. 11-14, parts of the straight line and the parabola are shown dotted, to indicate that the corresponding equations should not be used for slenderness ratios greater than certain limiting values. For very small values of $L/r$, the Gordon-Rankine curve approaches an upper horizontal asymptote; and for very large values of $L/r$, it approaches an Euler curve. Equation 11-15b should therefore be valid for the full range of slenderness ratios. For most materials, however, it is impossible to find values of $\sigma_l$ and $\phi$ that will enable the Gordon-Rankine formula to fit experimental data for all slenderness ratios.

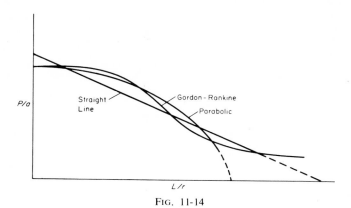

FIG. 11-14

After the constants in an empirical equation are determined to get the best fit of the resulting curve and the experimental points, a liberal factor of safety is usually applied to make the expression suitable for use as a design formula. The inherent unstable nature of the buckling phenomenon, as discussed earlier, causes a considerable scatter in test data for columns, even those from very carefully controlled laboratory tests. This condition necessitates much larger safety factors in columns than are usually necessary in most other structural elements.

## PROBLEMS

**11-1.** The bars $AB$ and $BC$ in part ($a$) of the illustration are considered to be weightless and infinitely rigid. A torsional spring connects the two bars and re-

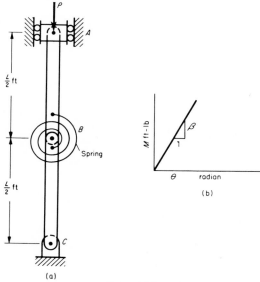

PROB. 11-1

sists rotation of the joint *B*.  Part (*b*) of the illustration gives the characteristics of the spring.  For what value of *P* will the system be in neutral equilibrium?

**11-2.**  The bar *AB* in the illustration is assumed to be weightless and infinitely rigid.  The helical spring is connected to the bar by means of a slider so that the spring can always remain vertical.  If the modulus of the spring is *K* lb/in., what is the value of the load *P* corresponding to the bifurcation point?

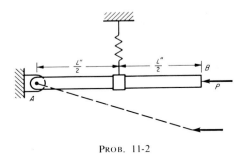

PROB. 11-2

**11-3.**  Bars *AB*, and *BC* are assumed to be weightless and rigid.  Find the critical force *P* if the spring constant is *k*.

**11-4.**  Find the critical load *P* for the setup in the illustration.  Both springs are linearly elastic with the constants as shown.

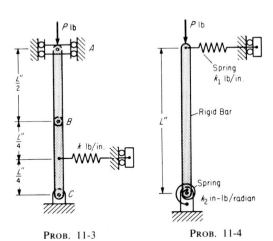

PROB. 11-3            PROB. 11-4

**11-5.**  How long is a round, hinged-ended column 3 in. in diameter, if the slenderness ratio is 100?

**11-6.**  What is the slenderness ratio of a column whose effective length is 22 ft?  The column is a 10WF60 structural steel section.

**11-7.**  What is the slenderness ratio of a hinged ended strut 14 in. long if the rectangular cross section is $1/2$ in. by 2 in.?

**11-8.** The cross section of a hinged-ended column 24 ft long is shown in the illustration. Find the slenderness ratio.

**11-9.** Two plates, each ³/₈ in. thick, are welded to two standard channels in the positions shown in the illustration so that the elements act as a unit. If the built-up member is used as a column, what must be the distance *d* in order that the column will have equal tendency to buckle about any axis of the cross section?

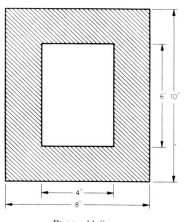

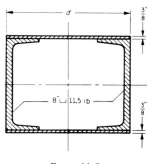

PROB. 11-8                 PROB. 11-9

**11-10.** What must be the wall thickness of a hollow tube whose outside diameter is 2.5 in., if it is 7 ft long and has a slenderness ratio of 110 when used as a hinged-ended column?

**11-11.** Show that a square cross section is a better column section than a circular one having the same area.

**11-12.** Find the theoretical buckling load of a wood yardstick ¹/₄ in. thick and 1 in. wide. Assume that the wood is linearly elastic and has a modulus of 1¹/₂ million psi, and that the ends are frictionless.

**11-13.** Find the maximum allowable load that a rectangular (1¹/₂ in. × ³/₄ in.) aluminum alloy hinged-ended column 4 ft long can carry, if the elastic modulus of the aluminum alloy is 10,400,000 psi. Assume linear behavior, and use a factor of safety of 2.

**11-14.** What axial compressive load will cause a 12 ft length of standard 2-in. steel pipe to buckle elastically? Actual dimensions of a 2-in. pipe are: O.D. = 2.375 in.; I.D. = 2.067 in. Assume frictionless ends.

**11-15.** Find the minimum required diameter of a solid, round structural steel strut that has hinged ends, is 60 in. long, and must support a compressive load of 8000 lb. Use a factor of safety of 2.

**11-16.** Find the dimensions of a hinged-ended timber column with a square cross section that is 20 ft long and is to carry an axial compressive load of 25,000 lb. The timber behaves linearly and has an elastic modulus of 1.8 million psi. Use a factor of safety of 4.

**11-17.** A hollow magnesium tube having an outside diameter of 2 in. and a wall thickness of $\frac{1}{8}$ in. is used for a compression member with hinged ends to support 3200 lb. How long can it be, assuming elastic behavior and a factor of safety of 1.5? Magnesium has a modulus of $6.5 \times 10^6$ psi.

**11-18.** What is the minimum slenderness ratio at which elastic buckling can occur for pivoted columns of the materials whose compression stress-strain curves are shown in the illustration?

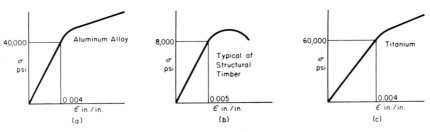

PROB. 11-18

**11-19.** Compare the Euler buckling loads of the following two very slender round columns: (1) manufactured plastic with modulus of 750,000 psi, cross-sectional area of $a$, length of $l$, fixed ends; and (2) clear spruce wood with modulus of 1,800,000 psi, cross-sectional area of $a$, length of $l$, rounded frictionless ends.

**11-20.** With a factor of safety of 4, how heavy a platform can a 180-lb flag-pole sitter place atop a 40-ft high, constant cross section flag pole made from a standard 4-in. steel pipe ($I = 7.233$ in.$^4$ and $r = 1.51$ in.)? One end of the pole is firmly fixed in the ground. The platform and the sitter are assumed to remain as axial loads. Neglect the weight of the pole itself, and assume elastic behavior.

**11-21.** A 160-lb student is climbing an aluminum alloy flag pole 80 ft tall to place a derby hat on its top. The pole has an inside diameter of 3.5 in. and an outside diameter of 3.80 in. Can he get to the top before the pole buckles? If not, how high can he go? Neglect the weight of the pole itself, and assume that he is a contortionist and is able to keep his center of gravity along the axis of the pole. The modulus of the aluminum is 10,000,000 psi.

**11-22.** An aluminum yard stick (rectangular cross section $\frac{1}{4}$ in. by 1 in.) having frictionless ball-and-socket joints at its ends is held between two rigid walls. If the yard stick is unstressed at room temperature, find the temperature change required to buckle the stick. It is assumed that the stick will buckle elastically when the buckling load is reached. Coefficient of thermal expansion for aluminum is $12.8 \times 10^{-6}$ in./in./°F, and the modulus is 10.6 million psi.

**11-23.** In the structure shown in part ($a$) of the illustration, the member $AB$ is made of a material which has the compression stress-strain curve shown in part ($b$). What is the minumum allowable area of a square cross section of member $AB$, if buckling is the mode of failure? Verify your answer.

**11-24.** As shown in the illustration, four high-strength steel wires $\frac{1}{8}$ sq in. in cross section and 10 ft long are used to guy the 8-ft mast whose cross section is

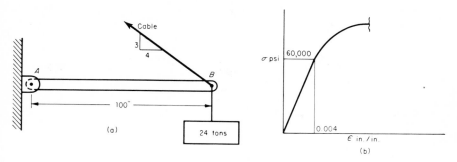

PROB. 11-23

$2'' \times 2''$ extruded magnesium. The turnbuckles on the guy wires are tightened so that the tension in each wire is negligible. The turnbuckles have a thread pitch of 32 threads per inch; $E$ for magnesium is $6.5 \times 10^6$. How many turns of each turnbuckle (simultaneously) will buckle the mast? Assume elastic behavior.

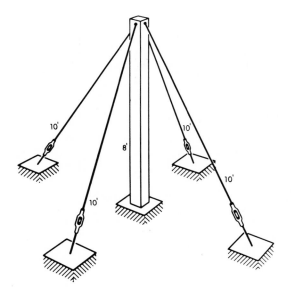

PROB. 11-24

**11-25.** The 90-ft steel mast with the 2000-lb platform on top is guyed at the third-points as shown in the illustration. If the cross section of the mast is solid and circular, what must be the diameter in order to prevent elastic buckling? Assume zero load in the guy wires.

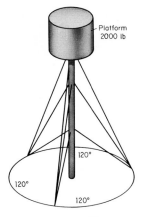

PROB. 11-25

**11-26.** A structural steel pivoted-ended column 12 ft long has two lateral supports at the mid-point, as shown in the illustration. If the cross section is as shown, what is the value of the critical buckling load. The elastic limit of the steel is 48,000 psi and $E = 30 \times 16^6$ psi.

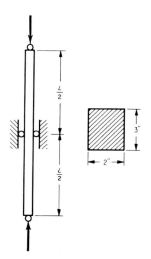

PROB. 11-26

**11-27.** The bar shown in the illustration has a slenderness ratio of 120 and a coefficient of thermal expansion of $6 \times 10^{-5}$ in./in./°F. At 70°F the bar is stress-free, while at 50°F a stress of 24,000 psi is present. Find the elastic modulus and the temperature at which the bar will buckle.

PROB. 11-27

**11-28.** An aluminum pole is 30 ft long and has a hollow circular cross section with a 6-in. outside diameter and a $5\frac{1}{2}$-in. inside diameter. One end of the pole is rigidly fastened in the ground, and the other end is free. What is the maximum height on this pole at which a 4-ton platform can be mounted without causing the pole to buckle? What is the weight of the heaviest platform which could be mounted on the top? Assume that the center of gravity of the platform is along the axis of the pole.

**11-29.** Determine the necessary minimum *width* and *thickness* of each of the two members *AB* in the illustration to prevent elastic buckling. Drum diameter is $1\frac{1}{2}$ ft; crank radius is 3 ft. Pull *P* exerted on the crank is 150 lb and the weight of the rock is negligible in comparison to the tension in the cord. The material is wood with a modulus of 2 million psi.

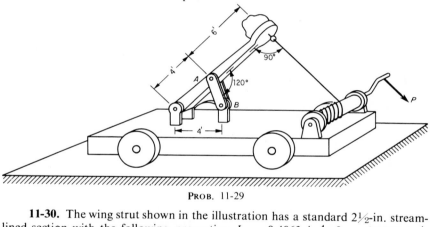

PROB. 11-29

**11-30.** The wing strut shown in the illustration has a standard $2\frac{1}{2}$-in. streamlined section with the following properties: $I_{11} = 0.4063$ in.$^4$; $I_{22} = 0.1018$ in.$^4$;

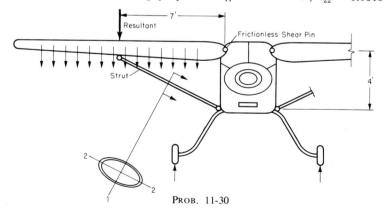

PROB. 11-30

$a = 0.3773$ in.$^2$  On a hard landing the resultant load is 1200 lb.  What factor of safety does the wing strut have?  The strut is pinned about the axis *2-2* and fixed about the axis *1-1*.  The strut is of alloy steel with a modulus of 30,000,000 psi.

**11-31.**  Four 6″ × 6″ × ¾″ equal-leg structural steel angles 38 ft long are riveted together to form a single unit, as shown in the illustration.  The ends are pivoted and frictionless.  What is the buckling load for this unit?  What is the buckling load if the rivets are removed and each angle acts as a separate column?

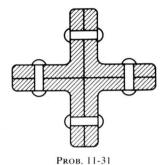

PROB. 11-31

**11-32.**  A slender, pivoted-ended column 8 ft long with a circular solid cross section is made of a material whose compression stress-strain curve is shown in the illustration.  Determine the minimum area this column must have so that it will not fail by elastic buckling.

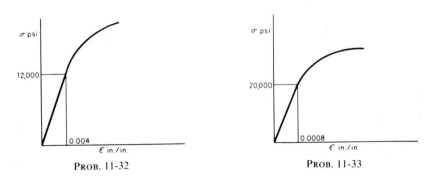

PROB. 11-32                    PROB. 11-33

**11-33.**  A pivoted-ended column 50 in. long is made of a material which has the compressive stress-strain relationship shown in the illustration.  The column has a square cross section 2 in. by 2 in.  *a)* Evaluate the slenderness ratio.  *b)* Does this column have an elastic buckling load?  If so, what is its value?  If not, why?

**11-34.**  Work Example 11-5 if the tube is fixed at both ends, and is 20 in. long.

**11-35.** Work Example 11-5 if the tube is 10 in. long instead of 8 in.

**11-36.** Work Example 11-5 if the cross section is solid and $3/4$ in. in diameter instead of hollow.

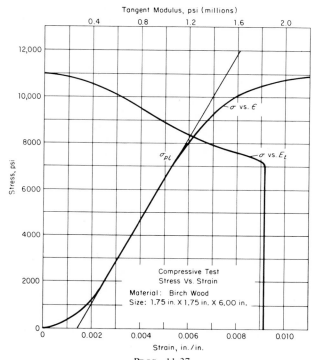

Prob. 11-37

**11-37.** How long can a 2 in. by 1 in. wood compression strut be, if it is to support an axial load of 10 tons and is made of the material whose stress-strain and tangent modulus curves are shown in the illustration. Assume fixed ends.

**11-38.** How much axial compression load can a wood dowel 2 in. in diameter and 10 in. long support, if made of the material in Prob. 11-37? One end is fixed and the other free.

**11-39.** How much axial compression load can a dowel 2 in. in diameter support, if the ends are pivoted and the length is 16 in.? The dowel is made of the material in Prob. 11-37.

**11-40.** Plot the curve of unit buckling load $(P/a)$ vs. slenderness ratio $(L/r)$ for pin-ended columns made from the 2024-T4 aluminum alloy having the properties given in the illustration for the range of $L/r$ from 20 to 200.

**11-41.** Plot the curve of unit buckling load $(P/a)$ vs. slenderness ratio $(L/r)$ for pin-ended columns made from the magnesium alloy having the properties given in the illustration for the range of $L/r$ from 20 to 200.

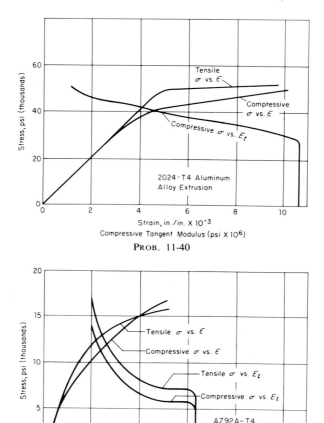

PROB. 11-40

PROB. 11-41

**11-42.** Plot the curve of unit buckling load ($P/a$) vs. slenderness ratio ($L/r$) for pin-ended columns made from the glass-resin laminate having the p•)perties shown in the illustration for the range of $L/r$ from 20 to 200. Assume that the warp of the fabric is parallel to the load axis.

**11-43.** A hollow member, 80 in. long and circular in cross section, is rigidly fixed at both ends and loaded in axial compression with a load of 375,000 lb. The member is to be proportioned so that its inside radius will be one-half of its outside radius. The compressive properties of the material are shown in the illustration for Prob. 11-40. Determine the minimum permissible dimensions of the member.

**11-44.** A pin-ended member is extruded from 2024-T4 aluminum alloy with a resulting cross section in the shape of an equilateral triangle having sides 1¼ in. wide. If it must support an axial compressive load of 32,000 lb, what is the greatest

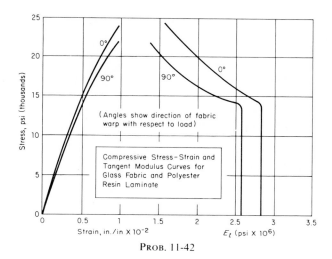

PROB. 11-42

length it can have? See the illustration for Prob. 11-40 for the compressive properties of the material.

**11-45.** A compression member is cast of AZ92A-T4 magnesium alloy (see the illustration for Prob. 11-41) with a box cross section having outside dimensions 4 in. by 2 in. and inside dimensions 3 in. by 1 in. If the member is completely free at one end and fixed at the other, and is 30 in. long, what is the maximum axial load it can withstand?

**11-46.** A fixed-ended strut 8 in. long is made of the glass-resin laminate whose properties are given in the illustration for Prob. 11-42. The warp of the fabric is perpendicular to the long axis of the column. The cross section is a rectangle 0.5 in. by 0.75 in. What is the axial buckling load?

**11-47.** The stress-strain curve for an aluminum alloy can be approximated as shown in the illustration. How much axial load can a 32-in. column made of

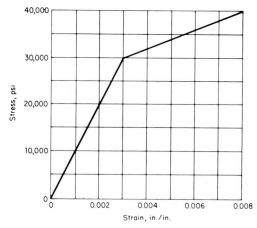

PROB. 11-47

the material support, if the cross section is 2 in. by 4 in. The ends are pivoted. *Verify the applicability of any formula used.*

**11-48.** A rounded-end column with a square cross section of 2 sq. in. is 20 in. long. If the material of the column has a stress-strain curve that can be approximated by the expression $\epsilon = \sigma^2/50{,}000{,}000$, what is the buckling load?

**11-49.** What is the buckling load of a $3'' \times 2'' \times \frac{3}{8}''$ angle (see Example 11-2 for properties) used as a compression strut 12 in. long with fixed ends, if the stress-strain law of the material is $\sigma = 6 \times 10^4 \epsilon^{1/3}$?

**11-50.** A slender member 2 in. square is pinned at both ends and loaded in axial compression. The member is 3 ft long and is made of an alloy whose properties are approximated in the illustration. Using a factor of safety of 3, find the maximum allowable load this member can safely carry before buckling.

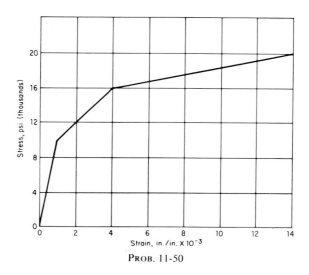

PROB. 11-50

**11-51.** a) Plot the experimental column test data given below for pin-ended $2'' \times 2''$ wood columns. b) Fit a straight line to these data. c) Derive the equation of the straight line. d) Apply a factor of safety of 2 to the derived equation to determine a working column formula for design.

| Slenderness Ratios | Buckling Loads (lb) | Slenderness Ratios | Buckling Loads (lb) |
|---|---|---|---|
| 15 | 22,000 | 50 | 17,100 |
| 15 | 23,100 | 60 | 13,000 |
| 20 | 23,400 | 70 | 12,100 |
| 20 | 21,800 | 80 | 8,900 |
| 30 | 19,900 | 80 | 10,000 |
| 30 | 21,600 | 90 | 8,000 |
| 40 | 17,000 | 100 | 5,000 |
| 50 | 16,300 | 100 | 5,100 |

**11-52.** The American Institute of Steel Construction (AISC) once specified the following parabolic formula for structural steel members used in buildings for which $l/r \le 120$:

$$\frac{P_D}{A} = 17{,}000 - 0.485\left(\frac{l}{r}\right)^2$$

where $P_D$ is the design load, *not* the critical buckling load. According to the AISC, how much axial compression load could a building column made from a 20 I 95.0 have supported if it is 15 ft long?

*part* **IV**

# Other Design Considerations

## chapter 12

# Additional Beam
# Topics

## 12-1. Introduction

In Chapter 10 we considered some of the basic ideas concerning the response of relatively long members subjected to transverse bending loads. In that presentation some rather restrictive assumptions were made in regard to the geometry of the member and the manner of loading. These will now be reviewed and discussed.

First, the members investigated in Chapter 10 were assumed to be straight, to have a constant cross section, and to be made of a homogeneous and isotropic material. In the later parts of that chapter the restriction of constant cross section was removed, and beams with gradually varying cross sections (tapered beams) and beams with reduced cross sections were considered. Also, the deflections of beams composed of several sections of different materials joined end to end were investigated by use of the moment-area method. However, all the beams in Chapter 10 were *straight*.

Second, the beams in Chapter 10 were assumed to have a longitudinal plane of symmetry, and the loads were assumed to lie in this plane of symmetry. With these restrictions, the neutral surface of the beam was found to be perpendicular to this plane of symmetry. Also, the symmetry of loading and cross section prevented the possibility that the beam would twist. The only time the requirement of symmetry was removed was in Sec. 10-6. It was shown there that if the loading were pure bending (no shear forces), the bending couple could be resolved into component moments about the principal axes of the cross section. However, at no time did we consider unsymmetrical problems in which shear forces were present.

Third, it was assumed in Chapter 10 that bending, rather than shearing or buckling, was the primary consideration in determining the kinematic response of the member. This assumption will be carried over to the analyses presented in this chapter.

In addition to these first three major restrictions, other special assumptions were made at various times throughout Chapter 10. It would be well for you to briefly review the important theory in that chapter before proceeding with the present chapter. The succeeding sections are devoted to types of members, such as beams with unsymmetrical shear loading, curved beams, and nonhomogeneous beams, which could not be discussed within the restrictions of Chapter 10.

## 12-2. Shear Center

It was shown in Chapter 10 that if a beam had a longitudinal plane of symmetry and if the loads were applied in this plane of symmetry, then the beam would not twist and the resulting deformation would be caused primarily by the bending effects. On this basis the flexure formula (Eq. 10-10) was derived and applied to beams subjected to transverse shear forces as well as to pure bending couples. We might now pose the problem of a beam which is subjected to transverse loads and has no longitudinal plane of symmetry; or even if the beam does have a longitudinal plane of symmetry, we might suppose that the loads do not lie in this plane. How then do we determine the stresses in the beam? This problem is much too general to be presented in a text of this level (indeed, it has been solved for relatively few cases). Instead, we shall concern ourselves with a problem which is solved much more easily but is still extremely useful and practical. In particular we shall consider a *beam whose cross section is relatively thin*, and shall determine where the transverse shear forces should be applied so that the beam does not twist. Since many standard structural beams have I-shaped, T-shaped, channel-shaped, and other types of thin cross sections, the solution of this problem has wide application.

The method of solution is conceptually simple. We merely try to locate the line of action of the resultant internal reaction shear forces assuming the beam bends but does not twist. Then for equilibrium the external resultant shear force must be colinear with the internal resultant shear force. Thus the problem is reduced to one of elementary statics.

Consider, for example, a thin channel-shaped section used as a cantilever beam and loaded as shown in Fig. 12-1(*a*). Since the bending moment varies throughout the length of the beam, we know that longitudinal shear forces will be present. One such force is $\Delta F_x$ in Fig. 12-1(*b*). (Refer to Sec. 10-4.) In a thin section of this type, it is assumed that the vertical web carries the entire vertical shear force and that the horizontal flanges do not carry any *vertical* shear forces. Thus, shear force "flows" through the cross section of the channel in a manner illustrated in Fig. 12-1(*c*). This condition is effectively a horizontal force in each of the flanges and a vertical force in the web, as in Fig. 12-1(*d*). From equilibrium requirements, this system of forces is statically equivalent to the single force shown in Fig. 12-1(*e*). The distance $s$ is such that the force $V$ must produce the same moment about any point on the center line of the vertical web as does the couple formed by the horizontal forces $F_1$ and $F_3$. Thus,

$$s = \frac{F_1 h}{V} \qquad (12\text{-}1)$$

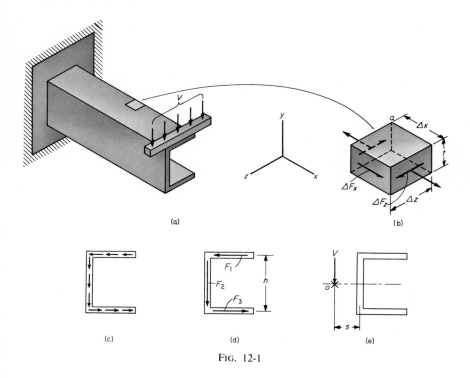

(a)

(b)

(c)

(d)

(e)

FIG. 12-1

If the line of action of the resultant external vertical force is located at the distance $s$ from the center of the web, there will be no twisting of the beam. If the line of action were in any other position, the beam would twist since the assumed internal reaction force distribution would not balance the external twisting effects.

For this example the distance $s$ will completely locate the desired line of action, because the only external resultant force on the beam is a vertical one. In general, when there is a horizontal as well as a vertical resultant shear force, it may be necessary to determine both a vertical and a horizontal distance $s$. The point located in this manner, such as $o$ in Fig. 12-1(e), is called the *shear center*. It locates a longitudinal axis through which the external resultant shear forces must pass in order that the beam will not twist. Notice, from the procedure used to determine the distance $s$, that if the thin cross section has an axis (or axes) of symmetry, the shear center will always lie on this axis. Thus, the intersection of any two axes of symmetry, if they exist, will locate the shear center.

Although Eq. 12-1 formally determines the distance $s$, the concept of shear flow is quite useful in actually evaluating $s$. (Refer to Secs. 9-8 and 10-4). The force $\Delta F_x$ in Fig. 12-1(b) can be expressed as $q\Delta x$, where $q$ is the shear force per unit length or the shear flow. Then by considering rotational equilibrium of an infinitesimal element $t\,dx\,dz$, we find that

$$dF_z = q\,dz$$

Hence, the total shear force in the flange is

$$F_z = \int_{\substack{\text{width of}\\\text{flange}}} q\,dz \qquad (12\text{-}2)$$

Similarly, the shear force in the lower flange and in the vertical web can be found by integrating the shear flows over their respective lengths.

In many cases, such as in the example involving the channel section, the integrating operation indicated in Eq. 12-2 can be accomplished by rather simple methods. For example, in Fig. 12-1 we already know from equilibrium requirements that the vertical force $F_2$ in the web is equal to the external shear force $V$. Also, since the beam bends but does not twist, we know that for elastic behavior the shear flow is given by Eq. 10-15. Thus,

$$q = \frac{VQ}{I} \qquad Q = \int_{\text{area}} y\,da$$

The integral $Q$ represents the first moment, about the centroidal bending axis, of the *area of that portion of the cross section which is considered to be separated from the remaining portion.* For instance, consider the channel section in Fig. 12-2($a$). In evaluating the shear flow at $c$-$c$ in the flange, $Q$ would be the first moment of all the area to the right of $c$-$c$. By similar evaluation throughout the entire cross section of the channel, you will find that $Q$, and hence the shear flow $q$, varies linearly in the flanges and parabolically in the web, as indicated in Fig. 12-2($b$). Thus, the horizontal force in either flange can be found by calculating the area under the

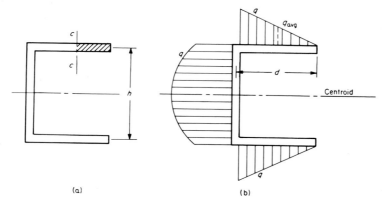

(a)                                                        (b)

Fɪɢ. 12-2

shear flow curve. The result is

$$F_1 = q_{avg}d = \frac{VQ_{avg}}{I}d$$

Substitution of this last result in Eq. 12-1 yields

$$s = \frac{VQ_{avg}}{I}d\frac{h}{V} = \frac{Q_{avg}(d)(h)}{I} \tag{12-3}$$

where $Q_{avg}$ is the first moment, about the centroid, of half the area of one of the flanges.

Although this last expression is strictly applicable only to a channel section, it illustrates the following important fact. *The location of the shear center depends only on the geometry of the cross section, and not on the loading.* The shear forces in the section and the shear flow used to calculate the location of the shear center were introduced merely to give a physical meaning to the shear center and to aid in finding its location. They did not influence the final result indicated by Eq. 12-3. This fact can be very useful in locating the shear center of an arbitrary cross section, since we are free to assume that the external shear force is acting in any convenient direction, such as perpendicular to an axis of symmetry or a principal axis. The following example illustrates this idea.

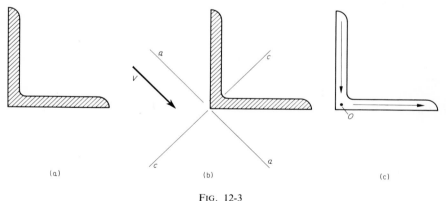

(a)                               (b)                               (c)

FIG. 12-3

EXAMPLE 12-1. Determine the location of the shear center for the 5″ × 5″ × ⅞″ equal-angle section of Fig. 12-3(a).

*Solution:*  First, the axis c-c in Fig. 12-3(b) is an axis of symmetry and thus the shear center must lie on this axis. We must now find where it is located on this axis. For this purpose, we shall find it convenient to assume that the applied external shear force $V$ is parallel to the axis a-a or perpendicular to the axis c-c. If the *external* force is applied in this manner, the *internal* shear forces will flow as illustrated in Fig. 12-3(c). Since these two forces intersect at point O, the moment of these internal forces with respect to O will be zero. Hence, the external force V must have zero moment about O, and thus this external force must pass through O. Therefore, O is the shear center.

NOTE: It is unnecessary to locate O any more accurately than to say that it lies slightly inside the corner of the angle section, since our shear flow theory is only an approximation based on the premise that the sections are very thin.

## 12-3. Bending of Curved Beams

Many structural elements are long curved members which carry some type of flexural loading. Examples are crane hooks, punch-press frames, C clamps, and rocking-chair rockers. For the most part this discussion will be concerned with determining the elastic longitudinal or circumferential flexural stresses[1] resulting from pure bending, although combined loading will be mentioned later.

**Kinematic Response.** Essentially the same assumptions regarding material, symmetry, and loading will be made here as were made in Chapter 10 in deriving the elastic flexure formula (Eq. 10-10) for straight beams. The material is assumed to be homogeneous, isotropic, and elastic. The beam has a longitudinal plane of symmetry. The resultant pure bending couples lie in this plane of symmetry. When we make these assumptions and use arguments similar to those in Sec. 10-2 for straight beams, we conclude that plane cross sections perpendicular to the longitudinal axis of the beam will remain plane after bending. Also, the beam will have a neutral surface throughout its length, and plane cross sections will remain perpendicular to this neutral surface after bending. A curved beam with a longitudinal plane of symmetry is shown in Fig. 12-4(a).

In Fig. 12-4(b) $\Delta\phi$ is the angle between two plane trapezoidal cross sections, a-b and c-d, *before* bending. During the application of a pure bending moment M, these two planes rotate relative to one another through a small angle $\Delta\theta$, as indicated by the dashed line c'-d'. The distance $\bar{r}$ and R are measured from the undeformed center of curvature to the centroidal axis and the neutral axis (wherever it may be), respectively. In the cross-sectional view in Fig. 12-4(c) a positive distance y is measured from the *neutral* surface *toward* the center of curvature. From the characteristics of the deformation and the geometry in Fig. 12-4(b) and (c), we see that the deformation at any distance y is given by the relationship

$$e_y = -y\,\Delta\theta$$

The minus sign indicates that those fibers on the concave side of the neutral surface are compressed and those fibers on the convex side are elongated.

[1]Although radial stresses do exist in curved flexural members, they are usually relatively small and become significant only in thin members such as shells.

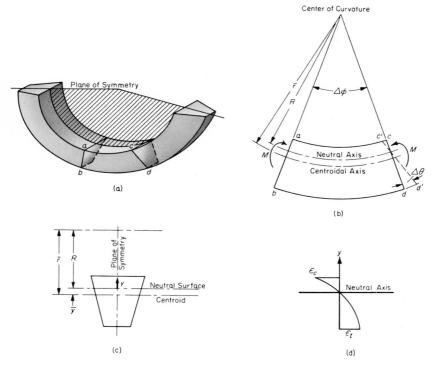

FIG. 12-4

Although the *deformation varies linearly* throughout the depth of the beam, the *strain does not vary linearly* because all the longitudinal fibers between section *a-b* and section *c-d* do not have the same initial length. The initial length of a fiber at a distance $y$ from the neutral surface is

$$\Delta L_y = (R - y)\Delta \phi$$

Then, by definition, the longitudinal strain in such a fiber is

$$\epsilon_y = \frac{e_y}{\Delta L_y} = \frac{-y\,\Delta \theta}{(R - y)\Delta \phi} \tag{12-4}$$

This last equation shows that the strain varies hyperbolically, as illustrated in Fig. 12-4(*d*). The maximum compressive strain $\epsilon_c$ occurs at the innermost concave side, and the maximum tensile strain $\epsilon_t$ occurs at the outermost convex side.

**Elastic Flexural Stresses.** Two questions need to be answered for a curved beam. Where is the location of the neutral surface, and what is the relation between the applied bending moment and the resulting flexural stresses? As was the case for a straight beam, these questions can be answered by utilizing the strain distribution given by Eq. 12-4 and the equilibrium requirements for the curved beam.

First, since the loading is assumed to cause pure bending, there is no resultant longitudinal normal force. Thus, the stress distribution must satisfy the relationship

$$\int_{\substack{\text{cross} \\ \text{section}}} \sigma \, da = 0 \tag{12-5}$$

For *elastic* behavior, $\sigma = E\epsilon$, where $E$ is the elastic modulus and is assumed to be the same for tension and compression. Then by Eqs. 12-4 and 12-5, we have

$$\int_{\substack{\text{cross} \\ \text{section}}} E\left(-\frac{y \, \Delta\theta}{(R - y)\Delta\phi}\right) da = 0$$

Since $E$, $\Delta\theta$, and $\Delta\phi$ are constants insofar as this integral is concerned,

$$\int_{\text{area}} \frac{y}{R - y} \, da = 0 \tag{12-6}$$

If we let $r$ be the distance from the center of curvature to any fiber of the beam, then $r = R - y$ and Eq. 12-6 becomes

$$\int_{\text{area}} \frac{R - r}{r} \, da = R \int_{\text{area}} \frac{da}{r} - \int_{\text{area}} da = 0$$

From this result we have

$$R = \frac{\displaystyle\int_{\text{area}} da}{\displaystyle\int_{\text{area}} \frac{da}{r}} \tag{12-7}$$

Recalling the definition of $R$, we have now determined (at least formally) the location of the neutral surface, which does *not* coincide with the centroid of the cross section.

In order to satisfy rotational equilibrium about the bending axis, the stress distribution must also satisfy the relationship

$$M_{\text{ext}} = -\int_{\substack{\text{cross} \\ \text{section}}} \sigma y \, da \tag{12-8}$$

The minus sign results from the assumed conditions of loading and coordinates, just as it did in deriving Eq. 10-2. Substitution in Eq. 12-8 of the expression for the flexural stress used previously yields

$$M_{ext} = - \int_{area} E\left(\frac{-y\,\Delta\theta}{(R-y)\Delta\phi}\right) y\,da$$

or

$$M_{ext} = E\frac{\Delta\theta}{\Delta\phi} \int_{area} \frac{y^2}{R-y}\,da \qquad (12\text{-}9)$$

Now, with the aid of Eq. 12-6, this last integral can be written as follows:

$$\int_{area} \frac{y^2}{R-y}\,da = \int_{area} \frac{Ry}{R-y}\,da - \int_{area} y\,da$$

$$= R(0) - a\bar{y}$$

or

$$\int_{area} \frac{y^2}{R-y}\,da = -a\bar{y} \qquad (12\text{-}10)$$

where $a$ is the area of the cross section and $\bar{y}$ is the distance between the neutral axis and the centroidal axis.

Notice that the integral on the left-hand side of Eq. 12-10 is inherently positive. Since $a$ is positive, $\bar{y}$ will necessarily be *negative*. From this we see that the neutral axis will always lie closer to the center of curvature than will the centroidal axis. Substitution of Eq. 12-10 in Eq. 12-9 yields

$$M_{ext} = -E\frac{\Delta\theta}{\Delta\phi}a\bar{y}$$

from which

$$\frac{\Delta\theta}{\Delta\phi} = -\frac{M}{Ea\bar{y}} \qquad (12\text{-}11)$$

Finally, from Eqs. 12-4 and 12-11, the relation between the external bending moment and the circumferential flexural stress is

$$\sigma = E\epsilon$$

$$= E\frac{-y}{R-y}\frac{\Delta\theta}{\Delta\phi}$$

or

$$\sigma = \frac{My}{(R-y)a\bar{y}} = \frac{My}{ra\bar{y}} \qquad (12\text{-}12)$$

Notice that a positive moment and a positive $y$ (inside fibers) produce a negative flexural stress (recall that $\bar{y}$ is negative). This result is consistent with the assumed deformation response.

Although Eq. 12-12 was derived on the basis that the loading was pure bending with no resultant axial force, within limitations the result can be applied to more general loading conditions that are usually encountered in engineering structures. It is merely necessary to transfer the line of action of any resultant axial force to the *centroid* of the cross section by use of the theory of transformation of a force into a force and a couple. Then the stress caused by the centroidal axial load and the bending couple can first be evaluated separately and later be superposed to give the total longitudinal stress in the beam (provided, of course, that the total stress is within the elastic limit).

As a final comment, we mention that the major difficulty in using Eq. 12-12 is in determining the value of $R$. Even for a relatively simple cross section, such as a circular or triangular section, the evaluation of Eq. 12-7 presents a formidable problem. Also, considerable accuracy in determining $R$ is required, because the value of $\bar{y}$ greatly influences the stress given by Eq. 12-12.

EXAMPLE 12-2. The curved steel member in Fig. 12-5($a$) has the cross section shown in Fig. 12-5($b$). $a$) Determine the maximum tensile and compressive stresses at section $c$-$c$. $b$) What would be the error in the flexure stresses if the elastic flexure formula for straight beams were used instead of Eq. 12-12?

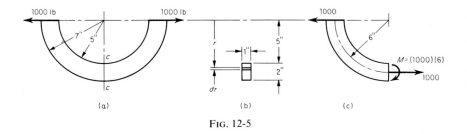

FIG. 12-5

*Solution:*   $a$) From Eq. 12-7,

$$R = \frac{\int da}{\int \dfrac{da}{r}} = \frac{2}{\displaystyle\int_5^7 \dfrac{{}^7(1)dr}{r}} = \frac{2}{\ln 7 - \ln 5}$$

$$= \frac{2}{\ln \dfrac{7}{5}} = 5.944$$

Therefore

$$\bar{y} = 5.944 - 6 = -0.056 \text{ in.}$$

From Fig. 12-5(c) and Eq. 12-12, the stress on the innermost fiber is

$$\sigma_i = \frac{P}{a} + \frac{My}{(R-y)a\bar{y}}$$

$$= \frac{1000}{2} + \frac{(-6000)(0.944)}{(5)(2)(-0.056)} = 500 + 10{,}120 = 10{,}620 \text{ psi}$$

The stress in the outermost fiber is

$$\sigma_o = \frac{1000}{2} + \frac{(-6000)(-1.056)}{(7)(2)(-0.056)} = 500 - 8090 = -7590 \text{ psi}$$

b) From the elastic flexure formula for straight beams,

$$\sigma = -\frac{My}{I} = -\frac{(-6000)(1)}{\dfrac{(1)(2^3)}{12}} = 9000 \text{ psi}$$

The error is

$$\left(\frac{10{,}120 - 9000}{10{,}120}\right)(100) = \left(\frac{1120}{10{,}120}\right)(100) = 10.6\%$$

## 12-4. Nonhomogeneous Beams

Thus far in this text we have dealt only with beams in which any cross section was composed of a single homogeneous material having the same modulus of elasticity for tension and compression. However, it is fairly common engineering practice to use beams in which a cross section is composed of two or more different materials. For example, steel plates are often used to reinforce timber beams, and steel rods are generally used to reinforce concrete beams. Also, several common structural materials, such as cast iron and concrete, do not have the same stiffness in compression as in tension and thus exhibit different elastic moduli in tension and compression.

One convenient method of analyzing the elastic stresses and deflections of a nonhomogeneous beam of the types mentioned above is to replace the actual beam by an *equivalent* beam composed of a single homogeneous material with the same modulus for tension and compression. By "equivalent" we mean that the new beam has the same kinematic response and internal *force* (not stress) distribution as does the original beam. Once the geometry of the equivalent beam has been determined, the elastic stresses and deflections can be found from the relations previously derived for beams made of a single homogeneous material.

Consider a straight, two-material beam whose symmetric cross section is shown in Fig. 12-6(a). The upper portion is made of a material with a modulus $E_1$, and the lower portion of a material with a modulus $E_2$. For convenience, we will assume that $E_1$ is greater than $E_2$. If the loading causes pure bending and is applied in the plane of symmetry, we can repeat the arguments of Sec. 10-2 for straight, homogeneous beams, and can conclude that plane sections remain plane after bending. The deformation, and hence the strain, will vary linearly from the neutral surface (wherever it may be), as indicated in Fig. 12-6($b$), and the stress is given by the relationship $\sigma = E\epsilon$. However, because of the different moduli, the stress distribution will be as shown in Fig. 12-6($c$), and an abrupt jump discontinuity will occur at the interface of the two materials.

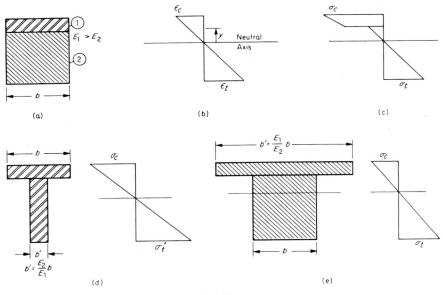

FIG. 12-6

As stated earlier, we wish to replace this beam having a nonhomogeneous cross section with an equivalent beam having a homogeneous cross section. First of all, we want the equivalent beam to have the same kinematic response. According to Eq. 10-1, the strain at any distance $y$ from the neutral surface is

$$\epsilon_y = -\frac{y}{\rho}$$

where $\rho$ is the radius of curvature of the bent beam. Thus, for any given radius of curvature, if the strain in a fiber of the equivalent beam is to be the same as the strain in the corresponding fiber of the actual beam, we must not change the distance $y$ from the neutral surface. Therefore, the *equivalent beam should have the same dimensions perpendicular to the neutral surface as the original beam.*

In addition to this kinematic equivalence, we also require static equivalence. That is, we require that the *force* on a fiber of the equivalent beam be the same as that on the corresponding fiber in the original beam. The force on any fiber at a distance $y$ from the neutral surface is

$$dF = \sigma_y \, da = E\epsilon_y \, da$$

$$= -E \frac{y}{\rho} \, da$$

Our requirement of static equivalence gives

$$dF' = dF$$

$$\sigma' \, da' = \sigma \, da$$

$$E' \frac{y}{\rho} \, da' = E \frac{y}{\rho} \, da$$

$$E' \, da' = E \, da \qquad (12\text{-}13)$$

where the primed quantities refer to the equivalent beam and the unprimed quantities to one of the materials of the actual beam.

Equation 12-13 shows that the elemental areas of the equivalent beam and the actual beam are inversely proportional to the moduli of the materials of the beams. Since we have said that we do not wish to change any height dimensions, we may only change the width of the elemental area. If the cross section has parallel straight sides,

$$da = b \, dy \qquad da' = b' \, dy$$

Hence, Eq. 12-13 yields

$$E'b' = Eb$$

from which

$$b' = \frac{E}{E'} b \qquad (12\text{-}14)$$

Thus, although the vertical dimensions of our equivalent beam will be the same as those of our original beam, the width of the equivalent beam is given by Eq. 12-14.

The equivalent beam is usually assumed to be made of one of the materials of the original beam. For example, if the beam of Fig. 12-6(a) is replaced by an equivalent beam made of a material with a modulus $E_1$, the equivalent beam will have the dimensions shown in Fig. 12-6(d). On the other hand, if the equivalent beam is made of a material with a modulus $E_2$, its dimensions will be those of Fig. 12-6(e). The choice has no effect on the actual stresses in the original beam, but it is usually convenient to assume that the equivalent beam is made from the material of that portion of the original beam in which you are most interested. That is, if you wish to know the stresses in the upper fibers, you would probably choose the equivalent beam of Fig. 12-6(*d*). However, it is not necessary to do so, since

$$\sigma' \, da' = \sigma \, da$$

$$\sigma = \frac{da'}{da} \, \sigma' = \frac{E}{E'} \, \sigma' \qquad (12\text{-}15)$$

Thus, once an equivalent beam has been found, all the flexural stresses in the entire original beam (straight or curved) can be determined.

As mentioned in the first paragraph of this section, probably the most widely used type of composite beam is one made of concrete reinforced with steel. Such a beam can be analyzed by the method just presented. However, concrete is so relatively weak in tension that it is usually assumed that the portion of the concrete in tension *does not exist*. That is, it is assumed that the original beam is composed of sections of concrete in compression and the steel rods which are in tension. This beam is then transformed into an equivalent beam by the usual method.

EXAMPLE 12-3. A beam with the rectangular cross section of Fig. 12-7(*a*) is made of a material that has a tensile modulus of $10 \times 10^6$ psi. and a compressive modulus of $15 \times 10^6$ psi. The beam is loaded in pure bending with a positive moment of 20,000 ft-lb. *a*) Find an equivalent beam for this section. *b*) Determine the maximum tensile and compressive flexural stresses in the original beam.

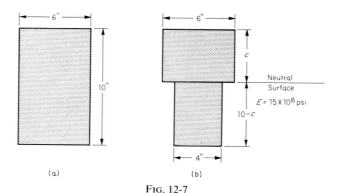

FIG. 12-7

*Solution:* *a*) The fibers above the neutral surface are in compression ($E = 15 \times 10^6$), while those below the axis are in tension ($E = 10 \times 10^6$). However, we do not as yet know the location of the neutral surface. Assume that it is $c$ inches from the top surface. Then by Eq. 12-14 an equivalent beam made of a material with a modulus of $E = 15 \times 10^6$ is shown in Fig. 12-7($b$). For elastic bending the neutral surface coincides with the centroid. Hence,

$$(6)(c)\left(\frac{c}{2}\right) = (10 - c)(4)\left(\frac{10 - c}{2}\right)$$

$$c^2 + 40c - 200 = 0$$

$$c = 4.5 \text{ in.}$$

*b*) The flexural stresses in the equivalent beam are given by the elastic flexure formula, which is

$$\sigma = -\frac{My}{I}$$

For the equivalent beam,

$$I = \frac{(6)(4.5^3)}{3} + \frac{(4)(5.5^3)}{3}$$

$$= 182 + 222 = 404 \text{ in.}^4$$

Then the stress in the uppermost fiber of the *actual* beam is

$$\sigma_u = -\frac{(20{,}000)(12)(4.5)}{404} = -2670 \text{ psi}$$

$$= 2670 \text{ psi (compressive)}$$

The stress in the lowermost fiber of the *equivalent* beam is

$$\sigma_L{}' = -\frac{(20{,}000)(12)(-5.5)}{404} = 3260 \text{ psi}$$

By Eq. 12-15 the stress in the lowermost fiber of the *actual* beam is

$$\sigma_L = \frac{E}{E'}\,\sigma_L{}' = \left(\frac{10 \times 10^6}{15 \times 10^6}\right)(3260) = 2175 \text{ psi (tensile)}$$

## PROBLEMS

**12-1 to 12-4.** For each of the thin sections shown in the illustrations, determine the location of the shear center.

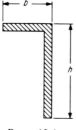

PROB. 12-1

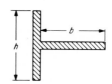

PROB. 12-2

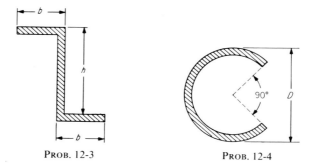

PROB. 12-3          PROB. 12-4

**12-5.** Place arrowheads on the two shear vectors shown for the removed element from the cantilever beam. The only load is vertically down through the shear center.

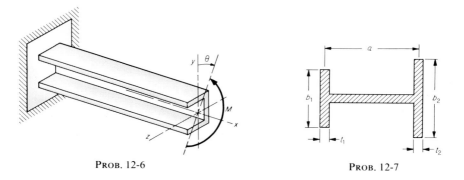

PROB. 12-5

**12-6.** The cantilever beam in the illustration is made of a thin channel section and is loaded with a pure couple $M$ as shown. For this loading condition, sketch the direction of the shear flow in the web and flanges.

**12-7.** The section shown in the illustration has a moment of inertia about the horizontal centroidal axis of $I$ in.[4] Locate the shear center in terms of $I$ and the dimensions given.

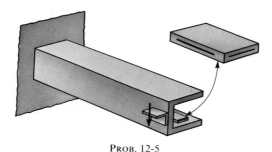

PROB. 12-6          PROB. 12-7

**12-8.** Locate the shear center for a 10⌴15.3 (American standard channel).

**12-9.** A cantilever beam with the cross section shown carries a vertical load of 6 tons. Sketch the direction of the shear flow throughout the cross section, and determine the location of the shear center. What is the value of the maximum horizontal shear stress?

**12-10.** Same as Prob. 12-9 but the cross section is shown in the illustration accompanying this problem.

**12-11 and 12-12.** Rework Example 12-2 for the cross sections shown in the accompanying illustrations.

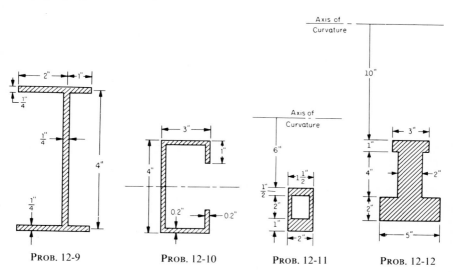

PROB. 12-9    PROB. 12-10    PROB. 12-11    PROB. 12-12

**12-13 and 12-14.** For curved beams loaded in pure bending, determine the value of the dimension $b$ of the cross section shown in each illustration which will make the maximum tensile and compressive bending stresses in the section numerically equal.

**12-15.** Determine the location of the neutral axis for a curved beam loaded in pure bending and having the cross section shown in the illustration.

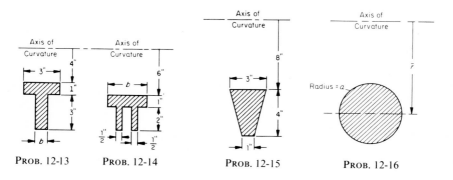

PROB. 12-13    PROB. 12-14    PROB. 12-15    PROB. 12-16

**12-16.** Show that the distance $R$ from the axis of curvature to the neutral axis for a curved beam with the cross section shown in the illustration is given by the relationship

$$R = \frac{\bar{r} + \sqrt{\bar{r}^2 - a^2}}{2}$$

(HINT: Use integral tables.)

**12-17.** Find the maximum normal bending stress in the member in the illustration.

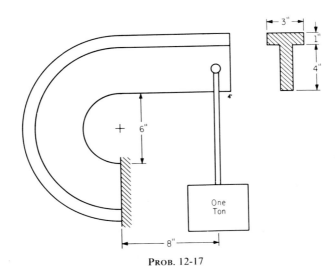

PROB. 12-17

**12-18.** For the composite steel-reinforced timber beam shown in the illustration, find the maximum tensile and compressive flexural stresses in the wood when the beam is subjected to a negative bending moment sufficient to cause a maximum flexural stress in the steel of 20,000 psi. The modulus of elasticity $E$ for the steel is $30 \times 10^6$ and that for the wood is $2 \times 10^6$ psi.

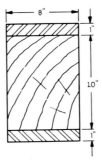

PROB. 12-18

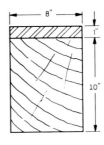

PROB. 12-19

**12-19.** The requirements are the same as for Prob. 12-18 but the beam has a cross section with a reinforcing plate only at the top as shown in the illustration.

**12-20.** The composite beam shown in the illustration is made of a timber reinforced with steel plates. The modulus of elasticity for the steel is $30 \times 10^6$ psi, and that for the wood is $2 \times 10^6$ psi. The allowable flexural stresses are 20,000 psi for the steel and 1000 psi for the wood. What is the maximum permissible bending moment about the axis *x-x*?

**12-21.** The cross section and the stresses are the same as for Prob. 12-20. What bending moment can be applied about the axis *y-y*?

**12-22.** The composite beam shown in the illustration is a steel-reinforced timber beam. The properties of the steel and wood are the same as those in Prob. 12-20. Determine the maximum permissible bending moment about a horizontal axis.

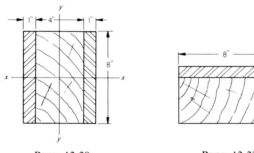

PROB. 12-20          PROB. 12-22

**12-23.** The steel and wood in the composite beam in the illustration for Prob. 12-22 have the properties given in Prob. 12-20. Determine the maximum permissible bending moment about a vertical axis. Where is the shear center for the beam?

**12-24.** A timber beam 6 in. high, 4 in. wide, and 22 ft long is to be simply supported at its ends and loaded with a uniformly distributed load of 200 lb/ft. The allowable bending stress in the wood is 1000 psi. *a*) If the timber is to be reinforced with two steel plates 4 in. wide on the top and bottom, what should be the thickness of the plates? *b*) What should be their width if steel plates $\frac{1}{4}$ in. thick are used? Use *E* for wood of $2 \times 10^6$ psi.

**12-25.** Select a timber beam approximately 4 in. wide to carry the loading in Prob. 12-24 over the same span. Compare the deflection of this timber beam to that of the reinforced beam in part (*a*) of Prob. 12-24. Discuss the relative merits of the two beams.

**12-26.** A positive bending moment of 30,000 ft-lb acts on a composite beam, the cross section of which is shown in the illustration. The steel and brass are bolted together. Determine the maximum flexural stresses in the materials if $E_{st} = 30 \times 10^6$ psi and $E_{br} = 20 \times 10^6$ psi.

**12-27.** A simply supported beam of reinforced concrete has the cross section shown in the illustration and carries a uniformly distributed load of 1800 lb/ft over a span of 20 ft. Evaluate the maximum flexural stress in the concrete and steel, assuming that $E_c = 2 \times 10^6$ psi and $E_s = 30 \times 10^6$ psi.

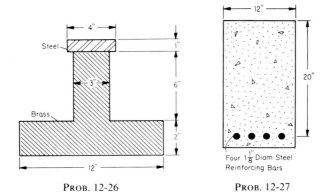

PROB. 12-26          PROB. 12-27

**12-28.** A reinforced concrete beam whose cross section is shown in the illustration is to carry a bending moment of 20,000 ft-lb. If the ratio of the modulus of elasticity of the reinforcing material to that of the concrete is 8, what are the maximum flexural stresses in the concrete and in the reinforcing bars?

**12-29.** The beam in the illustration is made of concrete reinforced with a material whose modulus of elasticity is 10 times that of the concrete. If the allowable bending stress in the concrete is 1200 psi and that in the reinforcing bars is 18,000 psi, what is the allowable bending moment?

**12-30.** In designing a steel-reinforced concrete beam it is possible to proportion the area of the steel to the area of the concrete so that the maximum permissible stresses in the steel and concrete occur simultaneously. Using the dimensions shown in the illustration, show that for balanced reinforcement

$$k = \cfrac{1}{1 + \cfrac{\sigma_s E_c}{\sigma_c E_s}}$$

where $\sigma_c$ and $\sigma_s$ are the permissible stresses in the concrete and steel, respectively.

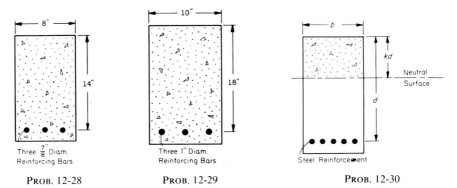

Three $\frac{7}{8}''$ Diam.
Reinforcing Bars

Three 1" Diam.
Reinforcing Bars

Steel Reinforcement

PROB. 12-28          PROB. 12-29          PROB. 12-30

**12-31.** Refer to Prob. 12-30. Determine the required reinforcing area for balanced reinforcement if $\sigma_c$ = 1000 psi, $\sigma_s$ = 20,000 psi, $b$ = 8 in., $d$ = 20 in., $E_s$ = 30 × 10⁶ psi, and $E_c$ = 2 × 10⁶ psi.

## chapter 13

# Strain Energy and Theories
# of Failure

## 13-1. Strain Energy and Failure

Part III of this text was devoted primarily to investigating the stresses, strains, and deformations in some of the more simple structural elements, such as shafts and beams. We saw that while it is possible for the state of stress in a member to be uniaxial, it is much more likely that a state of combined or biaxial stress exists throughout a loaded member. Consequently, when we talk about the stress in a member, we must necessarily be concerned with the entire *state of stress* or *state of strain* in the member.

Why were we interested in finding stresses or strains? The main reason was so that we could predict or determine the load-carrying abilities of the various members. By comparing the stresses or strains developed in the member with some preset standard or specification, such as the yield-point stress or the proportional-limit strain, we tried to make a judgment as to the cause and type of failure.

Failures can be observed and thereby classified into certain categories, such as fracture, general yielding, buckling, and others which were discussed in some detail in Chapter 7. But, what *causes* a material to fracture? Or to yield? There is probably no simple answer to either of these questions, and the best we can presently do is to propose *theories* as to why fracture or yielding occurs. Such theories are called "theories of failure," and they are just theories, not true facts or laws. In later sections of this chapter, we shall discuss and compare some of the more commonly accepted theories of failure and try to point out why in certain cases some of the theories may seem more valid than the others. But, remember that there is no one all-powerful, all-inclusive theory of failure, and that the engineer is therefore called upon to exercise his judgment as to which of the various theories is most applicable to a particular problem.

Why does a material fail? Because the largest stress exceeds some maximum permissible value? Because the largest strain exceeds some maximum permissible value? Both of these reasons seem plausible enough, but they do not provide the complete answer to the question of failure.

**329**

What then? Perhaps the individual values of the stresses and strains are not so important as their combined effect; that is, maybe the combination of smaller stresses with their accompanying strains is a more critical factor than any individual larger stress or strain. How then do we measure their combined effect? The concept of strain energy provides a possible answer to this question.

Strain energy was first introduced and defined in Sec. 4-4. However, that discussion was based mostly on the premise of a uniaxial state of stress which exists in a specimen during a uniaxial tension or compression test. The succeeding sections will be concerned with a more general development of strain energy in relation to biaxial and triaxial states of stress.

## 13-2. Total Strain Energy

Strain-energy density (hereafter referred to simply as strain energy) is the energy per unit volume which a material absorbs while undergoing forced deformations. For example, consider a small elemental volume $\Delta x \, \Delta y \, \Delta z$ in Fig. 13-1. If a centroidal force is applied parallel to the $x$ axis, deformations in the $x$, $y$, and $z$ directions will result. However, since

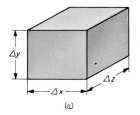 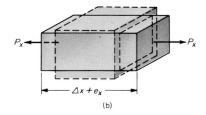

FIG. 13-1

the only force is parallel to the $x$ axis, the work done on the volume of material will depend only on this force and the deformation $e_x$ in the $x$ direction. Also, the magnitude of the force depends on the amount of deformation, and vice versa. Hence, the work per unit volume, or the strain-energy density, is

$$U = \frac{1}{V} \int_0^{e_x} P_x \, de_x = \int_0^{e_x} \frac{P_x}{\Delta z \, \Delta y} \frac{de_x}{\Delta x}$$

$$U = \int_0^{\epsilon_x} \sigma_x \, d\epsilon_x \tag{13-1}$$

In the last integral, the variable of integration is in terms of strain, and hence the limits are also in terms of strain.

The expression in Eq. 13-1 represents the strain energy (or stress work) for uniaxial loading only. Suppose, now, that the loading is not uniaxial, but that there are axial $y$ and $z$ forces as well as the $x$ force. Then the total work done on the elemental volume of material will be the work done by all of the forces acting through their respective displacements. Thus,

$$U = \int_0^{\epsilon_x} \sigma_x \, d\epsilon_x + \int_0^{\epsilon_y} \sigma_y \, d\epsilon_y + \int_0^{\epsilon_z} \sigma_z \, d\epsilon_z \qquad (13\text{-}2)$$

where the strains $\epsilon_x$, $\epsilon_y$, and $\epsilon_z$ are the *total* strains in their respective directions. Remember that each of the individual total strains is influenced by each of the individual axial stresses, and vice versa. Hence, in order that the integrals in Eq. 13-2 may be evaluated, the generalized stress-strain law for the material must be known. In particular, if the material is homogeneous, elastic, and isotropic, so that the stress-strain law is the Generalized Hooke's Law, the evaluation of these integrals is considerably simplified as will be seen in the next section.

Instead of axial loads, suppose that the forces acting on the elemental volume in Fig. 13-2(*a*) are shear forces which produce the deformation of Fig. 13-2(*b*). Then, from our definition of work, and assuming

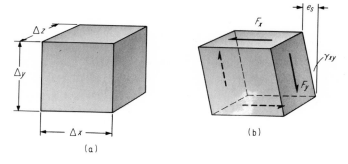

Fig. 13-2

the deformations are small, the strain-energy density for this system becomes

$$U = \frac{1}{V} \int_0^{e_s} F_x \, de_s = \int_0^{e_s} \frac{F_x}{\Delta x \, \Delta z} \frac{de_s}{\Delta y}$$

$$U = \int_0^{\gamma_{xy}} \tau_{yx} \, d\gamma_{xy} = \int_0^{\gamma_{xy}} \tau_{xy} \, d\gamma_{xy} \qquad (13\text{-}3)$$

Similarly, if all possible shear forces are taken into consideration, the strain energy due to these forces will be

$$U = \int_0^{\gamma_{xy}} \tau_{xy} \, d\gamma_{xy} + \int_0^{\gamma_{xz}} \tau_{xz} \, d\gamma_{xz} + \int_0^{\gamma_{yz}} \tau_{yz} \, d\gamma_{yz} \qquad (13\text{-}4)$$

Finally, in the most general case of a triaxial state of stress, the total strain energy will be that due to the three normal stresses and the three shear stresses, or the sum of Eqs. 13-2 and 13-4.

Before concluding this section, we shall make one very important observation. Work, and therefore strain energy, is a scalar quantity. That is, it is independent of the coordinate system used to identify the various forces, displacements, stresses, and strains. With this in mind, the calculation of the total strain energy for a combined state of stress can be considerably simplified if a principal coordinate system of stress and strain is used. For this case, there are no shear stresses or shear strains, and the total strain energy is then represented by Eq. 13-2 alone, where the stresses and strains are the principal values.

## 13-3. Elastic Strain Energy

Of particular importance, insofar as theories of failure are concerned, is the *elastic* strain energy associated with a combined state of stress. For a linear, homogeneous, isotropic material, elastic behavior means that the stress-strain law is the Generalized Hooke's Law (refer to Sec. 3-5). Also, since elastic strains and deformations are usually quite small, the convenient method of superposition may be used in calculating the elastic strain energy. When elastic stresses are superposed, the order in which the stresses are applied to the elemental volume is immaterial.

Before we proceed with the calculation of elastic strain energy, a comment on notation is worthwhile. Since each of the normal stresses influences each of the axial strains, a distinction will be made as to which stress causes what strain. The strain in a particular direction produced by the stress in that same direction will be denoted by $\epsilon_{xx}$, $\epsilon_{yy}$, and $\epsilon_{zz}$. The strain in one direction produced by the Poisson effect of a normal stress in another direction will be denoted by $\epsilon_{xy}$ and $\epsilon_{xz}$; $\epsilon_{yx}$ and $\epsilon_{yz}$; and $\epsilon_{zx}$ and $\epsilon_{zy}$. Thus, with this notation, the total strain in the $x$ direction, for example, is

$$\epsilon_{x(\text{total})} = \epsilon_{xx} + \epsilon_{xy} + \epsilon_{xz}$$

$$\epsilon_{x(\text{total})} = \frac{\sigma_x}{E} - \mu \frac{\sigma_y}{E} - \mu \frac{\sigma_z}{E} \tag{13-5}$$

Similar expressions can be obtained for the $y$ and $z$ directions as well.

In using the method of superposition to calculate elastic strain energy, we shall assume arbitrarily that the normal stresses are applied in the sequence $\sigma_x$, $\sigma_y$, and $\sigma_z$. You should satisfy yourself that the same result is obtained for any other sequence you may happen to select. Now, if the stress $\sigma_x$ is the first one applied to an elemental volume, strains will be produced in the $x$, $y$, and $z$ directions which are denoted by $\epsilon_{xx}$, $\epsilon_{yx}$, and $\epsilon_{zx}$, respectively. However, since for this condition the loading is uniaxial, the strain energy is given by Eq. 13-1 and is

$$U = \int_0^{\epsilon_{xx}} \sigma_x \, d\epsilon_{xx}$$

For uniaxial loading, Hooke's law is simply

$$\epsilon_{xx} = \frac{\sigma_x}{E}$$

and the elastic strain energy at this stage can be written as

$$U = \int_0^{\epsilon_{xx}} E\epsilon_{xx}\,d\epsilon_{xx} = \frac{1}{2}E\epsilon_{xx}^2 = \frac{1}{2}\sigma_x\epsilon_{xx} = \frac{\sigma_x^2}{2E} \tag{a}$$

Next, the stress $\sigma_y$ is applied. This stress also produces a strain in each of the directions; and these strains are denoted by $\epsilon_{yy}$, $\epsilon_{xy}$, and $\epsilon_{zy}$. During this deformation process there are two stresses acting on the body, namely, $\sigma_y$ and $\sigma_x$. The work done by the stress $\sigma_y$ is

$$U = \int_0^{\epsilon_{yy}} \sigma_y\,d\epsilon_{yy} \tag{b}$$

since this stress varies from zero to its maximum value during this straining process. On the other hand, during the same process the stress $\sigma_x$ retains a constant value. Thus, the work done by the stress $\sigma_x$ during this deformation process is

$$U = \int_0^{\epsilon_{xy}} \sigma_x\,d\epsilon_{xy} = \sigma_x\int_0^{\epsilon_{xy}} d\epsilon_{xy} = \sigma_x\epsilon_{xy} \tag{c}$$

Now, from Hooke's law,

$$\epsilon_{yy} = \frac{\sigma_y}{E} \qquad \epsilon_{xy} = -\mu\frac{\sigma_y}{E}$$

Substituting these values in equations (b) and (c), we have

$$U = \int_0^{\epsilon_{yy}} \sigma_y\,d\epsilon_{yy} + \sigma_x\epsilon_{xy}$$

$$U = \frac{\sigma_y^2}{2E} - \mu\frac{\sigma_x\sigma_y}{E} \tag{d}$$

Finally, the stress $\sigma_z$ is applied, causing the strains $\epsilon_{zz}$, $\epsilon_{xz}$, and $\epsilon_{yz}$. During this process three stresses are acting on the element. The magnitudes of the work done by the stresses are, respectively,

$$\int_0^{\epsilon_{zz}} \sigma_z\,d\epsilon_{zz} \qquad \int_0^{\epsilon_{xz}} \sigma_x\,d\epsilon_{xz} \qquad \int_0^{\epsilon_{yz}} \sigma_y\,d\epsilon_{yz}$$

where $\sigma_z$ varies while $\sigma_x$ and $\sigma_y$ are constant. By Hooke's law,

$$\epsilon_{zz} = \frac{\sigma_z}{E} \qquad \epsilon_{xz} = -\mu\frac{\sigma_z}{E} \qquad \epsilon_{yz} = -\mu\frac{\sigma_z}{E}$$

So we find that the work during the process is

$$U = \int_0^{\epsilon_{zz}} \sigma_z \, d\epsilon_{zz} + \sigma_x \int_0^{\epsilon_{xz}} d\epsilon_{xz} + \sigma_y \int_0^{\epsilon_{yz}} d\epsilon_{yz}$$

$$U = \frac{\sigma_z^2}{2E} - \mu \frac{\sigma_x \sigma_z}{E} - \mu \frac{\sigma_y \sigma_z}{E} \tag{e}$$

Thus, after the application of all three normal stresses, the final strain energy per unit volume is the algebraic sum of the work in equations (a), (d), and (e). This sum is

$$U = \frac{1}{2E} (\sigma_x^2 + \sigma_y^2 + \sigma_z^2) - \frac{\mu}{E} (\sigma_x \sigma_y + \sigma_x \sigma_z + \sigma_y \sigma_z) \tag{13-6}$$

This is a quadratic expression in the stresses $\sigma_x$, $\sigma_y$, and $\sigma_z$ and represents the elastic strain energy due to the three normal stresses. By use of Generalized Hooke's Law, Eq. 13-5, the expression for the strain energy can also be written in the bilinear form

$$U = \frac{1}{2} (\sigma_x \epsilon_x + \sigma_y \epsilon_y + \sigma_z \epsilon_z) \tag{13-6a}$$

where $\epsilon_x$, $\epsilon_y$, and $\epsilon_z$ are the *total* strains in the x, y, and z directions.

If, in addition to the three normal stresses, shear stresses also act on the element, then the total elastic strain energy must include the energy due to the shear stresses. However, for an isotropic material a shear stress does not influence either the axial strains or the other shear strains. So we can compute the strain energy produced by each of the shear stresses individually and then add their contributions algebraically. By Hooke's law for elastic shear stresses,

$$\gamma_{xy} = \frac{\tau_{xy}}{G} \qquad \gamma_{xz} = \frac{\tau_{xz}}{G} \qquad \gamma_{yz} = \frac{\tau_{yz}}{G}$$

From Eq. 13-4 the elastic strain energy due to the shear stresses is

$$U = \int_0^{\gamma_{xy}} \tau_{xy} \, d\gamma_{xy} + \int_0^{\gamma_{xz}} \tau_{xz} \, d\gamma_{xz} + \int_0^{\gamma_{yz}} \tau_{yz} \, d\gamma_{yz}$$

$$U = \frac{1}{2G} (\tau_{xy}^2 + \tau_{xz}^2 + \tau_{yz}^2) \tag{13-7}$$

or

$$U = \frac{1}{2} (\tau_{xy} \gamma_{xy} + \tau_{xz} \gamma_{xz} + \tau_{yz} \gamma_{yz}) \tag{13-7a}$$

The *total* elastic strain energy per unit volume for a general state of triaxial stress is the sum of the energies due to the normal and shear stresses, or the sum of Eqs. 13-6 and 13-7. This sum is

$$U = \frac{1}{2E}(\sigma_x^2 + \sigma_y^2 + \sigma_z^2) - \frac{\mu}{E}(\sigma_x\sigma_y + \sigma_x\sigma_z + \sigma_y\sigma_z)$$

$$+ \frac{1}{2G}(\tau_{xy}^2 + \tau_{xz}^2 + \tau_{yz}^2) \qquad (13\text{-}8)$$

or

$$U = \frac{1}{2}(\sigma_x\epsilon_x + \sigma_y\epsilon_y + \sigma_z\epsilon_z + \tau_{xy}\gamma_{xy} + \tau_{xz}\gamma_{xz} + \tau_{yz}\gamma_{yz}) \qquad (13\text{-}8a)$$

If the stresses and strains are measured in the principal directions, Eq. 13-8 reduces to Eq. 13-6, because the shear stresses and strains are zero for the principal directions. As a result, it is often more convenient to determine the strain energy after the principal stresses and strains have been found.

## 13-4. Volumetric and Distortion Components of Elastic Energy

Although the total elastic strain energy may be an important criterion in determining the cause of failure of a material, an even more important consideration is *how* the strain energy input affects the material. One way of investigating the effect which the energy input has on the elemental volume of material is to divide the total energy into two components. One component is the strain energy associated with the change in volume of the element. The other component is the strain energy associated with the distortion which the element undergoes. Later we shall see what roles the two component energies play in the theories of failure.

First of all, we know that any generalized triaxial state of stress may be resolved into a system of normal stresses and a system of shear stresses, as in Fig. 13-3. Also, in Sec. 3-6 it was shown that the change in volume per unit volume (cubical dilatation) $\epsilon_V$ was the sum of the axial strains, or $\epsilon_x + \epsilon_y + \epsilon_z$. Hence, the shear stresses and strains do not influence the volume change which may occur during a deformation process, and only the normal stresses and strains influence the volume change.

Let us now further decompose the system of normal stresses into two component systems of normal stresses. In one of these component stress

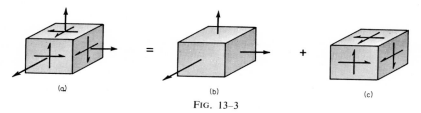

(a)          =          (b)          +          (c)

FIG. 13-3

systems, the stress will be the same in all directions, and this stress will be equal to the average value of the original system of normal stresses; thus, $\sigma_{avg} = \frac{1}{3}(\sigma_x + \sigma_y + \sigma_z)$. The other component stress system will consist of the remainders of the original normal stresses, as illustrated in Fig. 13-4.

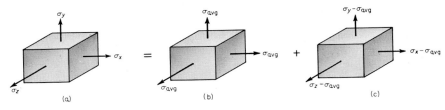

FIG. 13-4

The axial strain of the element in Fig. 13-4(b) will be the same in each direction and denoted by $\epsilon_{avg}$. From Hooke's law, we have

$$\epsilon_{avg} = \frac{\sigma_{avg}}{E} - \mu\frac{\sigma_{avg}}{E} - \mu\frac{\sigma_{avg}}{E} = \frac{\sigma_{avg}}{E}(1 - 2\mu)$$

$$\epsilon_{avg} = \frac{\sigma_x + \sigma_y + \sigma_z}{3E}(1 - 2\mu) = \frac{1}{3}(\epsilon_x + \epsilon_y + \epsilon_z) = \frac{1}{3}\epsilon_V$$

Thus, the element in (b) undergoes a unit volume change of

$$3\epsilon_{avg} = \epsilon_x + \epsilon_y + \epsilon_z = \epsilon_V$$

This is also the unit volume change of the element in Fig. 13-4(a). Hence, the element in Fig. 13-4(c) actually undergoes no volume change, and the energy associated with the change of volume of the element in Fig. 13-4(a) is the energy produced by the system of stresses in (b). Using Eq. 13-6a we find this volumetric energy to be

$$U_V = 3\left(\frac{1}{2}\sigma_{avg}\epsilon_{avg}\right) = 3\left(\frac{1}{2}\sigma_{avg}\frac{1}{3}\epsilon_V\right) = \frac{1}{2}\sigma_{avg}\epsilon_V \qquad (13\text{-}9)$$

Using the definition of bulk modulus (see Sec. 3-6),

$$K = \frac{\sigma_{avg}}{\epsilon_V} = \frac{E}{3(1 - 2\mu)} \qquad [3\text{-}19]$$

the volumetric energy can also be written as

$$U_V = \frac{1}{2}K\epsilon_V^2 = \frac{\sigma_{avg}^2}{2K} \qquad (13\text{-}9a)$$

The total energy produced by the original system of normal and shear stresses is found by Eq. 13-8. Thus

$$U_{total} = \frac{1}{2E}(\sigma_x^2 + \sigma_y^2 + \sigma_z^2) - \frac{\mu}{E}(\sigma_x\sigma_y + \sigma_x\sigma_z + \sigma_y\sigma_z)$$

$$+ \frac{1}{2G}(\tau_{xy}^2 + \tau_{xz}^2 + \tau_{yz}^2) \qquad [13\text{-}8]$$

If this expression represents the total elastic strain energy and that in Eq. 13-9a represents the volumetric strain energy, the remaining strain energy, or the energy of distortion, is

$$U_d = U_{total} - U_V$$

Using Eqs. 13-8, 13-9a, and 3-19, and simplifying, we obtain

$$U_d = \frac{1 + \mu}{6E}[(\sigma_x - \sigma_y)^2 + (\sigma_x - \sigma_z)^2 + (\sigma_y - \sigma_z)^2]$$

$$+ \frac{1}{2G}(\tau_{xy}^2 + \tau_{xz}^2 + \tau_{yz}^2)$$

Since $G = [E/2(1 + \mu)]$, the energy of distortion is

$$U_d = \frac{1}{12G}[(\sigma_x - \sigma_y)^2 + (\sigma_x - \sigma_z)^2 + (\sigma_y - \sigma_z)^2]$$

$$+ \frac{1}{2G}(\tau_{xy}^2 + \tau_{xz}^2 + \tau_{yz}^2) \qquad (13\text{-}10)$$

This energy of distortion is the energy that causes the change in shape, or distortion, of the original elemental volume. It does not cause any volume change. As we shall soon see, energy of distortion plays an important role in the theories of failure. Note that Eq. 13-10 becomes simplified somewhat if the principal directions are used. In this case the shear stresses are zero and the normal stresses are the principal values.

EXAMPLE 13-1. Compute the total elastic strain energy, the energy of distortion, and the energy of volume change for a biaxial state of pure shear.

*Solution:* For a biaxial state of pure shear with a maximum shear stress $\tau$, the *principal* stresses are $\sigma_1 = \tau$, $\sigma_2 = -\tau$, and $\sigma_3 = 0$. Hence, if a principal coordinate system is used, the total strain energy is

$$U_{total} = \frac{1}{2E}(\sigma_1^2 + \sigma_2^2 + \sigma_3^2) - \frac{\mu}{E}(\sigma_1\sigma_2 + \sigma_1\sigma_3 + \sigma_2\sigma_3)$$

$$= \frac{1}{2E}[\tau^2 + (-\tau)^2 + 0^2] - \frac{\mu}{E}[\tau(-\tau) + \tau(0) + (-\tau)0]$$

or

$$U_{\text{total}} = \frac{1 + \mu}{E} \tau^2$$

The energy of distortion is

$$U_d = \frac{1}{12G} [(\sigma_1 - \sigma_2)^2 + (\sigma_1 - \sigma_3)^2 + (\sigma_2 - \sigma_3)^2]$$

$$= \frac{1}{12G} \{[\tau - (-\tau)]^2 + (\tau - 0)^2 + (-\tau - 0)^2\}$$

$$= \frac{1}{2G} \tau^2$$

The energy of volume change is

$$U_V = \frac{1}{2K} \sigma_{\text{avg}}^2 = \frac{1}{2K} \left( \frac{\sigma_1 + \sigma_2 + \sigma_3}{3} \right)^2$$

$$= \frac{1}{2K} \left( \frac{\tau - \tau + 0}{3} \right) = 0$$

This last result could have been found from $U_{\text{total}}$ and $U_d$ by realizing that

$$G = \frac{E}{2(1 + \mu)}$$

Hence, for pure shear, $U_{\text{total}} = U_d$. This means that all the energy goes to distorting the element, and no volume change occurs.

## 13-5. Theories of Failure

As noted in Sec. 13-1, the phenomena of fracture and yielding can be observed and described, but the reasons (or reason) for their occurrence are as yet not completely understood. From a practical point of view, the designer or engineer would like to have some single criterion on which he can base his designs. That is, he would like to be able to say that failure in the form of fracture or yielding will occur when a certain critical quantity (whatever it may be) reaches a limiting value. But what is the critical quantity and what is its limiting value? These are the questions to which the theories of failure attempt to provide answers. In effect, a theory of failure establishes some quantity (or quantities) as a criterion for predicting failure.

As an example, consider a uniaxially loaded member. How can we establish a critical quantity for failure? In this case, comparison with experimental data could determine a fairly reliable criterion, such as the axial stress, axial strain, or strain energy. If plastic deformation were the mode of failure, the yield-point stress or yield strength could be used as the limiting value for the permissible axial stress. If fracture were the mode of failure, the ultimate strength would probably be used as the limiting value for the stress. Of course, the criterion need not be a stress. The elastic-limit strain might just as well be used as the criterion for plastic behavior, and the modulus of toughness could be used for fracture.

How is a critical quantity determined if the loading is not uniaxial and produces some general biaxial or triaxial state of stress? Certainly it is not feasible to attempt to conduct experiments for all possible combinations of stresses. In fact, it is difficult, and in some cases impossible, to produce controlled states of triaxial stress. It is at this stage that the need for a *theory of failure* for any arbitrary state of stress arises. The following paragraphs describe some of the more commonly accepted theories of failure.

**Maximum Normal Stress Theory.** This theory asserts that failure occurs at some point in a body only when the maximum principal normal stress at that point reaches some limiting value. According to this theory, only the magnitudes of the principal stresses are important, and the principal stress having the greatest magnitude governs the failure of the material.

This theory appears to give reasonable results for fracture of brittle materials. However, its limitations are rather self-evident, in that it does not take into account either the type of normal stress (tensile or compressive) or the orientation of the principal planes. Thus, this theory of failure is strictly applicable only to isotropic materials; and even for such materials it still does not take into account the effects of states of combined tensile and compressive stresses. For a ductile material, a biaxial combination of one tensile stress and one compressive stress can be more critical than the magnitude of either of the stresses individually.

**Maximum Shear Stress Theory.** This theory predicts that failure will occur at some point in a loaded body when the maximum shear stress at that point reaches a certain limiting value. If the principal stresses are $\sigma_1$, $\sigma_2$, and $\sigma_3$, and if $\sigma_1 > \sigma_2 > \sigma_3$, then we recall from Chapter 1 that the maximum shear stress is

$$\tau_{max} = \frac{\sigma_1 - \sigma_3}{2}$$

Thus, if a limiting value of $\tau_{max}$ is assigned or known, this equation represents the mathematical formulation of the maximum shear stress theory.

This theory seems to be applicable to members made of ductile materials in which relatively large shear stresses are developed. However, as in the case of the maximum normal stress theory, this theory fails to take account of the orientation of the maximum shear stress. Hence, it is strictly applicable only to isotropic materials. Also, according to this theory, an element subjected to a state of triaxial hydrostatic stress (three equal principal stresses) would never fail; of course, such a conclusion is erroneous. Nevertheless, this theory is widely used in the design of steel members, since shear stresses play a significant role in the phenomenon of yielding.

**Maximum Axial Strain Theory.** This theory proposes that axial strain, rather than stress, is the criterion for failure of a material. In particular, it proposes that failure depends on the magnitude of the largest principal axial strain, regardless of the combination of stresses causing the strains. If this theory is to be used, the generalized stress-strain law for the material must be known so that the principal strains can be evaluated.

As with the first two theories, this theory does not take into account the directions of the principal strains. Hence, this theory is strictly applicable only to isotropic materials. Also, since Hooke's law is valid only for elastic behavior, it becomes difficult to apply this theory if inelastic behavior is permitted to occur. For these and other reasons, this theory is rarely used in modern design.

**Maximum Total Energy Theory.** This theory states that failure occurs when the total energy per unit volume reaches some predetermined limiting value. This theory has one obvious advantage over the previous three theories in that the strain energy is independent of the orientations of the stresses and strains involved. For this reason an energy theory of this type can be more readily applied to nonisotropic materials. As indicated in Sec. 13-2, the problem of evaluating the total strain energy for an arbitrary state of stress can be extremely difficult. This is especially true if inelastic behavior occurs, because the generalized stress-strain law is then quite complex and may possibly be unknown. For this reason this theory is almost never used to predict failure by fracture for a ductile material.

If the total *elastic* strain energy is the criterion for failure, then the total elastic strain energy for an isotropic material can be evaluated by the equations derived in Sec. 13-3, which only require that the elastic constants of the material be known. Thus, this theory can be used to predict fracture of a very brittle material or the initiation of yielding in a ductile material. However, the next theory has been found to be more accurate in predicting the initiation of yielding and the maximum normal stress theory is easier to use in predicting fracture of brittle materials. For these reasons the total strain energy theory has been virtually abandoned.

**Energy of Distortion Theory.** This theory, often called the Huber-Hencky-von Mises theory or the octahedral shear stress theory, predicts that failure will occur when an equivalent stress $\sigma_e$ defined by

$$\sigma_e^2 = [(\sigma_1 - \sigma_2)^2 + (\sigma_1 - \sigma_3)^2 + (\sigma_2 - \sigma_3)^2] \tag{13-11}$$

where $\sigma_1$, $\sigma_2$, $\sigma_3$ are the principal stresses, reaches a predetermined limiting value.

One development of this theory is based on the elastic energy of distortion, which was discussed in the Sec. 13-4. It was shown that any arbitrary state of stress can be decomposed into two systems of stresses. One system, consisting of equal normal stresses, produces only volume change, and the other system of stresses causes the distortion. Experimental evidence indicates that a state of stress due to hydrostatic compression has very little tendency to cause yielding in a ductile material. On this motivation, the energy of distortion theory states that yielding in a ductile material is independent of the energy of volume change and is *dependent* only on the energy of distortion. If the principal stress system is used, we see from Eq. 13-10 that, except for the elastic constant multiplier, the energy of distortion is proportional to the quantity indicated in Eq. 13-11. Hence, that quantity is used as the criterion for this theory of failure.

This theory of failure is distinctly different from the total energy theory. For example, if it were possible to construct a state of stress due to hydrostatic tension (three equal tensile principal stresses), there definitely would be strain energy but there would be no energy of distortion. Hence, according to the distortion theory of failure, yielding should never occur. However, except for the rare possibility of hydrostatic stress, the energy of distortion theory seems to be the most accurate for predicting failure by yielding in ductile materials.

## 13-6. Use of the Theories of Failure

Each of the preceding theories of failure establishes a criterion for predicting failure of a material. Of course, by their very nature, some of the theories are more applicable to predicting failure by fracture, and others to predicting failure by yielding. Consequently, one might say that generally the maximum normal stress theory and the maximum axial strain theory are more valid for design of members made of brittle materials, and the maximum shear stress theory and the energy of distortion theory are more valid for design of ductile members. But, regardless of which theory of failure is employed, the question still remaining is, what number is to be used as the limiting value of the established criterion? From an engineering point of view, this value should be based on some experimental evidence. The most easily controlled states of stress are those associated with loading conditions causing uniaxial stress, torsion, or bending. Hence, the limiting value for the critical quantity is usually based on the value obtained from one or more of these simple loading conditions.

If all theories of failure (or even any one theory) were really correct, the value of the critical quantity would not depend on what experiment was used to determine it. Unfortunately, this is not true, as can be seen by a simple example. Suppose that the maximum shear stress theory is to be used to predict the initiation of yielding in a ductile material. Recalling

Mohr's circle, for a uniaxial tensile test we see that the maximum shear stress is equal to one-half of the maximum tensile stress. Thus, the shear yield point $\tau_{yp}$ is one-half the tensile yield point $\sigma_{yp}$. However, it has been found that in a torsion test where the state of stress is pure shear, the value of the shear stress at yielding is approximately $0.57\ \sigma_{yp}$. Which of these values is to be used as the limiting value of the maximum shear stress? The answer depends on the particular problem and the experience of the engineer. The value obtained from a uniaxial tension test is probably more conservative than that obtained from a torsion test. It is left as an exercise for the student (Prob. 13-13) to show that if the energy of distortion theory is used, and if this theory is correct for both uniaxial and torsion loading, the relation between $\sigma_{yp}$ and $\tau_{yp}$ should be

$$\tau_{yp} = \frac{\sigma_{yp}}{\sqrt{3}} = 0.577\ \sigma_{yp}$$

which, as previously mentioned, agrees with experimental evidence for ductile materials.

Remember that nobody really knows what factor or combination of factors causes yielding or fracture to occur. A theory of failure does not try to explain why these phenomena occur, but merely attempts to establish some way of predicting when a certain phenomenon should occur. Thus, the theories of failure are really just reasonable guesses. Which theory should be used in a particular engineering problem? Here, again, an intelligent guess is involved. Before making this decision, the engineer or designer must assimilate the available information in regard to loading conditions, use of the structure, material, reliability of experimental data, and other influencing factors. Finally, the engineer must call on his own experience and good judgment.

EXAMPLE 13-2. Data from a uniaxial tensile test are to be used to determine the limiting value for failure. If the mode of failure is the initiation of inelastic action, determine the limiting value for each of the theories of failure in terms of the available data.

*Solution:* From the uniaxial test, the values of $E$, $\mu$, and the proportional-limit stress $\sigma_{PL}$ can be found. We can then determine the limiting value for each critical quantity in terms of these properties.

a) *Maximum normal stress theory.* The critical quantity is the largest principal stress. Thus

$$\sigma_{cr} = \sigma_{PL}\ \text{psi}$$

b) *Maximum shear stress theory.* The critical quantity is the maximum shear stress. Thus

$$\tau_{cr} = \frac{\sigma_{PL}}{2}\ \text{psi} \quad \text{(since the test loading is uniaxial)}$$

c) *Maximum axial strain theory.* The critical quantity is the largest principal axial strain. Thus

$$\epsilon_{cr} = \frac{\sigma_{PL}}{E} \text{ in./in. \quad (since the test loading is uniaxial)}$$

d) *Maximum total energy theory.* The critical quantity is the total elastic strain energy. Thus

$$U_{cr} = \frac{1}{2} \frac{\sigma_{PL}^2}{E} \frac{\text{in-lb}}{\text{in.}^3} \quad \text{(from Eq. 13-6 or 13-8)}$$

e) *Maximum energy of distortion theory.* The critical quantity is the effective stress given by Eq. 13-11. Thus

$$\sigma_{cr} = \sigma_e = \sqrt{\sigma_{PL}^2 + \sigma_{PL}^2} = \sqrt{2}\,\sigma_{PL} \text{ psi}$$

## PROBLEMS

**13-1.** Compute the total elastic energy, the energy of distortion, and the energy of volume change for a uniaxial state of stress.

**13-2.** Compute the total elastic energy, the energy of distortion, and the energy of volume change for a general state of plane stress.

**13-3.** Compute the total elastic energy, the energy of distortion, and the energy of volume change for a state of plane strain for which $\epsilon_z = 0$. Express your answer in terms of the stresses $\sigma_x$, $\sigma_y$, $\sigma_z$, and $\tau_{xy}$.

**13-4.** Carry out the details in the derivation of Eq. 13-10.

**13-5.** Calculate the elastic volumetric strain energy for the element shown.

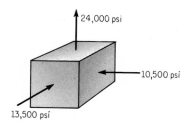

24,000 psi

10,500 psi

13,500 psi

PROB. 13-5

**13-6.** Data from a pure torsion test are to be used to determine the limiting value for failure. If the mode of failure is the initiation of inelastic action, determine the limiting value for each of the theories of failure in terms of the available data. The values of $E$, $G$, $\mu$, and $\tau$ can be determined from an elastic torsion test.

**13-7.** A thin-walled spherical pressure vessel is made of a ductile material and the parts are riveted together. If the only significant loading is the internal pressure, discuss fully the relative merits of each of the theories of failure. Which is the most conservative? Which is the least conservative?

**13-8.** A thin-walled, riveted cylindrical pressure vessel is made of a ductile material. The longitudinal axis of the vessel is horizontal, and its ends are simply supported. If the only significant loading is the internal pressure, discuss fully the relative merits of each of the theories of failure. If the weight of the vessel causes significant bending effects, how will the applicability of each theory of failure be affected?

**13-9.** The member shown in the illustration is made of mild steel. The member will cease to function properly if the load causes excessive inelastic deformation. Derive a design formula for the required radius of the member in terms of the dimensions $L$ and $b$, the load $P$, and some limiting quantity. Also, specify how the value of the limiting quantity should be determined.

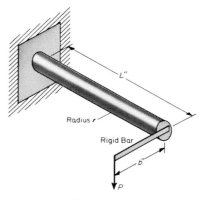

PROB. 13-9

**13-10.** The member shown in the illustration is made of a brittle material. Derive an expression for the allowable value of $P$ in terms of the radius of the member, some limiting quantity, and the ratio $T/P$. Also, specify how the value of the limiting quantity should be determined.

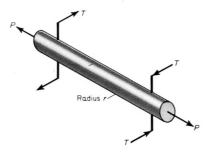

PROB. 13-10

**13-11.** The member shown in the illustration for Prob. 13-9 is made of a linearly brittle material. The only available properties for the material are those from a uniaxial tensile test which indicated that $E = 18 \times 10^6$ psi, $\mu = 0.30$, and $\sigma_{ult} = 60,000$ psi. If the radius of the member is 1 in. and the dimensions $L$ and $b$ are 15 in. and 8 in., respectively, determine the maximum permissible load $P$ based on a safety factor of 2. Use the most conservative of the applicable theories of failure.

**13-12.** The member shown in the illustration for Prob. 13-10 is made of a ductile material. As determined from a uniaxial test, the yield strength (based on 0.2% offset) is $\sigma_{ys} = 50,000$ psi, $E = 24 \times 10^6$ psi, and $\mu = \frac{1}{4}$. Although data from a torsion test might be desirable, it is rather difficult to determine inelastic stresses in a torsion specimen. Hence, the uniaxial data must be used in the design of the member. If failure is due to excessive inelastic behavior, what theory or theories of failure are most applicable and why? If $T = 60,000$ in-lb and $P = 20,000$ lb, what minimum radius is required based on a safety factor of 1.5?

**13-13.** In a ductile material failure is usually based on the initiation of inelastic action. Compare the limiting values for each of the theories of failure obtained from a uniaxial test with those obtained from a torsion test. (HINT: See Example 13-2 and Prob. 13-6.)

## chapter 14

# Work and Energy Methods

## 14-1. Introduction

We have seen in Chapter 2 how the concept of strain energy can be used as a material parameter; i.e., the modulus of resilience and modulus of toughness. In Chapter 13 we showed how strain energy per unit volume can be used as a criterion for failure particularly in situations in which a multi-axial state of stress exists. We now turn to yet another, and perhaps the most widely used and practical, application of energy concepts, namely, the determination of loads and/or deformations from strain energy considerations.

## 14-2. Work

Whenever external forces are applied to structural members such as beams, shafts, etc., these members deform and the forces do work, the work of each force being equal to the product of the force and the displacement parallel to the force. In most deformation processes the applied forces vary from zero to their final values as the process progresses. Therefore, the work of each force would have to be measured continuously during the process in order to know the total amount of work done. Theoretically, if we knew how each force varied with the deformation, the work could be computed as

$$W_{total} = \sum_{i=1}^{n} \int_{0}^{s_i} P_i \, ds_i$$

where $n$ is the number of forces on the body, $P_i$ is the $i$th force and $s_i$ is the displacement of the $i$th force parallel to that force. In general, the obtainment of all the necessary information to compute the work would be a difficult and tedious task.

The essential characteristic of a linearly elastic body is that the deformation is directly proportional to the force producing that deformation, or, vice versa. That is

$$P = ke$$

For this situation the work done by a force as it is applied to an elastic body is

$$W = \int_0^e P\, de = \int_0^e ke\, de = k\frac{e^2}{2} = \frac{1}{2}Pe$$

or                                                                                        (14-1)

$$= \int_0^e P de = \int_0^P P\frac{dP}{k} = \frac{1}{2}\frac{P^2}{k} = \frac{1}{2}Pe$$

In either expression we see that the work depends only on the proportionality constant $k$ and the final value of the load $P$ and/or the deformation $e$. Thus we need not know the entire history of the loading and deformation process in order to compute the work done by the force. The work done by forces producing elastic deformations can be computed by measuring the forces and the final deformation of the body at the points of application without regard as to the order in which the forces were applied.

EXAMPLE 14-1. Compute the work done by a torque $T$ applied to a solid circular shaft of length $L$ and constant radius $r$ loaded in pure torsion. Assume elastic behavior.

*Solution:*  The work done by a couple $C$ is given by the integral

$$W = \int_0^\theta C d\theta$$

when $\theta$ is the angular displacement of the couple. From Chapter 9 we recall that the torque-angular deformation relationship for an elastic shaft is

$$\theta = \frac{TL}{JG} \qquad \text{or} \qquad T = \frac{JG\theta}{L}$$

Hence

$$W = \int_0^\theta \frac{JG\theta}{L} d\theta = \frac{JG}{L}\frac{1}{2}\theta^2 = \frac{1}{2}T\theta$$

or

$$W = \int_0^T T\frac{L\, dT}{JG} = \frac{L}{JG}\frac{1}{2}T^2 = \frac{1}{2}T\theta$$

EXAMPLE 14-2. A two section bar shown in Fig. 14-1($a$) is attached at the top to a rigid fixture. A load of 100 lbs is applied as shown in Fig. 14-1($b$) causing section A to stretch 0.1 inch. Then a second 100 lb load is applied as shown in Fig. 14-1($c$) causing section A to stretch an additional 0.1 inch and section B to stretch 0.25 inch. What was the work done during this entire process if the bar is elastic?

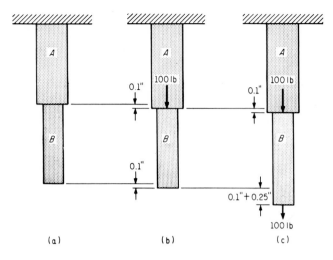

$$\text{Fig. } 14\text{-}1$$

*Solution:* During the application of the first force the work is by Eq. 14-1

$$W = \frac{1}{2} Pe = \frac{1}{2} (100)(0.1) = 5 \text{ in-lb}$$

During the application of the second force the first force is constant while undergoing a displacement of an additional 0.1 inch. Hence the work of force one is an additional

$$W = Ps = (100)(0.1) = 10 \text{ in-lb}$$

while the work of force two is

$$W = \frac{1}{2} Pe = \frac{1}{2} (100)(0.1 + 0.25) = 17.5 \text{ in-lb}$$

since this second force produces a total elongation of 0.1 + 0.25 at the point of application.
Thus the total work is

$$5 + 10 + 17.5 = 32.5 \text{ in-lb}$$

*Alternate Solution:* The total deformation of the point of application of force one is $(0.1 + 0.1) = 0.2$ and that of the point of application of force two is $(0.1) + (0.1 + 0.25) = 0.45$. Thus the total work is by Eq. 14-1

$$W = \frac{1}{2} P_1 e_1 + \frac{1}{2} P_2 e_2 = \frac{1}{2} (100)(0.20) + \frac{1}{2} (100)(0.45)$$

$$= 10 + 22.5 = 32.5 \text{ in-lb}$$

Re-study this example to be sure that you understand the ideas behind both procedures.

## 14-3. Elastic Strain Energy of Structural Members

In Chapter 13 it was shown that the *work* done on a unit cube of *elastic* material by a general triaxial system of stresses during a quasi-static process was given by the elastic strain energy density (Eq. 13-8)

$$U = \frac{1}{2E}(\sigma_x{}^2 + \sigma_y{}^2 + \sigma_z{}^2) - \frac{\mu}{E}(\sigma_x\sigma_y + \sigma_x\sigma_z + \sigma_y\sigma_z)$$

$$+ \frac{1}{2G}(\tau_{xy}{}^2 + \tau_{xz}{}^2 + \tau_{yz}{}^2) \tag{13-8}$$

Therefore, the total work done on an elastic structural member during a quasi-static process can be obtained by integrating the elastic strain energy per unit volume over the entire volume of the member.

$$W = \int_{\text{Volume}} U\, dV = \text{Total Elastic Strain Energy of the Body} \tag{14-2}$$

For the more simple structural members, the stresses can be expressed in terms of the applied loads and the resulting integrals for $W$ easily evaluated. Thus for such members Eq. 14-2 offers us a convenient method for determining the work done on an elastic member during quasi-static processes.

EXAMPLE 14-3. Using Eq. 14-2, re-work Example 14-1.

*Solution:* For a solid circular shaft subjected to pure torsion the shear stress is given by Eq. 9-3

$$\tau_\rho = \frac{T\rho}{J}$$

and all other stresses are zero.

Hence from Eqs. 13-8 and 14-2

$$W = \int_V \frac{1}{2G}(\tau^2)dV = \frac{1}{2G}\int_V \frac{T^2\rho^2}{J^2}\,dV = \frac{T^2}{2J^2G}\int_V \rho^2\,dV$$

Now an elemental volume $dV$ can be written as

$$dV = da\, dL$$

when $da$ is an elemental area and $dL$ an elemental length parallel to the axis of the shaft. Then

$$\int_V \rho^2\, dV = \int_{\text{length}} dL \int_{\text{area}} \rho^2\, da$$

But

$$\int_{\text{area}} \rho^2\, da = J \text{ polar moment of inertia}$$

Thus

$$\int_V \rho^2 \, dV = \int_{length} J \, dL = JL$$

provided the shaft is not tapered. Hence

$$W = \frac{T^2}{2J^2G} JL = \frac{T^2 L}{2JG} = \frac{1}{2} T\theta \qquad (14\text{-}3)$$

which is precisely that of Example 14-1, as expected.

In a manner similar to that employed in the previous examples, you can show (Probs. 14-1 and 14-2) that the work done during an elastic quasi-static process on an axially loaded member is given by

$$W = \frac{P^2 L}{2aE} \qquad (14\text{-}4)$$

and for a straight beam loaded in pure bending

$$W = \frac{1}{2} \frac{M^2 L}{EI} \qquad (14\text{-}5)$$

where $M$ is the bending moment and $L$ the length of the beam.

NOTE: In an arbitrary deformation process, the work done by external forces contributes to changes of the kinetic energy of the body of material as well as to the production of strain energy. However, for processes in which little or no changes in kinetic energy occur, the entire work of the external forces can be assumed to be equal to the stress work or strain energy as we have done here and in Chapter 13.

## 14-4. Castigliano's Theorem

We stated in the introduction of this chapter that energy concepts can be used to determine loads and deformations of structural members. In the previous sections we have seen some samples of how the total elastic strain energy of a member can be related to the applied forces or vice versa. We now wish to make a more general study of the relation between the applied loads and the elastic strain energy.

For convenience, consider a uniaxially loaded member and suppose we record the force deformation data as it is loaded up to some load $P$ as shown in Fig. 14-2. As we know, the work done by the force $P$ would be the area *under* the load deformation curve which we call $W$. The shaded area *above* the curve we will call the complementary work $W_c$. Although no simple physical interpretation can be given this quantity, we find it convenient to use in our discussions. Suppose now that $P$ is given a very slight increase $\delta P$ which results in a slight increase in deformation $\delta e$ and accompanying increases in $W$ and $W_c$ of $\delta W$ and $\delta W_c$, respectively. To first order terms

$$\delta W = P \, \delta e \qquad \delta W_c = e \, \delta P$$

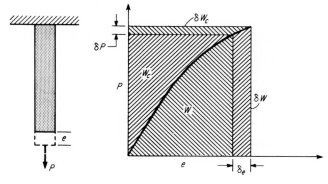

FIG. 14-2

or (14-6)

$$P = \frac{\delta W}{\delta e} \qquad e = \frac{\delta W_c}{\delta P}$$

We can interpret these equations as stating that the rate of change of the work with respect to the deformation is equal to the load and that the rate of change of the complementary work with respect to the load is equal to the deformation. If the member had been *linearly elastic so that the load-deformation curve was a straight line* then $W = W_c$ and we would have

$$\delta W = \delta W_c$$

and (14-7)

$$e = \frac{\delta W}{\delta P}$$

which is a simple example of Castigliano's Theorem. A more general statement of this important and widely used theorem may be paraphrased as follows: *For an elastic body under equilibriated loads, the rate of change of work (or total elastic strain energy) with respect to any statically applied force P gives the deformation of the point of application of the force in the direction of the force.* We will not attempt to derive the theorem in its full generality but will demonstrate its derivation to point out its significant features.

Consider a simply supported beam with two concentrated loads $P_1$ and $P_2$ as shown in Fig. 14-3. We assume *elastic* behavior so that the figure is considerably exaggerated. Recalling Eq. 14-1, the work done by these forces in producing the elastic deflection of the beam is

$$W(P_1 P_2) = \frac{1}{2} P_1 y_1 + \frac{1}{2} P_2 y_2 \qquad (14-8)$$

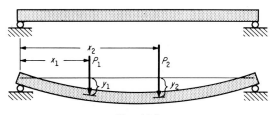

FIG. 14-3

Now, apply an additional force at $x_1$, $\Delta P_1$, which causes a change in deflection $\Delta y_1$ at $x_1$ and $\Delta y_2$ at $x_2$. The work done during this change is

$$\Delta W = P_1 \Delta y_1 + P_2 \Delta y_2 + \frac{1}{2} \Delta P_1 \Delta y_1 \qquad (14\text{-}9)$$

since $P_1$ and $P_2$ were constant during the application of $\Delta P_1$. The total work of the entire system is then

$$W = W(P_1 P_2) + \Delta W \qquad (14\text{-}10)$$

Suppose we had applied these same forces but in reverse order; that is, we apply $\Delta P_1$ then $P_1$ and $P_2$. In this case the work during the application of $\Delta P_1$ would be

$$\frac{1}{2} \Delta P_1 \Delta y_1$$

while the work done during the application of $P_1$ and $P_2$ would be

$$\frac{1}{2} P_1 y_1 + \frac{1}{2} P_2 y_2 + \Delta P_1 y_1$$

since $\Delta P_1$ is constant during this application. According to our discussion in Sec. 14-2, the work on an elastic body is independent of the order of application of the forces and so the total work for the first order should be the same as that for the second order. Therefore

$$\Delta P_1 y_1 = P_1 \Delta y_1 + P_2 \Delta y_2$$

and Eq. 14-9 becomes

$$\Delta W = \Delta P_1 y_1 + \frac{1}{2} \Delta P_1 \Delta y_1 \qquad (14\text{-}11)$$

Since the beam is linearly elastic $\Delta y_1 = k \Delta P_1$, and

$$\Delta W = \Delta P_1 y_1 + \frac{1}{2} k (\Delta P_1)^2$$

Dividing through by $\Delta P_1$ and letting $\Delta P_1 \to 0$ we obtain

$$\frac{\partial W}{\partial P_1} = y_1 \qquad (14\text{-}12)$$

The partial notation must be employed since $W$, you recall, was dependent upon $P_1$ and $P_2$. Equation 14-12 is the mathematical equivalent of Castigliano's Theorem stated earlier. In a similar manner, by considering couples $C$ and their resulting angular deformations $\theta$ we can obtain the result

$$\frac{\partial W}{\partial C} = \theta \qquad (14\text{-}13)$$

You must realize that in order to use Eq. 14-12 or 14-13, the work (or total elastic strain energy) must be expressed as a function of the applied loads. The practical usefulness of this theorem depends upon how easy or difficult it is to obtain the necessary differentiable expression for the work.

We now have a new method for computing the deformations at points of application of concentrated loads or concentrated couples. However, suppose we wish to determine the deformation at a point where no load is applied? We get around this dilemma by realizing that a zero load is the same as no load at all. That is, we can place a *ficticious load* at the point in question, then carry out the formal calculations as if that load were actually there, and finally *after differentiation* set this load equal to zero. We demonstrate these ideas in the following examples.

EXAMPLE 14-4. What is the maximum deflection of a straight, simply supported elastic beam loaded with a single load $P$ at its midspan?

*Solution:* From symmetry, the maximum deflection will occur directly under the load. Hence

$$y_{max} = \frac{\partial W}{\partial P} \qquad (a)$$

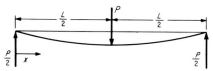

FIG. 14-4

For an elastic beam the work (strain energy) is given by

$$W = \int_{Volume} U dV = \int_V \frac{\sigma^2}{2E} dV = \int_V \left(\frac{My}{I}\right)^2 \frac{dV}{2E}$$

$$= \int_{length} dx \int_{area} \frac{M^2 y^2 da}{2EI^2} = \int_{length} \frac{M^2 dx}{2EI}$$

Before this expression can be integrated, we must obtain an expression for the bending moment $M$ in terms of the loads on the beam. In this case

$$M = \frac{P}{2} x \qquad 0 \leq x \leq \frac{L}{2}$$

$$= \frac{P}{2} x - P\left(x - \frac{L}{2}\right) \qquad \frac{L}{2} \leq x \leq L$$

Thus, since $E$ and $I$ are constant along the length of the beam

$$W = \frac{1}{2EI} \int_0^{L/2} \left(\frac{P}{2} x\right)^2 dx + \frac{1}{2EI} \int_{L/2}^L \left(\frac{-Px}{2} + \frac{PL}{2}\right)^2 dx \qquad \text{(b)}$$

Expression (b) is representative of the type that is usually encountered in using energy methods to solve beam deflection problems. Since $P$ is constant insofar as the variable of integration $x$ is concerned we may, if we wish, perform the differentiation required in (a) before integrating (b). This procedure is often economical and saves some unnecessary integration operations. In this case

$$\frac{\partial W}{\partial P} = \frac{1}{2EI} \int_0^{L/2} \frac{P}{2} x^2 dx + \frac{1}{2EI} \int_{L/2}^L 2P\left(\frac{-x}{2} + \frac{L}{2}\right)^2 dx$$

$$= \frac{1}{2EI} \frac{P}{6} \left(\frac{L}{2}\right)^3 + \frac{2P}{2EI} \frac{L^3}{96}$$

$$= \frac{PL^3}{48EI}$$

Note: Because of the symmetry in this problem, the strain energy in the left half of the beam will be one half of the total strain energy. Thus, we could have written

$$W = 2 \int_0^{L/2} \frac{M^2}{2EI} dx = \frac{1}{EI} \int_0^{L/2} \left(\frac{P}{2} x\right)^2 dx$$

EXAMPLE 14-5. Assuming elastic behavior for the beam shown in Fig. 14-5(a), determine the deflection of the free end.

*Solution:* Although the free end has no load applied there, we will put a ficticious load of $P$ at the left end as shown in Fig. 14-5(b). Then the moment equations are

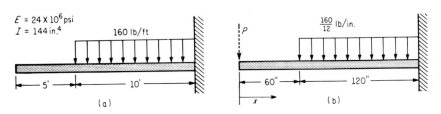

FIG. 14-5

$$M = -Px \qquad 0 \le x \le 60$$

$$= -Px - \frac{160}{12} \frac{(x-60)^2}{2} \qquad 60 \le x \le 180$$

From the previous example

$$W = \int_0^L \frac{M^2}{2EI} \, dx$$

$$= \frac{1}{2EI} \int_0^{60} (-Px)^2 \, dx + \frac{1}{2EI} \int_{60}^{180} \left[ -Px - \frac{160}{24}(x-60)^2 \right]^2 dx$$

Since we want the deflection at the left end

$$y_L = \frac{\partial W}{\partial P} = \frac{1}{2EI} \int_0^{60} 2Px^2 \, dx$$

$$+ \frac{1}{2EI} \int_{60}^{180} 2 \left[ -Px - \frac{160}{24}(x-60)^2 \right](-x) \, dx$$

Since $P$ is actually equal to zero, we need only evaluate the second term in the latter integral and thus

$$y_L = \frac{1}{EI} \frac{160}{24} \left[ \frac{x^4}{4} - 40x^3 + (60)^2 \frac{x^2}{2} \right]_{60}^{180}$$

$$= \frac{1}{(24)(10^6)(144)} \frac{160}{24} (400)(60)^3 = \frac{1}{6} \quad \text{inch}$$

Compare this example with Example 10-5.

In both of the previous examples singularity functions (Secs. 6-5 and 10-10) could have been employed in writing the moment equations although in these examples there was no particular advantage in doing so.

We close this chapter with a final example illustrating how the energy method can be used in composite structures.

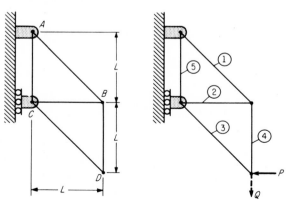

Fig. 14-6

EXAMPLE 14-6. All the members in the pin connected structure shown have the same cross section and elastic modulus. Find the movement of point $D$ caused by the application of a horizontal force $P$ at point $D$.

*Solution:* Even though only a horizontal force is applied at $D$, chances are that point $D$ will move vertically and horizontally. Therefore, we will also apply a ficticious vertical force $Q$ at $D$.

The total elastic strain energy in the structure will be the sum of the energy in each of the members. Since each member is an axially loaded one, by Eq. 14-4

$$U = \sum_{i=1}^{5} \frac{P_i^2 L_i}{2a_i E_i}$$

Now

$$L_1 = \sqrt{2}L \qquad L_2 = L \qquad L_3 = \sqrt{2}L \qquad L_4 = L \qquad L_5 = L$$

By the method of joints (or sections)

$$P_1 = \sqrt{2}(P + Q) \text{ tension} \qquad\qquad P_2 = (P + Q) \text{ compression}$$

$$P_3 = \sqrt{2}\, P \text{ compression} \qquad\qquad\qquad P_4 = (P + Q) \text{ tension}$$

$$P_5 = P \text{ compression}$$

Therefore

$$U = \frac{1}{2aE} \left[ 2(P + Q)^2 \sqrt{2}L + (P + Q)^2 L + 2P^2 \sqrt{2}L + (P + Q)^2 L + P^2 L \right]$$

From Castigliano's Theorem, the horizontal displacement will be

$$\delta_H = \frac{\partial U}{\partial P} = \frac{1}{aE} [2(P + Q)\sqrt{2}L + (P + Q)L + 2P\sqrt{2}L + (P + Q)L + PL]$$

For $Q = 0$

$$\delta_H = \frac{PL}{aE} [2\sqrt{2} + 1 + 2\sqrt{2} + 1 + 1] = 8.6 \frac{PL}{aE} \longleftarrow$$

Similarly, the vertical displacement will be

$$\delta_V = \frac{\partial U}{\partial Q} = \frac{1}{aE} [2(P + Q)\sqrt{2}L + (P + Q)L + (P + Q)L]$$

For $Q = 0$

$$\delta_V = \frac{PL}{aE} [2\sqrt{2} + 1 + 1] = 4.8 \frac{PL}{aE} \downarrow$$

## PROBLEMS

**14-1.** Carry out the details by two methods in the derivation of Eq. 14-4 for a straight elastic member loaded in pure tension.

**14-2.** Carry out the details by two methods in the derivation of Eq. 14-5 for a straight elastic beam loaded in pure bending.

**14-3.** Two elastic cylindrical bars of the same material and shown in the figure are to absorb the same total amount of elastic strain energy delivered by uniaxial loads applied at their ends. Neglecting stress concentrations, compare the stresses in the two bars.

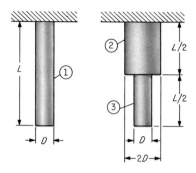

PROB. 14-3

**14-4.** Two solid circular elastic shafts of the same material are to absorb the same amount of total elastic strain energy delivered by pure torques applied at their ends. Neglecting stress concentrations, compare the shear stresses in the two shafts.

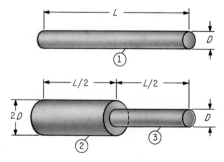

PROB. 14-4

**14-5.** Derive an expression for the total elastic strain energy of a solid homogeneous bar of constant cross-sectional area $A$, and length $L$, and modulus $E$ due to its own weight $W$ when suspended from one end.

**14-6.** Using the procedure of Prob. 14-5 and a ficticious load at the lower end of the bar, determine the total elastic deformation of the bar due to its own weight.

**14-7.** What is the total elastic strain energy for a straight, circular, hollow steel shaft with outside diameter 2 in., inside diameter 1 in. and length 6 ft subjected to a pure torque of 100 ft-lbs.

**14-8.** The elastic circular shaft in the illustration carries a uniformly distributed torque of $T$ in.-lb/in. along its entire length. Derive an expression for the total elastic strain energy.

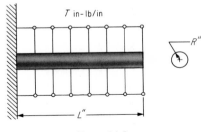

$T$ in-lb/in

$R''$

$L''$

PROB. 14-8

**14-9.** Using the procedure of Prob. 14-8 and a ficticious torque at the right end of the shaft, determine the total elastic angular twist of the shaft in Prob. 14-8.

**14-10.** A simply supported elastic beam of length $L$, second moment $I$, and modulus $E$ is loaded with a uniformly distributed load of $w$ lb/in. along its entire length. Derive an expression for the maximum deflection of the beam using a ficticious load at its midspan.

**14-11.** Using Castigliano's Theorem, determine the wall reactions on the beam shown in the illustration.

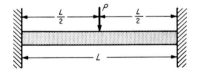

$P$

$\frac{L}{2}$   $\frac{L}{2}$

$L$

PROB. 14-11

**14-12.** Work Prob. 10-47 using the energy methods of this chapter.
**14-13.** Work Prob. 10-49 using the energy methods of this chapter.
**14-14.** Work Prob. 10-50 using the energy methods of this chapter.
**14-15.** Work Prob. 10-51 using the energy methods of this chapter.
**14-16.** Work Prob. 10-57 using the energy methods of this chapter.
**14-17.** Work Prob. 10-59 using the energy methods of this chapter.
**14-18.** Work Prob. 10-66 using the energy methods of this chapter.
**14-19.** Work Prob. 10-67 using the energy methods of this chapter.
**14-20.** Work Prob. 10-68 using the energy methods of this chapter.
**14-21.** Work Prob. 10-69 using the energy methods of this chapter.
**14-22.** Work Prob. 10-70 using the energy methods of this chapter.
**14-23.** Work Prob. 9-54 using the energy methods of this chapter.
**14-24 to 14-27.** The pin-connected trusses shown in the accompanying illustrations are made of mild structural steel. The cross-sectional area of each member is 2 sq in. Determine the displacement of point $B$ in each case.

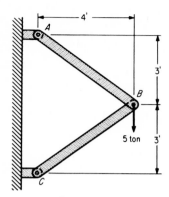

PROB. 14-24

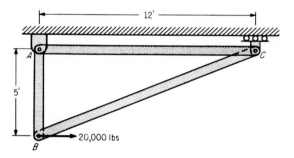

PROB. 14-25

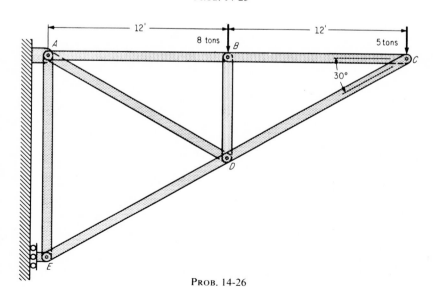

PROB. 14-26

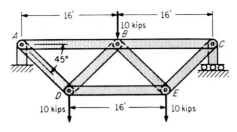

PROB. 14-27

## chapter 15

# Nonstatic Loads, Strain Concentrations, and Time Dependent Properties

## 15-1. Introduction

Thus far we have been concerned primarily with the response of engineering members to static loads. We have not considered in any detail the influence of time effects in loading, and we have not considered abrupt geometry changes in members which cause concentrations of strain and stress. Concentrations resulting from abrupt geometry changes and material imperfections become increasingly important under repeated loading conditions (fatigue). It is important and appropriate to discuss these two topics, concentrations and repeated loading, in the same chapter so that their relations with each other can be emphasized. These and several other topics, including impact and time dependent properties, are discussed in the following sections.

## 15-2. Stress Concentrations

The "flow" of force through a member will be disrupted by an abrupt change in the geometry of the member, and a concentration of the lines of force will result. For example, the lines of force in a simple plate loaded uniaxially are shown in Fig. 15-1(a), and Fig. 15-1(b) suggests how the lines of force become concentrated around a hole in the plate.

The lengths of the vectors in Fig. 15-1(c) and (d) indicate the magnitudes of the stresses at various points along plane A-A. In (c) the stress is uniform and undisturbed in the portion away from the end effects of the loads. In (d), where a hole is introduced, the strains and therefore the stresses are more concentrated at the hole. The peak stress on plane A-A is adjacent to the hole and has a much larger value than the average stress in a cross section without a hole. The peak stresses in structures at points of abrupt changes in geometry are extremely important in the design of members of any material subjected to repeated loads and in the design of members made of brittle material subjected to loading of any type.

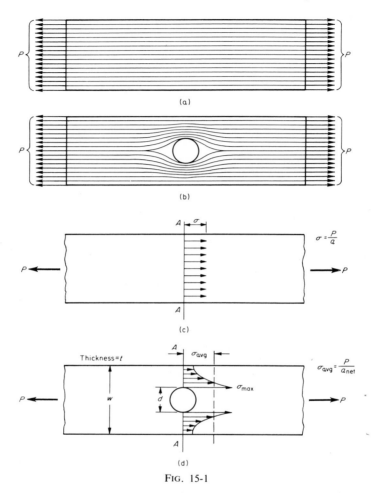

FIG. 15-1

It is often difficult to determine the maximum value of the stress at a concentration by theoretical analysis, but designers can usually determine the maximum stress by applying a factor to an average stress obtained from the elementary load-stress relationship. Such factors are called *stress concentration factors* and have been determined experimentally for many common geometry changes, such as holes, grooves, notches, and fillets in plates and shafts subjected to axial, bending, and torsion stresses.

The stress concentration factor, denoted here by $k$, is defined as the ratio of the maximum stress to the average stress. That is,

$$k = \frac{\sigma_{\max}}{\sigma_{\text{avg}}}$$

The average stress is usually determined by using the net area of the cross section. For example, in Fig. 15-1 (*d*),

$$\sigma_{max} = k \frac{P}{a_{net}}$$

where $a_{net}$ is $(w - d)t$, $t$ being the plate thickness. The stress concentration factor, except in very unusual instances, is greater than unity. For a uniaxially loaded plate with a hole, the factor approaches a theoretical value of 3 as the plate width becomes extremely large in comparison with the diameter of the hole.

Many curves and tables showing values of stress concentration factors for various geometries and loadings are given in technical references. One of the best sources of such information is *Stress Concentration Design Factors,* by R. E. Peterson (John Wiley & Sons, Inc., 1953).

Generally a stress concentration factor is determined experimentally by using models (such as photoelastic models discussed in the next chapter). As a result, the factor is based only on geometrical considerations (shape) and does not take into account the effect of the material. A factor that is based only on geometry is called a theoretical stress concentration factor and is denoted by $k_t$. Factors obtained by comparing the stress distributions determined from the mathematical theory of elasticity with the average values determined from elementary relationships also do not consider material effects and are considered theoretical stress concentration factors.

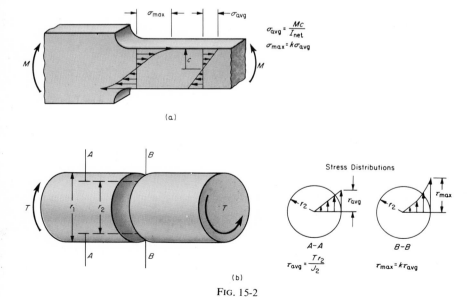

FIG. 15-2

If the factor is obtained by comparing the actual stress distribution (including the effects of the behavior of the specific material on the distribution) with the average stress determined from elementary theory, it is known as an effective stress concentration factor and is denoted by $k_e$. Effective factors can differ from theoretical factors because of such things as inelastic action, nonhomogeneity due to large grain size, and anisotropy.

In Fig. 15-2 (*a*) and (*b*) are shown the effects of a typical geometry change on the stress distributions in bending and torsion members, respectively. The curves in Fig. 15-3 and Fig. 15-4 give some values of the

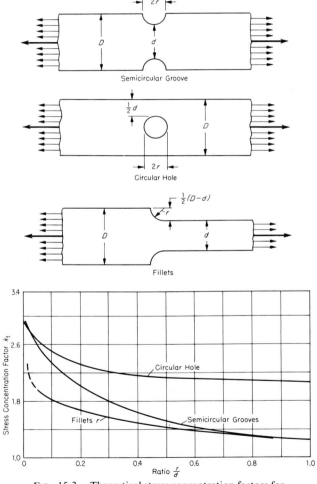

FIG. 15-3. Theoretical stress concentration factors for axial loading obtained by photoelasticity. (After Frocht)

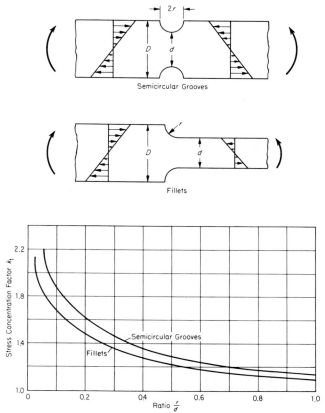

FIG. 15-4. Theoretical stress concentration factors for
bending obtained by photoelasticity. (After Frocht)

theoretical stress concentration factors for plates with axial and bending
loads and show their variation with various dimension parameters. The
values were determined by photoelastic methods.

For a member made of a very brittle material the "peaked" stress
distribution will be present at any load level, and a rupture will ordinarily
start at the point of maximum stress when that stress reaches the value of
the ultimate stress. For this reason, abrupt geometry changes which cause
stress concentrations must always be carefully considered, and avoided if
possible, in the design of members of brittle materials.

In a member made of a ductile material the "peaked" stress distribution, or stress concentration, will be reduced as yielding occurs. The yielding will start at the point of maximum stress. If the material has a large
inelastic range, as do several common structural materials (e.g., mild steel),
the material can yield in the region of the stress concentration without

danger of rupture. This is one of the greatest advantages of ductility in a material. In designing a member made of a ductile material for static loads, stress concentration factors are seldom considered. However, if repeated loads are expected, even a ductile material may fail as a result of the propagation of cracks which start at concentrations of stress. Repeated loads are discussed in more detail in the next section.

Figure 15-5 illustrates the reduction of the stress concentration factor and the redistribution of stress for a notched plate of a very ductile ma-

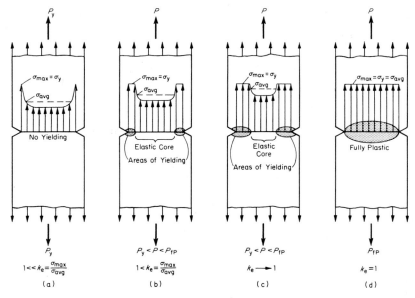

FIG. 15-5

terial. It is assumed that the material has a flat-topped stress-strain diagram. Notice the progression of yielding and how it is accompanied by a redistribution of stress to the extent that the stress finally becomes uniform over the area of the plate. The effect of the abrupt change in geometry is nullified and the effective stress concentration factor is reduced to unity.

EXAMPLE 15-1. The member in Fig. 15-6 is made from a brittle plastic whose stress-strain relationship is linear to rupture at 8000 psi. Using a factor of safety of 3, find the allowable load.

*Solution:* For a brittle material the stress concentrations must be considered. For the section at the hole,

$$\sigma_{max} = k_t \sigma_{avg} = k_t \frac{P_f}{a_{net}}$$

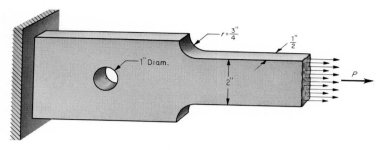

FIG. 15-6

The failure load is $P_f = NP_a$, where $P_a$ is the allowable load and $N$ is the factor of safety. Therefore

$$P_a = \frac{\sigma_{max} \, a_{net}}{N \, k_t}$$

Referring to Fig. 15-3, where $D = 3\frac{1}{2}$ in., $d = 2\frac{1}{2}$ in., $r = \frac{1}{2}$ in., and $r/d = 0.20$, we get $k_t = 2.3$. Therefore, the allowable load based on the stress at the hole is

$$P_a = \frac{(8000)\,(2\frac{1}{2})\,(\frac{1}{2})}{(3)\,(2.3)} = 1449 \text{ lb}$$

For the section at the fillet, $r/d = 0.375$. From Fig. 15-3, $k_t = 1.5$. Hence, the allowable load based on the stress here is

$$P_a = \frac{(8000)\,(2)\,(\frac{1}{2})}{(3)\,(1.5)} = 1778 \text{ lb}$$

The allowable load for the member is the smaller of the two, or 1449 lb.

## 15-3. Fatigue

**Description of Failure.** The word *fatigue* is commonly used in engineering to describe repeated-load phenomena. The word is a misnomer in the lay sense, in that the materials do not get "tired." A majority of fractures that occur in engineering members are due to repeated loads or fatigue, and it is therefore important that engineers have some understanding of the phenomena and how to deal with fatigue. When we speak of repeated loads we usually are referring to moving parts or members which are found, for example, in engines, turbines, pumps, motors, and other machinery with moving parts. Failure by fatigue is a progressive cracking and, unless detected, this cracking leads to a rupture which is often catastrophic.

The exact mechanism of the initiation of a fatigue failure is complex and is not completely understood. Detailed discussion of the theories of its initiation and initial propagation will not be given here. It is more important at this stage to be able to determine where, when, and under what conditions such a fatigue crack is likely to start. If a repeated load is large enough to cause a fatigue crack, the crack will start at a point of maximum stress. This maximum stress is usually due to a stress concentration (often called a stress raiser). Stress concentrations can occur in the interior of a member as a result of the inclusion of foreign matter or voids in the material. They can occur on the exterior surface of the member because of scratches, rust pits, machining marks, or the more obvious sharp corners or other abrupt geometry changes in the designs.

If a stress raiser occurs in a region of high overall stress (e.g., at a section of maximum bending moment) and if the effect of the stress raiser is superimposed on this already high stress, then the chances of a fatigue crack developing under repeated loading are greatly increased. Since a fatigue crack or failure is undesirable, it is important for the engineer to take great care to eliminate or reduce the possibility of conditions which might lead to the initiation of a fatigue crack. High quality control may reduce chances of interior imperfections. Polishing the surface in critical areas may be necessary. Careful designs can also reduce stress concentrations. The idea is not to give a fatigue crack a place to start!

In the previous sections we mentioned that stress concentrations are sometimes ignored in ductile materials subjected to *static* loads. Under repeated loading ductile materials, as well as brittle materials, are susceptible to fatigue failures. In a fatigue failure of a ductile material there is no large amount of plastic deformation, or "necking," as in a static-load rupture of the same material. The fatigue failure of a ductile material appears somewhat as would the failure in a brittle material.

After a fatigue crack is initiated at some microscopic or macroscopic stress raiser, the crack itself acts as an additional stress raiser. The crack grows with each repetition of the load until the effective cross section is reduced to such an extent that the remaining portion will fail with the next application of the load. It may take many thousands, or even millions, of stress repetitions for a fatigue crack to grow to such an extent that it causes rupture. The required number of repetitions depends on the magnitude of the load (and stress) and other related factors.

The appearance of most of the fatigue-ruptured surface of a metal is rather rough and crystalline, and it is often erroneously thought by laymen that the metal has crystallized in the process of breaking. Since the solid metal is crystalline, and has been since it was transformed from a liquid, the appearance is perfectly understandable. What is interesting about the appearance is the fact that you can usually detect the point of initiation of the crack. The progression of the crack from this point causes the appearance of rings about the point somewhat like growth rings in timber. This portion of the fracture is rather smooth because of the rubbing action in the development of the crack.

**The *S-N* Curve.**   Now that we have some idea where and how a fatigue failure might occur, we must try to determine *when* it is likely to occur.   Since susceptibility of a member to fatigue failure is greatly influenced by the material from which it is made, we must depend on experimental tests of the material in order to design against failure by fatigue. The most common test used to evaluate fatigue properties of a material is a rotating-beam test, in which the number of completely reversed cycles of bending stress required to cause failure (this is called the fatigue life) is measured at various stress levels.   By a completely reversed cycle of stress we mean that in one complete cycle the stress goes from some maximum tensile stress, to zero, to a maximum compressive stress of the same magnitude as the maximum tensile stress, and then back to zero and on to the original tensile stress again.   These test data are usually plotted on semilog paper, and the resulting plot is referred to as an *S-N* curve.   Some typical curves are shown in Fig. 15-7.

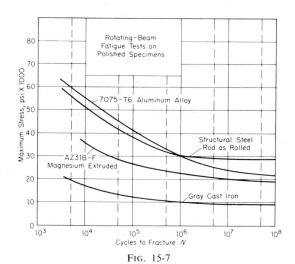

Fig. 15-7

An individual test in which the specimen is fractured is required for each experimental point on an *S-N* curve.   The inherent nature of fatigue tests gives rise to a great deal of scatter in the data.   That is, if several specimens that have been carefully machined and polished (as all fatigue-test specimens should be) are tested at the same stress level, it is not unusual to have a variation of 10 to 20 percent in their fatigue life measured in terms of the number of loading cycles at which the specimen ruptures.   It therefore requires quite a few tests to correctly identify an *S-N* curve for a material.

The property called *fatigue strength* refers to the stress level that can be applied a specific number of times before failure. For any certain life or number of stress repetitions, there is a corresponding fatigue strength. This strength can be taken from the *S-N* curve.

**Endurance Limit.** As specimens are tested at lower stress levels, more cycles or repetitions of this stress are required to cause failure. Notice that each *S-N* curve of Fig. 15-7 has an approximately horizontal line as a lower limit. When the stress level for a specimen is at or below this limit, which is called the *endurance limit* or *fatigue limit,* the specimen does not fail. The machine is stopped and the unbroken specimen is removed usually after about $10^8$ (one hundred million) cycles. Such a test is called a "run out," and the corresponding point is usually plotted on the diagram with an arrow extending from it to the right. The endurance limit for a material is defined as the stress below which an indefinite number of stress repetitions may be applied without a fatigue failure.

It is important for the designer to know what type of fatigue test was used to determine the fatigue properties of a material. Although data from rotating-beam fatigue tests are perhaps the most common, other types of fatigue testing are in common use. Some of them involve bending of specimens of thin plates, uniaxial repeated loading, and torsion loading.

The endurance limit for most engineering materials is less than the yield strength, elastic limit, or proportional limit. It is important for you to realize that a member can fracture when loaded at a stress level in the elastic range if the stress is high enough and is repeated enough times. It has been proved that a member is damaged to some degree after very few repetitions of stress. This damage is permanent and remains after the member is unloaded and left stress-free. If at any future time the member is again stressed repeatedly, the damage progresses from where it left off. Hence, we see that stress history is important in materials. Generally, tensile stresses are more damaging in fatigue than are compressive stresses.

Effective stress concentration factors for use in the design of members subjected to repeated loads are commonly determined from fatigue tests. To determine the factor in such a case the endurance limit for a material with the stress concentration in question (for example, a grooved or notched specimen may be used) is divided into the endurance limit for the same material obtained from specimens free of the stress concentration.

Fatigue properties for materials are often determined at elevated temperatures and also in various corrosive environments. Temperature and/or environment can drastically influence the fatigue properties. Fatigue machines can be equipped for such special tests as these. Also, many fatigue machines are designed to subject the specimen to a varying deflection of some pre-set amplitude, whereas others impose a varying load with some pre-set maximum.

EXAMPLE 15-2. A part of the landing gear of a military airplane is made from 7075-T6 aluminum alloy. The landing gear is to be designed for a life of 10,000 landings. A factor of safety of 1.5 is to be used. The part is cylindrical and is subjected to a reversed bending moment of 800 ft-lb as the plane lands. Find the required diameter.

*Solution:* From the *S-N* curve of Fig. 15-7 the fatigue strength for a life of 10,000 cycles is 54,000 psi. For bending we assume that the flexure formula is applicable. The 800 ft-lb, or $(800)(12)$ in-lb, is the working moment, and the failure moment is $(800)(12)(1.5)$ in-lb. Hence,

$$\sigma = \frac{M_f c}{I}$$

where $c = d/2$ and $I = \pi d^4/64$,

$$54,000 = \frac{(800)(12)(1.5)(d/2)}{\pi d^4/64}$$

$$d = 1.4 \text{ in.}$$

EXAMPLE 15-3. An engine part made from gray cast iron must be designed for an indefinite fatigue life. A factor of safety of 4 is needed because of the quality of the material, the consequences of a failure, and the fact that weight is not important. If the cross section is $\frac{1}{2}$ in. square, how much bending moment may be applied to the part?

*Solution:* Since the life is indefinite, the endurance-limit stress will be used. As obtained from Fig. 15-7, it is 9000 psi. Assuming that the flexure formula is applicable, we have

$$M_f = \frac{\sigma I}{c} \quad \text{and} \quad M_w = \frac{\sigma I}{Nc}$$

$$M_w = \frac{(9000)(\frac{1}{12})(\frac{1}{2})^4}{(4)(\frac{1}{4})} = 47 \text{ in-lb}$$

**Nonreversed Cycles of Loading.** Up to this point we have discussed only completely reversed cycles of stress in relation to fatigue failures. In many cases fluctuating loads do not approximate complete reversals. They do, however, in many instances approximate complete reversal superimposed on a uniform mean stress. This is often true where a load varies from some high limit to some low limit and both limits are tensile or both are compressive. To determine the allowable limits for such a range of stress, it is desirable to have experimental data from fatigue tests which cover the range of interest. In the absence of such test data, there are several approximate and conservative criteria which have been proposed for design under conditions of fluctuating stress.

Two approximate criteria are shown in Fig. 15-8(*a*), where the initial mean stress is plotted against the alternating stress. If the criterion of failure is yielding (slip), the so-called Soderberg straight line is assumed. Its limits are the endurance limit for zero mean stress and the yield point

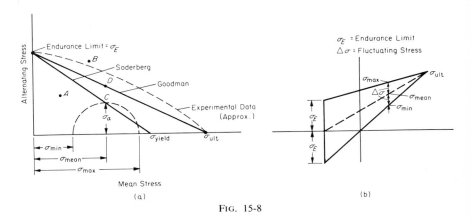

Fig. 15-8

or yield strength for the case of no alternating load. If the criterion for failure is fracture, the so-called Goodman diagram is commonly used. It allows a higher value of alternating stress for a given value of mean stress, or vice versa.

For any combination of alternating stress and mean stress which plots as a point below the line, such as point *A* in Fig. 15-8(*a*), there would be no failure. For a combination of stresses which plots outside the line, as at point *B*, it is supposed that failure will occur.

For a mean stress $\sigma_{mean}$ when the criterion of failure is yielding, the maximum alternating stress that can be superposed on the mean stress is $\sigma_a$, up to point *C*. If the fracture criterion is assumed, an alternating stress up to the value corresponding to point *D* can be superposed on the mean stress. Actual test data have verified the fact that the Soderberg and Goodman diagrams are on the safe or conservative side. For the fracture criterion, for example, points plotted from the experimental test data fall outside the straight Goodman line about as indicated by the dashed line. Thus, if adequate test data were available, we would not need the Soderberg or the Goodman diagram.

Another common way of presenting the Goodman criterion is shown in Fig. 15-8(*b*). The fluctuating stress $\Delta\sigma$ that can be superposed on a steady or mean stress is the distance between either sloping solid line and the sloping dashed line indicating the mean stress. If the mean stress is equal to the ultimate stress (right-hand end of diagram), no fluctuating stress may be added. If the mean stress is zero (left-hand end of diagram), the fluctuating stress may be equal to the endurance limit $\sigma_E$.

EXAMPLE 15-4. What amount of fluctuating stress would cause rupture if superposed on a maximum steady bending tensile stress of 8000 psi in a machine part of AZ31B-F magnesium whose ultimate static tensile strength is 35,000 psi and whose *S-N* curve is given in Fig. 15-7?

*Solution:* Since the mode of failure is rupture, we will use the Goodman diagram. The endurance limit from Fig. 15-7 is 18,000 psi, and the ultimate stress is given as 35,000 psi. The Goodman diagram is plotted in Fig. 15-9. The equation of the inclined line is

$$\sigma_a = 18,000 - \frac{18,000}{35,000}\sigma_m$$

For the mean stress of 8000 psi, the superimposed alternating stress which would cause rupture is

$$\sigma_a = 18,000 - \frac{18,000}{35,000}\, 8000$$

$$= 13,880 \text{ psi}$$

The maximum and minimum values of stress which would lead to rupture, based on a steady tensile stress of 8000 psi, are

$$\sigma_{max} = 8000 + 13,880 = 21,880 \text{ psi, tension}$$

$$\sigma_{min} = 8000 - 13,880 = \phantom{2}5880 \text{ psi, compression}$$

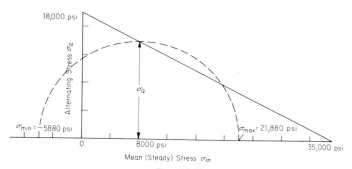

FIG. 15-9

EXAMPLE 15-5. A machine part of AZ31B-F magnesium is subjected to a varying load that causes the maximum bending stress to vary from 8000 psi, tension, to 20,000 psi, tension. The ultimate tensile strength of the material is 35,000 psi. What is the ratio of the fluctuating stress that will cause rupture (failure stress) to the actual fluctuating stress on the member?

*Solution:* Since the mode of failure is rupture, we will use the Goodman diagram. The endurance limit from Fig. 15-7 is 18,000 psi, and the ultimate stress is given as 35,000 psi. From the Goodman diagram in Fig. 15-10, it is seen by similar triangles that the failure stress $\sigma_f$ is

$$\frac{\sigma_f}{35,000 - \sigma_m} = \frac{18,000}{35,000}$$

where

$$\sigma_m = \frac{20,000 + 8000}{2} = 14,000 \text{ psi}$$

$$\sigma_f = 10,800 \text{ psi}$$

The actual fluctuating stress is

$$\frac{20,000 - 8000}{2} = 6000 \text{ psi}$$

Hence, the ratio of failure stress to actual stress is $10,800/6000 = 1.77$. This ratio might be interpreted as a margin of safety.

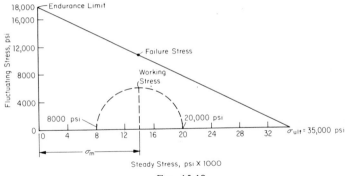

Fig. 15-10

## 15-4. Creep and Relaxation

**Creep.** As defined in Chapter 4, creep is the behavior wherein the strain in a member undergoes changes while the loading remains essentially unchanged. Most engineering materials, particularly metals, do not creep appreciably at ordinary temperatures. Creep is a function of stress and time, as well as temperature and a material is more likely to creep at a higher stress level or after a long period of time even if the temperature is not elevated. However, a great majority of creep problems in engineering materials arise from conditions of high temperatures. For example, the chemical engineer must often design containers or vessels for a process that takes place at very high pressures and temperatures. Mechanical engineers must consider creep problems in tubes of high-pressure boilers, in steam-turbine blades, in jet-engine blades, and in other applications where high stresses and high temperatures must be sustained for long periods of time.

The engineer must design the part so that the stress is low enough to limit creep to an allowable value. This maximum permissible stress is often called the *creep strength* or *creep limit*. The creep strength is defined as the highest stress that a material can stand for a specified time without excessive deformation. The *creep-rupture strength*, sometimes called the *rupture strength*, is defined as the highest stress a material can stand for a specified time without rupture. Creep properties are determined primarily from experimental tests on the materials of interest. For the most common type of creep test a tensile specimen is loaded uniaxially with a dead load (constant stress) at a constant temperature while the data for strain vs. time are recorded. The general shape of the creep curve for many materials is illustrated in Fig. 15-11.

Creep behavior for most materials exhibits the three stages of creep shown in the curve of Fig. 15-11. After the "instantaneous" strain at the time of application of the load, the *rate* of strain in most cases (depending on test conditions) gradually decreases to a more or less constant value. In this second stage the material is almost entirely viscous. The constant slope, in this stage, defined as the creep rate, is the minimum strain rate

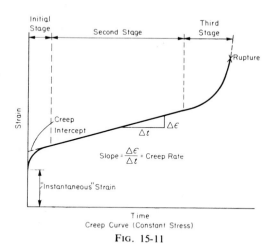

Creep Curve (Constant Stress)

FIG. 15-11

encountered in the test. In the third stage, called the *tertiary creep* range, the strain rate increases abruptly primarily because of the necking of the specimen, and rupture follows. In Fig. 15-12(*a*) is shown a family of actual creep curves for a constant temperature and three stress levels. In Fig. 15-12(*b*) is shown another common way of presenting creep data. Here the

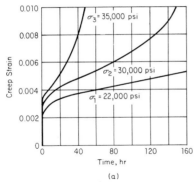
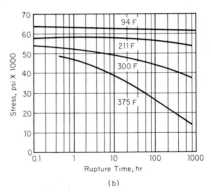

FIG. 15-12. Creep curves for Alclad 2024-T3 aluminum alloy sheet (*a*) at 400 F and (*b*) at a thickness of 0.040 in. (after Flanagan, Tedsen, and Dorn).

time to rupture is plotted against stress for various temperatures. Notice that only one creep test is required to obtain each curve in Fig. 15-12(*a*), whereas a separate creep test is required for each experimental point of each curve in Fig. 15-12(*b*).

EXAMPLE 15-6. A cylindrical space capsule 48 in. in diameter must resist a pressure differential of 8 psi for 36 hr at 400 F. To prevent leakage at the joints, a strain of 0.006 in./in. must not be exceeded. How thick must the wall be if it is made from 2024-T3 Alclad aluminum alloy? Assume a stress concentration factor of 2.0 at the joint and a factor of safety of 1.5. Assume that the curves of Fig. 15-12(*a*) apply for the biaxial stress condition of this wall.

*Solution:* By interpolating in Fig. 15-12(*a*) for 36 hr and a strain of 0.006 in./in., a constant stress of approximately 32,000 psi is obtained. This is the failure, or maximum, stress. The average permissible stress is the maximum peak stress of 32,000 psi divided by the stress concentration factor of 2. Thus,

$$\sigma_{avg} = \frac{36,000}{2} = 16,000 \text{ psi}$$

The design pressure is 8 psi, and the failure pressure is therefore $8N = (8)(1.5) = 12$ psi. For a thin-walled pressure vessel,

$$\sigma = \frac{pD}{2t}$$

$$t = \frac{pD}{2\sigma} = \frac{(12)(48)}{(2)(16,000)} = 0.018 \text{ in.}$$

**Relaxation.** Relaxation is defined as the relief of stress by *internal* creep. In other words, it is the time-dependent decrease in stress in a member which is held in a relatively fixed external configuration. A material which is susceptible to creep is also susceptible to relaxation. The same mechanisms within the material are involved in both behaviors. In creep, external or over-all deformation occurs while the member is subjected to a constant load; whereas in relaxation, internal creep causes a decrease and redistribution of the stresses *without* significant deformation of the boundaries.

Bolts and other connectors used at joints of such elements as pipes, valves, pressure vessels, turbine casings, and cylinder heads are often subjected to high temperatures and initial stresses. In these instances the boundary deformation remains essentially constant but, as internal creep occurs, the bolt relaxes. To analyze this relaxation behavior for constant strain, a plot of the reduction of stress with time can be made as in Fig. 15-13.

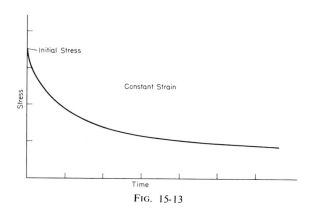

FIG. 15-13

Data for plots of stress vs. time to determine relaxation properties are usually obtained by holding tensile specimens at a constant temperature while the load (stress) is reduced with time so that the strain remains essentially constant. As with creep, relaxation is more prevalent and occurs at a faster rate at elevated temperatures.

## 15-5. Impact Loads

Creep and relaxation concern time-dependent behavior from the standpoint of what happens over relatively long periods of time. Now let us consider briefly material and structural behavior when loads are applied suddenly or are varied rapidly for very short periods of time.

Loads which are applied for extremely short periods of time are called impact loads, or shock loads, and a detailed analysis of their effects is very complex. Strain waves are set up in the material, and the magnitudes and distributions of the resulting stresses and strains depend on the velocity of propagation of these strain waves, as well as on the usual material properties, dimensions, and loading. If there is only elastic action, the member will vibrate until it comes back to equilibrium. If the time of application of the load is short compared to the natural period of vibration of the member, the load is usually said to be high-velocity impact.

Since theoretical analyses are difficult for impact loading, empirical test results are commonly used to get an index of a material's ability to resist high-velocity impact loads. A pendulum-type impact test machine is commonly used for this purpose. The pendulum, which has a known kinetic energy at impact, strikes a 10mm × 10mm × 50mm standard notched specimen supported as a simple beam (Charpy test) or as a cantilever beam (Izod test), and the energy required to rupture the specimen is recorded by measuring the loss of energy of the pendulum. Such tests are good for comparing materials, but this test does not give quantitative data that can be used directly in design. Also, the results do not correlate well with those of tests for other mechanical properties.

One interesting and important phenomenon illustrated by impact tests is the sudden ductile-brittle transition which is exhibited by some materials as the temperature is lowered. Recall from Chapter 4 that the amount of energy absorbed to rupture can be used as a measure of the ductility (or brittleness) of a material. As illustrated in Fig. 15-14, some materials (particularly steels) lose their ductility (as measured by the impact test)

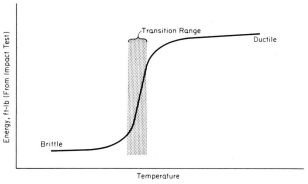

FIG. 15-14

and become brittle very abruptly at a low temperature. The center of this transition range can vary for common engineering steels from about $+50\,F$ to $-100\,F$, the exact value for a particular material depending on the specific metallurgical structure.

When the velocity of impact of loads is not too large, that is, when the time of application of the load is greater than several times the lowest natural frequency of the loaded member, and when the load causes only elastic behavior, the engineer can usually make simplifying assumptions which will enable him to develop a reasonably simple rational analysis of the problem. Under these conditions, the time of loading is low enough to enable the resisting body to behave in the same way as it would under a static load. The inertia of the resisting member or system is usually neglected, and the deformation of the member is directly proportional to the magnitude of the applied force as is the case in elastic static loading. *The energy imparted to the member by the low-velocity impact load is equal to the work done by some equivalent static load causing the same deformation.* Thus,

Energy imparted = work of equivalent static load

This approach is often called the equivalent static load method.

The energy imparted may be, for example, the potential energy of a weight $W$ to be dropped from a height $h$ onto a member; or the imparted energy may be kinetic energy of a translating mass imparted to the resisting member. In the first case, the energy imparted would be $W(h + \Delta)$, where $\Delta$ is the vertical deformation of the member. In the second case, the imparted energy would be $Wv^2/2g$, where $W$ and $v$ are the weight and velocity of the translating mass, respectively.

Recall that for a quasi-static deformation process the work done by a static load is

$$\text{Work} = \int_0^e P\,de \quad \text{or} \quad \int_0^\theta T\,d\theta$$

The first integral represents the work done by a force $P$ acting through a collinear deformation $e$, and the second integral represents the work done by a moment or torque $T$ acting through an angular deformation $\theta$ (measured in radians). For linear elastic behavior, the deformations are directly proportional to the loads causing them. Hence

$$\text{Work} = \frac{P_{max}}{2}e \quad \text{or} \quad \frac{T_{max}}{2}\theta$$

Now, by equating the imparted energy to the work done by an equivalent static load, the equivalent static load can be determined. This equivalent load can then be used to investigate the stresses and deformations in the member.

EXAMPLE 15-7. The cantilever beam shown in Fig. 15-15 is made of mild steel. Determine the maximum stress which occurs in the beam when the 5-lb weight is dropped on it.

*Solution:* The 5-lb weight will move a distance of 2 in. plus the beam deflection $\Delta$, and the energy imparted to the resisting beam will be equal to $W(2 + \Delta)$. The static deflection $\Delta$ of a cantilever beam with a concentrated load at the free end is $Pl^3/(3EI)$. In this case $E = 30 \times 10^6$ psi and $I = (4)(0.5)^3/12 = 1/24$, and $l = 24$ in. Therefore,

$$\text{Energy imparted} = \text{work of equivalent load } P$$

$$5(2 + \Delta) = \frac{P}{2}\,\Delta$$

$$5\left(2 + \frac{Pl^3}{3EI}\right) = \left(\frac{P}{2}\right)\left(\frac{Pl^3}{3EI}\right)$$

$$(10) + \frac{(5)(P)(24^3)}{(3)(30)(10^6)\,(\frac{1}{24})} = \frac{P^2(24^3)}{(6)(30)(10^6)\,(\frac{1}{24})}$$

$$P = 5 \pm \sqrt{25 + 5420}$$

$$P = +79 \text{ lb or} -69 \text{ lb}$$

The positive root is the value of the force required to cause the maximum downward deflection $\Delta$ (the negative root would be the force required to cause the maximum upward deflection which would result if the beam vibrated back upward).

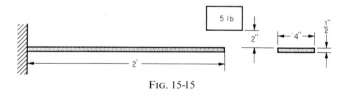

FIG. 15-15

The maximum stress would occur in the outer fibers where the maximum moment occurs. Neglecting stress concentrations and using the elastic flexure formula, we find that the maximum stress $\sigma$ is

$$\sigma = \frac{Mc}{I} = \frac{(79)(24)(\frac{1}{4})}{(\frac{1}{24})} = 11{,}370 \text{ psi}$$

## PROBLEMS

**15-1.** Find the maximum stress which occurs in an axially loaded plate whose cross section is $3\frac{1}{2}$ in. by $\frac{1}{4}$ in. with a hole $\frac{3}{4}$ in. in diameter in the center, if the plate is made of a linearly elastic material and the axial load is 8 kips.

**15-2.** How much load can be applied to the end of the cantilever beam with a circular cross section shown in the illustration, if a flexural stress of 18,000 psi must not be exceeded? Assume elastic action.

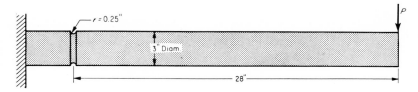

PROB. 15-2

**15-3.** A long plate whose cross section is 1 in. by $\frac{1}{8}$ in. is loaded with an axial static load of 1500 lb. A hole $\frac{1}{4}$ in. in diameter is drilled in the center of the plate. *a*) Find the maximum stress in the plate if the material is mild steel with a yield-point stress of 35,000 psi and an ultimate stress of 65,000 psi. *b*) Find the maximum stress if the plate is made from a brittle material with a linear stress-strain curve to its ultimate stress of 40,000 psi.

**15-4.** The 5 in. × 1 in. bar shown in the illustration is made from a brittle material with an ultimate stress of 12,000 psi. How narrow can it be cut down (find *d*) and still carry the 10,000-lb load? Assume a factor of safety of 2.4 based on failure by rupture. (HINT: Use trial-and-error solution.)

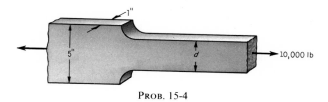

PROB. 15-4

**15-5.** Where does the maximum stress occur, and what is its magnitude, for the beam shown in the illustration? Assume elastic action.

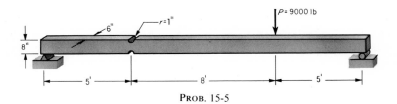

PROB. 15-5

**15-6.** It is estimated that the service life of a machine part made from extruded magnesium alloy AZ31B-F is 500,000 cycles. Determine a suitable working stress, using a factor of safety of 2.5.

**15-7.** What is the magnitude of a completely reversed load that can be applied to the free end of a cantilever beam of structural steel which is 36 in. long and has a cross section 2 in. by 4 in.? It bends about the axis of least moment of inertia, and the load is to be applied an indefinite number of times. A factor of safety of 2 is required.

**15-8.** The beam shown in the illustration is made of 7075-T6 aluminum alloy. How many times can the load be completely reversed before you can expect a failure? Neglect stress concentrations at the connections.

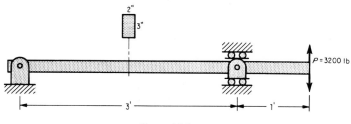

PROB. 15-8

**15-9.** At a certain engine speed a thin plate of 7075-T6 aluminum alloy in an airplane vibrates to the extent that a completely reversed stress of 30,000 psi is caused. After how many hours of flying time at this engine speed would you expect a failure, if the frequency of vibration is 40 cycles per second? Needless to say, the pilot will avoid this critical engine speed!

**15-10.** A circular shaft has a hollow cross section, and the outside diameter is twice the inside diameter. It is subjected to a completely reversed bending moment of 6000 in-lb. If the material is gray cast iron and the shaft must have an indefinite fatigue life, determine the outside diameter. Use a factor of safety of 3.

**15-11.** An automobile torsion bar is subjected to completely reversed cycles of torsional load. The torsional fatigue strength at the design life is 18,000 psi, and the diameter of the shaft is 1 in. Determine the torsional load, using a factor of safety of 2. Assume elastic behavior.

**15-12.** The cantilever beam shown in the illustration is made of 7075-T6 aluminum alloy and is to have a life of $10^6$ cycles of completely reversed bending under a load $P$ of 8000 lb. What radius $r$ of fillet should be used if a factor of safety of 1.5 is required? Assume that the theoretical stress concentration factors of Fig. 15-4 are applicable here.

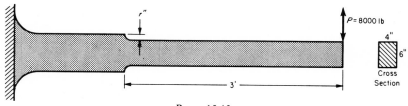

PROB. 15-12

**15-13.** What maximum value can $P$ have if the material is 7075-T6 aluminum alloy. The load is to be repeated only 5000 times during the life of the part. A factor of safety of 1.2 is to be used.

**15-14.** A machine part of AZ31B-F magnesium is subjected to a varying load that causes the maximum bending stress to vary from 6000 psi, tension, to

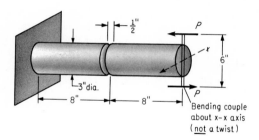

18,000 psi, tension. The yield strength of the material is 24,000 psi. What is the margin of safety based on failure by yielding?

**15-15.** A simple beam 12 ft long has a downward concentrated load at the center that varies from 500 to 4000 lb. The ultimate strength and the endurance limit of the material are 55,000 psi and 28,000 psi, respectively. If the depth of the beam is twice the width, what are the minimum dimensions needed to prevent rupture?

**15-16.** Determine the factor of safety with respect to failure by yielding for a structural steel machine part for which the stress varies repeatedly from 15,000 psi tension to 5,000 psi compression.

**15-17.** A circular cantilever beam $1\frac{1}{2}$ in. in diameter is subjected to a transverse load (on its free end) which varies from 800 lb upward to 300 lb downward. The material is 7075-T6 aluminum alloy (see Fig. 15-7 and the appendix for properties). Assuming a yielding mode of failure, determine the permissible length of the beam.

**15-18.** From Fig. 15-12(a) determine the minimum creep rate in inches per inch per hour for the material at the 30,000-psi stress level.

**15-19.** Determine the rupture time for a uniaxial tensile specimen of Alclad 2024-T3 aluminum alloy 2 in. wide and 0.025 in. thick, if it is subjected to a constant load of 1600 lb and a constant temperature of 375 F.

**15-20.** The service life of an oven part is 400 hr at 300 F. If it is made of Alclad 2024-T3 aluminum alloy, sustains a constant bending moment of 300 in-lb, and is 0.5 in. wide, what must be the height of its cross section? Assume a factor of safety of 3.

**15-21.** A cantilever beam of Alclad 2024-T3 aluminum alloy is 0.063 in. deep, 0.750 in. wide, and 12 in. long. It supports a dead load of 2 lb on the free end. If the temperature of the beam is raised from 70 F to 300 F, will the beam fail? If so, how long will it take for failure to occur after the temperature is raised? Neglect stress concentrations.

**15-22.** A load of 30 lb is dropped from a height of 2 in. onto a helical spring. The modulus of the spring is 120 lb per in. Find the maximum deflection of the spring.

**15-23.** A simply supported beam 12 ft long is 2 in. wide and 1 in. deep and is made of an aluminum alloy which has a modulus of 10,000,000 psi. A weight of 50 lb is dropped a distance of 4 in. onto the center of the beam. Will any yielding occur in the beam, if the proportional-limit stress is 30,000 psi?

**15-24.** A 160-lb student jumps onto the diving board shown in the illustration from a height of 3 ft. The board is hickory with a modulus of 1.8 million psi. How much will the board deflect, and what maximum stress will it be subjected to? How much would the board deflect if the student walked out slowly and stood still on the end?

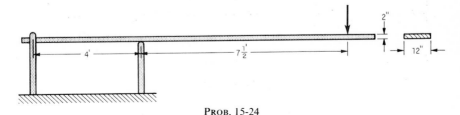

PROB. 15-24

**15-25.** When the block shown in the illustration strikes the round steel bar, which is 1 in. in diameter, assume that all of the energy of the block is transferred to the bar. Do you think the bar will buckle? If so, why? If not, why not?

**15-26.** When the 30-lb weight shown in the illustration is dropped and hits the stop, all of its energy is transferred into the steel bar. What is the maximum height through which the weight can be dropped without the stress exceeding 20,000 psi in the bar?

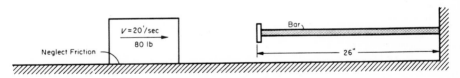

PROB. 15-25

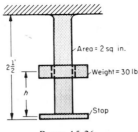

PROB. 15-26

**15-27.** A hollow aluminum-alloy shaft 36 in. long is made from a tube with a wall thickness of $\frac{1}{8}$ in. and an average diameter of 2 in. How much shear stress will result in the shaft from a pure torsional energy load of $20/\pi$ in-lb?

**15-28.** For the conditions shown in the illustration, from what height $h$ must the weight be dropped to cause the beam to deflect three times what it would if the weight were applied very slowly? The spring is force-free before the weight hits.

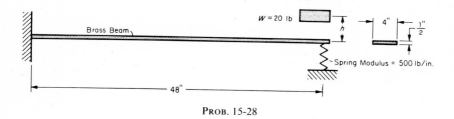

PROB. 15-28

## chapter 16

# Experimental Mechanics

## 16-1. Introduction

Suppose it is imperative for an aero-space engineer to know the magnitude and direction of the maximum stress which occurs at a critical point in a rocket structure for less than a second during the acceleration of the rocket when it is several miles from the earth. Perhaps it is necessary to know the stress in a blade of a jet-engine turbine while it is turning at 10,000 rpm at a temperature of 1800 F and an altitude of 60,000 ft. Or it may be necessary to know the forces acting on a drill bit rotating in an oil well several miles below the earth's surface, or to know the complete state of stress in the wall of a nuclear reactor in a highly radioactive environment. Can the answers to such difficult problems be obtained with experimental techniques available today? Generally, the answer is yes. The answer 20 years ago was no. Great advances have been made in the field of *experimental* mechanics in recent years. This field of study is commonly called experimental stress analysis, although experimental strain analysis is perhaps a better name. We compromise with "experimental mechanics."

This chapter is a brief introduction to experimental mechanics. It outlines some of the more important principles, techniques, equipment, and limitations of the state of the art today. Because of the tremendous advancements now being made in this field, such a presentation may be somewhat obsolete before this text is available to the student, particularly in regard to the equipment used. However, the underlying principles remain basically the same, regardless of new refinements in techniques and equipment.

The methods of experimental mechanics to be discussed in the following sections are based on the use of electrical resistance strain gages, photoelasticity (including birefringent coatings), and brittle lacquer coatings.

Experimental analyses of stresses and strains, and/or their distribution, by actual measurements are important in *supplementing* theoretical analyses, as well as substituting for them. In many cases the geometry or loading of a member or structure may be so complex that a theoretical solution is too cumbersome or uneconomical, or impossible. In such a case, the engineer may have no choice but to resort to experimental methods. However, experimental analyses are often used to establish sufficient boundary conditions (usually surface conditions) to enable the completion of a theoretical analysis. Experimental measurements are also used many times to verify theoretical predictions. Also it is possible to use an experimental approach to determine the stresses and strains, or their distributions, when the loading condition or load distribution is unknown.

Experimental stress or strain analysis is based on *measurements*. Measurement of what? We learned in Chapters 1 and 2 that stress and strain are mathematical quantities. *We cannot measure stress or strain directly.* We must measure some quantity which is related to stress or strain. We can measure force, length, and time.

Perhaps the most common measurement in experimental mechanics is the *surface* deformation in a small gage length. Generally, any device used to measure surface deformations is called a strain gage. A suitable mechanical device, such as a tensile extensometer on a test specimen, can be used to translate this deformation into some other form of language, such as a dial reading, an electrical signal by the proportional movement of a miniature transformer core, or a light-spot movement greatly magnified by a proportional mirror rotation. Much sophistication and precision can be built into these essentially mechanical devices, but they are inherently rather clumsy compared to electrical resistance strain gages.

## 16-2. Electrical Resistance Strain Gages

Strictly speaking, any strain gage—whether it be mechanical, electrical, optical, or of some other type—does not measure strain. Strain gages are deformation-sensitive; that is, they are able to sense and respond to a deformation, such as a *change* of a finite length. Therefore, the "measured" quantity obtained from a strain gage is *proportional* to an *average* strain in the gage length of the gage. In a region of a high strain gradient, an average reading of a strain gage may be appreciably less than the maximum occurring at some specific point in the gage length. This difficulty may be overcome to a great degree by using a strain gage with a very small gage length, but there is a physical limitation on the size of the smallest gage that can be made. At present, electrical resistance gages can be made with a gage length as small as 0.015 in., although the most common convenient gage length for such a gage is about ½ in.

The most common electrical resistance strain gages used universally are *bonded* gages. That is, the gage is intimately bonded to the surface on which the strain is desired, and it is therefore deformed along with the surface. Essentially, a wire gage consists of a length of very fine wire (about 1 mil in diameter) which is looped into a pattern, as in Fig. 16-1(*a*) or (*b*). A foil gage is made by etching a pattern on very thin metal foil (about 0.0001 in. thick), as in Fig. 16-1(*b*) or (*c*). The wire or foil is usually bonded to a thin base of paper or plastic. When in use, the bonded gage is cemented firmly to the member under investigation, with the wire or foil side out.

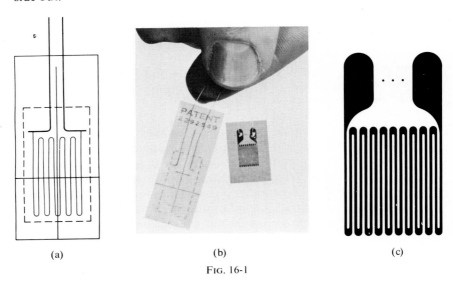

(a)            (b)            (c)

Fig. 16-1

An electric current is passed through the wire or foil. The resistance of the element (wire or foil) changes as the surface under it (and therefore the gage) is strained. The basic principle involved is simply Lord Kelvin's discovery that a wire changes its electrical resistance when deformed. Since the wire or foil is bonded throughout its length, the gage is able to sense a compressive strain as well as a tensile strain. The resistance change, which is accurately proportional to the strain, is measured by appropriate instruments.

Resistance gages are made in a variety of shapes, sizes, and types. Figure 16-2 is a photograph of various gages.

A resistance gage can be used on the surface of almost any solid material important in engineering, such as metal, plastic, concrete, wood, glass, and paper. In some special cases the gage can be embedded in the interior of a cast material such as concrete. The weight it adds to the member is usually negligible, and it can be used for static or dynamic strains. Also, it does not have to be read "on the spot." That is, a remote indicating device can be used to record the output information from the gage. Gages are sensitive to strains as small as 5 millionths of an inch per inch and usually have a range up to strains of 1 to 2 percent (10,000 to 20,000 microinches per inch). Special gages have ranges up to a strain of 10 percent. If properly protected, the typical gages can be used in various

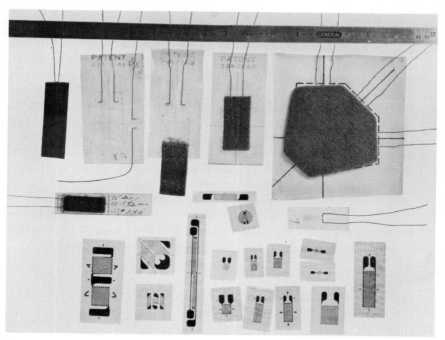

Fig. 16-2

environments, such as underwater and when exposed to temperatures of about 300 to 400 F. At the present time, special high-temperature gages can be used up to 2000 F, and this limit is sure to go higher with the tremendous interest and research in this direction. The cost of a gage varies from about $1.50 for a common general-purpose wire gage with a $\frac{1}{2}$-in. gage length to about $10.00 to $15.00 for a three-element foil rosette gage.

Previously we stated that the resistance change of the gage was proportional to strain. How are these two quantities related? Each gage has a gage factor, which we will denote by GF. The gage factor is defined as the ratio of the unit change in resistance to the unit change in length. That is,

$$\text{GF} = \frac{\Delta R/R}{\Delta L/L} = \frac{\Delta R/R}{\epsilon_{\text{avg}}} \tag{16-1}$$

where $\Delta R$ is the total change in resistance and $\Delta L$ is the total change in length.

The gage factor is an index of the strain sensitivity of a gage and is a constant for the small range of resistance changes and strains usually encountered. The higher the gage factor, the more sensitive is the strain gage. The gage factor is a function of the material of the wire or foil. The most typical gages have a gage factor of about 2. Although it is possible to use materials which will give higher gage factors, such materials usually have other undesirable properties that offset their high gage factor.

The success of a bonded strain gage depends to a great extent on the quality of the bond between it and the member on which it is used. It is of the utmost importance to use great care in cleaning and preparing the surface, applying the adhesive, and curing the adhesive to make sure that an intimate bond is obtained between the gage and the test piece. Nitro-cellulose cement (such as ordinary Duco household cement) is commonly used to bond standard paper-back gages. Epoxy cement may be required for plastic-backed gages. Eastman 910 Contact Cement is commonly used for a very fast installation because no cure time is then required.

Electrical resistance strain gages of all types are often referred to as SR-4 strain gages. The SR-4 is a trade name of the Baldwin-Lima-Hamilton Corporation, one of the leading manufacturers of strain gages today and for many years the only source of electrical resistance gages.

## 16-3. Strain-Gage Circuitry

How is the resistance change due to strain in a gage measured? Can we use an ordinary ohmmeter? To answer these questions, we must first determine the order of magnitude of the resistance change. A standard gage has a resistance of 120 ohms, and its gage factor is 2.0. If we bond such a gage to a bar of aluminum with a modulus of 10 million psi, and subject it to a uniaxial elastic stress of 500 psi in the direction of the gage axis, we can determine the strain and resistance change from Hooke's law and Eq. 16-1. Thus,

$$\epsilon = \frac{\sigma}{E} = \frac{500}{10,000,000} = 0.00005 \text{ in./in.}$$

$$\Delta R = \epsilon(GF)R$$

$$\Delta R = (0.00005)(2.0)(120) = 0.0120 \text{ ohm}$$

Resistance changes of this magnitude cannot be measured accurately with an ordinary ohmmeter. A Wheatstone-bridge circuit is usually used, in which one or more of the four arms of the bridge are strain gages. Following is a brief discussion of the use of the Wheatstone bridge for strain gages.

The basic Wheatstone-bridge circuit is shown in Fig. 16-3(a), and the basic relationship for a balanced bridge with zero output is

$$R_A \times R_D = R_B \times R_C \tag{16-2}$$

If a strain gage is placed in one arm of the bridge to replace the fixed resistance $R_A$, as in Fig. 16-3(b), when the gage is applied to a member and subjected to a strain, say tensile, the magnitude of $R_A$ in the basic equation would be increased, causing the equation to become unbalanced. *The magnitude of this unbalance is measured as output of the bridge and is proportional to the strain* in the member. Thus, because of unbalance due to $\Delta R_A$,

$$(R_A + \Delta R_A) \times R_D \neq R_B \times R_C \tag{16-2a}$$

We are here and hereafter assuming that initially all the resistances $R_A$, $R_B$, $R_C$, and $R_D$ are equal and that they change only as indicated.

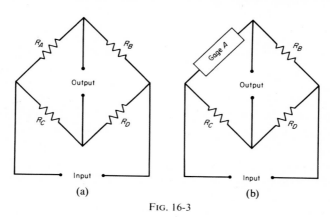

FIG. 16-3

The fact that the strain gage is very sensitive to temperature prevents the circuit in Fig. 16-3(*b*) from being entirely satisfactory. If the temperature of the member changes even slightly, this change will manifest itself as a "strain" or resistance change in the gage $A$. Therefore, the output or equation unbalance measured in such a circuit would actually be due to temperature strain as well as strain due to the loading. The correct equation, including unbalance due to $\Delta R_A$ and $\Delta R_T$, is

$$(R_A + \Delta R_A + \Delta R_T) \times R_D \neq R_B \times R_C \tag{16-2b}$$

By replacing $R_B$ with another gage $B$, identical to gage $A$, mounting it on a similar but unstressed material, and placing it in the immediate vicinity of gage $A$, then gage $B$ will be subjected to the same temperature influences as gage $A$. This circuit is shown in Fig. 16-4(*a*), and the basic equation for it would be

$$(R_A + \Delta R_A + \Delta R_T) \times R_D \neq (R_B + \Delta R_T) \times R_C \tag{16-2c}$$

Notice that since $R_B$ (gage $B$) is on the opposite side of the equation from $R_A$ (gage $A$), the terms $\Delta R_T$ due to temperature are cancelled out, and the only unbalance of the equation (output of the bridge) is due to the applied strain in the gage $A$ which is desired. For this reason the gage $B$ is often referred to as a temperature-compensating gage or dummy gage.

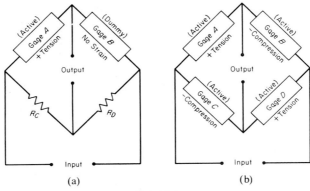

(a)                                                                     (b)

Fig. 16-4

## 16-4. Measuring Instruments

**Static Measurement.** The most commonly used strain-gage indicator for static measurement is the null-balance indicator, various types of which are shown in Fig. 16-5. After the active and compensating gages are connected to the indicator and the gage-factor adjustment is made, the indicator is adjusted until the circuit is balanced, as indicated by a zero deflection of the meter dial. A reading is then taken. After the active gage undergoes a strain, the circuit is again balanced and a new reading is taken. The difference in readings is the change in strain directly in microinches per inch.

This type of indicator is usually battery-operated and is therefore completely portable. Power for the bridge and amplification of the signal is provided within the indicator. Specifically, such an indicator usually consists of an audio-frequency oscillator for the bridge power, an amplifier, a detection circuit, and an indicating meter. Such an indicator can also be

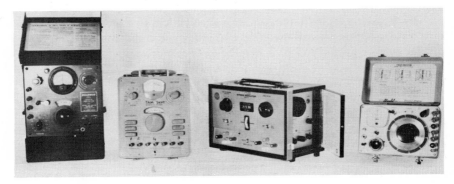

Fig. 16-5

used as an amplifier in connection with a cathode-ray oscilloscope for dynamic work up to frequencies of about 100 cps. With a slight external adjustment, the null indicator can readily be used with four external active gages (full external bridge). More will be said about the value of four active gages in the following section on transducers.

Frequently it is necessary to have many gages on a single structure subjected to static loads. Since each of these gases must be read individually, various types of switching units, or switching-and-balancing units, are available to handle such situations with a minimum of time and maximum convenience. In Fig. 16-6(*a*) is shown a commonly used 20-channel switching-and-balancing unit linked with a null indicator. Each channel has a balance potentiometer which makes it possible, in most cases, for a convenient zero reading to be the same for all gages. For any load on the structure the gages are read, one by one, as each gage is selected on the switch box and balanced on the null indicator. If there is not too much difference in the strain readings between channels, each gage can be read in about 15 to 30 seconds with a little practice. For higher speed and less chance of human reading error, an automatic recording system, such as that in Fig. 16-6(*b*), is required. Here the switching is automatic, the balancing is accomplished with a servo (rather than by hand), and the readings and recordings are made by an automatic printer. After the instrument has been properly adjusted for initial balance, it is only necessary to push one

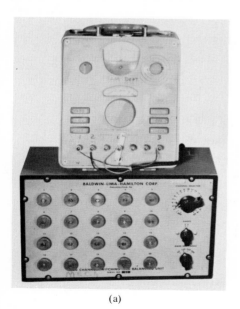

(a)

(b)

Fig. 16-6

button to obtain a complete and permanent record of strain in each gage from the printer in a few seconds. The automatic system in Fig. 16-6(*b*) is shown with only one 10-channel switching-and-balancing unit. Any number of additional 10-channel units may be provided as needed.

**Dynamic Measurement.** For applications where a strain gage is applied to a member in which the strain is changing rapidly, many types of

(a)

(b)

(c)

FIG. 16-7

recording instruments are available. In practically every case it is necessary to have means for powering the bridge and amplifying the output signal so that this signal can drive some type of readout device.

Several commonly used types of readout recording instruments are shown in Fig. 16-7. The portable instrument in Fig. 16-7(*a*) is a 2-channel recording oscillograph which plots output vs time. The plotting is done with a mechanical pen. The inertia of the pen mechanism limits this common type device to frequencies below approximately 100 Hz. For higher frequency response a recorder such as the "light beam" or "mirror galvanometer" type of oscillograph shown in Fig. 16-7(*b*) is commonly used. Here a low-inertia galvanometer with an attached mirror deflects a light beam in proportion to the signal, and the deflection is recorded as a trace on photosensitive paper. Since the paper is moving with a constant speed, a plot of output vs time is obtained. The particular model shown is 14 channels with 8 amplifiers just to its left and 6 in the rack above. Galvanometers of various frequency response (up to approximately 12,000 Hz) are available for these type instruments.

A cathode-ray oscilloscope may also be used as the readout instrument for strain signals. One type used for strain-gage work is shown in Fig. 16-7(*c*), along with a camera attachment at the top which clamps over the tube face. This instrument can be used to photograph a steady-state dynamic signal or a transient signal. Since the CRO uses an essentially inertialess electron beam for indication, it is suitable for any range of frequency up to the limit of the amplifier.

This section is a rather condensed survey of strain-gage instrumentation. It is highly recommended that the student read Chapter 5 in *The Strain Gage Primer,* second edition, by Perry and Lissner (McGraw-Hill Book Co., Inc., 1962) or some other reference on instrumentation of strain gages. Other chapters of *The Strain Gage Primer* are also of great value for the beginning student in strain-gage work.

## 16-5. Strain-Gage Transducers

A strain-gage transducer is a device which uses a strain gage to produce an electrical signal that is proportional to some mechanical phenomenon. This is done by making the strain in some part of the device proportional to the phenomenon to be measured. Strain-gage transducers are used to measure force (tension or compression), displacement, pressure, torque, or acceleration. These devices can usually be made very small, rugged, and simple, and they have the advantages which are inherent in electrical strain gages; that is, they are very accurate, are reliable, and can be used for static or dynamic phenomena, and the readout instrument can be remote from the sensing device. Many types of very elegant and sophisticated strain-gage transducers can be purchased commercially. The prices are quite high but reliable transducers, such as a torque pickup with slip rings or an accurately calibrated and damped accelerometer, are worth a high price. A simple (but reliable) "homemade" load, displacement, or pressure transducer can be constructed from a few dollars' worth of gages and materials. A few basic transducers will be discussed later in this section.

Generally, when a transducer is used, it is highly desirable to increase the output from the Wheatstone bridge as much as possible, in order to increase the sensitivity of the transducer. To increase the output, we must increase the unbalance of the basic bridge equation (Eq. 16-2). If tension in gage $A$ in one arm of the bridge caused a certain unbalanced resistance $(+\Delta R_A)$ on the *left-hand* side of the equation, as in Eq. 16-2a, then the unbalance and hence the output could be *doubled* if at the same time the gage $B$ in another arm could be subjected to an equal compression which would cause an equal but opposite change $(-\Delta R_B)$ on the *right-hand* side of the equation. This condition is often referred to as having two "active" gages or arms of the bridge, or an "active half bridge." The equation for the circuit would then be

$$(R_A + \Delta R_A + \Delta R_T) \times R_D \neq (R_B - \Delta R_B + \Delta R_T) \times R_C \quad (16\text{-}2d)$$

In this case, the output (unbalance due to $+\Delta R_A$ and $-\Delta R_B$) is doubled while the temperature effects are still cancelled out. The most common example of doubling the output by the use of two active gages is the application of putting gages "back to back in bending," as in the cantilever beam in Fig. 16-8($a$). The circuit is represented in Fig. 16-8($b$).

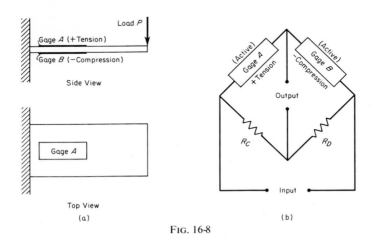

FIG. 16-8

By the same reasoning as above, to *quadruple* (multiply 4 times) the output, it is possible also to replace the other two fixed resistances $R_C$

and $R_D$ in Fig. 16-8(b) with gages, as shown in Fig. 16-4(b) or Fig. 16-9(b). Then if gage D is subjected to tension since it is on the left-hand side of Eq. 16-2, and if gage C is subjected to compression since it is on the right-hand side of the equation, the unbalance of the equation (the output of the bridge) will be four times as great as it was in Eq. 16-2a or 16-2c.

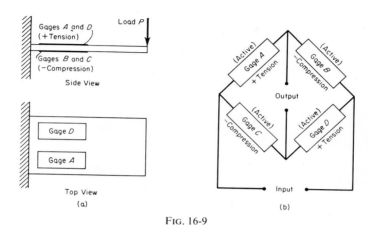

FIG. 16-9

This condition is often referred to as having four active gages or a "full bridge." The equation for a full bridge is

$$(R_A + \Delta R_A + \Delta R_T) \times (R_D + \Delta R_D + \Delta R_T)$$

$$\neq (R_B - \Delta R_B + \Delta R_T) \times (R_C - \Delta R_C + \Delta R_T) \qquad (16\text{-}2e)$$

It can be seen that the output (unbalance due to $+\Delta R_A$, $+\Delta R_D$, $-\Delta R_B$, and $-\Delta R_C$) is quadrupled while the significant (measurable) temperature effects are still cancelled out.

The determination of the unbalance for a full bridge does not consider the unbalance due to the products of the $\Delta$ terms (higher-order terms), which are negligible and not measurable. An example of this arrangement, which is often used in the sensing elements of transducers, is shown in Fig. 16-9(a). The circuit is represented in Fig. 16-9(b). This is the most desirable arrangement for transducers, because it yields the greatest output from the bridge and therefore the greatest sensitivity to the transducer.

Sketches of several types of simple transducers, which can be "home-made," are shown in Fig. 16-10. Notice how the strain gages are used in a full bridge. They are placed back to back in bending wherever possible. In Figs. 16-10(b), (c), (e), (f), and (g), four fully active gages give the

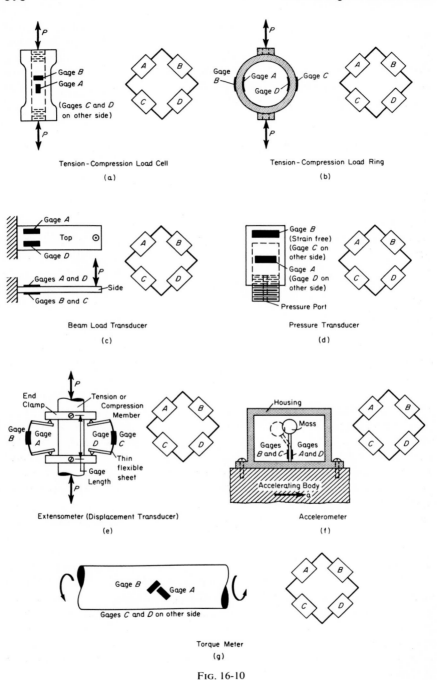

Tension-Compression Load Cell
(a)

Tension-Compression Load Ring
(b)

Beam Load Transducer
(c)

Pressure Transducer
(d)

Extensometer (Displacement Transducer)
(e)

Accelerometer
(f)

Torque Meter
(g)

Fig. 16-10

maximum sensitivity possible. In (*a*), the lateral gages *B* and *C* increase the output somewhat as a result of the Poisson effect. They, of course, also compensate for temperature. Only in the pressure transducer of Fig. 16-10(*d*) is it impossible to find places for gages *B* and *C* where they could read strains that are opposite to those read by gages *A* and *D*. Here gages *B* and *C* only compensate for temperature.

The load and torque transducers may easily be calibrated with dead weights or with a reliable test machine. A displacement transducer may usually be calibrated with a good micrometer, and a pressure transducer with a good pressure source and a reliable pressure gage or with a dead-weight pressure tester. Can you recommend a way to calibrate the accelerometer?

Electrical resistance gages have revolutionized experimental mechanics in the last 25 years. They have tremendous advantages over their predecessors, but they are not without some limitations. A resistance strain gage, except in a few special applications, measures only the *average surface* strain over the finite length of the gage *in the direction of the gage axis*. Rosette gages are useful in that they enable the determination of the maximum strain at a "point" (spot or small area is a better word). The important thing strain gages cannot do is to tell you *where* the maximum strain occurs. At exactly what point should the gage be bonded and, if a single gage is used, how should it be orientated? This question has worried many an engineer. The methods discussed in the next three sections help, in many cases, to answer the question.

## 16-6. Photoelasticity

Photoelasticity is a method of stress analysis in which a model of the member or structure of interest is used. The model is made of special plastic which possesses desirable strain-optical properties. Since the model is geometrically similar to the actual structure, the stress distribution in the model indicates the effects of the geometry on the stress distribution in the actual structure. However, the mechanical properties of the model and structure may differ considerably, and the model may therefore give a false impression of the actual behavior of the structure.

This method differs from strain-gage analysis in that it provides an over-all determination of the stresses at all points, both surface and interior, and not just a measure of the stress at one specific point. Photoelasticity is primarily a two-dimensional method in that the models used have a constant thickness. In advanced techniques, stresses are "frozen" in three-dimensional models and then two-dimensional slices are analyzed.

A photoelastic model is analyzed by use of an optical apparatus called a polariscope. It consists of a source of polarized light which passes through the model, a means of loading the model, and some means of analyzing the resulting phenomena (viewing screen or camera). A typical polariscope is shown in Fig. 16-11. Photoelasticity is based on the phenomenon which occurs when polarized light passes through the special plastic of the model. The plastic is *birefringent* when loaded. This means that it has unique properties which cause a ray of polarized light passing through the thickness at any point to be resolved into two components along the two principal stress directions at the point. One of these com-

FIG. 16-11

ponents is retarded, relative to the other, as they pass through the thickness of the stressed plastic. In other words, one component needs more time to go through than the other. This relative retardation (time difference) is proportional to the difference of principal stresses at the point. Thus,

$$\sigma_1 - \sigma_2 = \frac{C}{t} f \qquad (16\text{-}3)$$

where $C$ is the photoelastic constant for the material, $t$ is the model thickness, and $f$ is the number of full wave lengths of relative retardation and is called the fringe order. In the usual polariscope set-up, if the relative retardation at some point of the model is exactly one full wave length or any multiple thereof, the light is extinguished in the optical system and a black spot results on the viewing screen or camera plate for that point of the model. If the retardation is a half wave length or any multiple thereof, then a maximum amount of light is passed by the optical system and a bright spot on the screen represents that point in the model.

Since the model represents an infinite number of points, an infinite number of light rays pass through it, and the image seen on the screen is an infinite number of light, dark, and shady spots which merge to form a pattern of light and dark bands. Light very close to one frequency (monochromatic light) is usually used, so that the bands are sharper and more nearly black and white, rather than colored. A band, also called a fringe or isochromatic, represents the locus of points having the same difference of principal stresses. A qualitative study of the viewing screen will give an immediate and over-all "picture" of the stress distribution in the model. Points of maximum stress can be detected, and the effect of the model geometry can readily be seen.

A photograph of the isochromatic fringes is usually made for future quantitative study. Figure 16-12 is such a photograph. Special devices or techniques of compensation are available which make possible the determination of the stress differences at points of partial fringe orders. By using white light, removing certain elements of the optical system, and manipulating other elements, the locus of points having the same directions of principal stress can be sketched on the screen. Several such lines

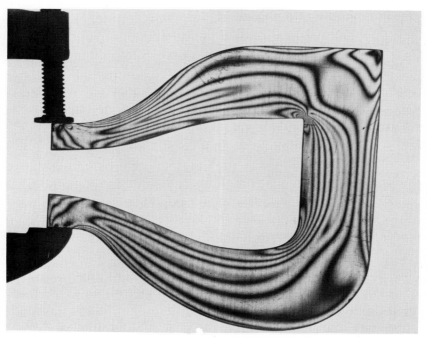

Fig. 16-12

can be drawn. These lines, called isoclinics, permit the determination of the directions of the principal stresses at any point.

The isochromatic fringe patterns represent the *differences* of the principal stresses at a point. Since $\sigma_1 - \sigma_2 = 2\tau$, we see that the fringe patterns actually give an over-all picture of the *shear-stress* distributions. Along the edges of the model, where the maximum stress usually occurs, one principal stress is zero. There the difference of the principal stresses, or $\sigma_1 - \sigma_2$, reduces directly to one stress. Elsewhere, the principal stresses can usually be separated only by rather long, tedious numerical computation.

A photoelastic material is usually chosen for its high value of $C$ (see Eq. 16-3), ease of machining, high proportional limit, optical transparency, good creep properties, and reasonable price. Two of the most commonly used materials are a type of Bakelite, usually referred to by the number BT-61-893, and a Columbia Resin called CR-39.

Before it is used, a material must be "calibrated." That is, the constant $C$ must be determined. This is usually done by loading a simple tensile-test specimen in a polariscope and plotting the axial load versus the fringe order. The slope of this curve is proportional to $C$. The other common method is to photograph the isochromatic pattern for a simple beam with a constant bending-moment loading. The slope of the plotted stress distribution is proportional to $C$.

## 16-7. Birefringent Coatings

In the past few years, with the development of excellent adhesives such as those of the epoxy family, photoelastic coatings (sometimes called birefringent coatings) have come into wide use. Techniques and instrumentation for their use have become highly developed. In this method of analysis the birefringent plastic is bonded directly to the surface of the actual structure to be stress-analyzed. When the surface of the structure undergoes a strain, the bond causes the plastic coating to undergo the same strain. The resulting isochromatic fringes and isoclinics of the plastic can be seen, sketched, or photographed, and they can be interpreted by using the principles of ordinary photoelasticity. Provision must be made to assure that the light directed through the plastic will be properly reflected for analysis. In this case, the light passes through the plastic twice. Instruments called reflective polariscopes are required for this technique. Some are extremely simple, while others are rather complex and expensive. For example, extremely valuable qualitative analyses are often made with a small piece of polaroid plastic used as a reflective polariscope.

Special techniques and equipment are available for dynamic work, for application to curved surfaces, for easy determination of separate principal stresses, for taking account of the stiffening effect of the plastic on thin sections, for use on nonmetallic surfaces, and for many other special applications. Although conventional photoelastic analysis of models is considered the more quantitatively accurate method, the photoelastic-coating method has the one big advantage of allowing an analysis of the actual member or structure of interest, rather than requiring the analysis of a model.

By necessity this text can include only a very brief survey of the photoelastic method. You are encouraged to refer to a text, such as *Photoelasticity*, vol. I, by M. Frocht (John Wiley & Sons, Inc., 1941) or *Introduction to Photomechanics*, by Durelli and Riley (Prentice-Hall Inc., 1965) for a detailed coverage of the basic principles of photoelasticity and the techniques of classical photoelastic model analysis. For more information about photoelastic-coating techniques, refer to the more recent issues of the *Proceedings of the Society for Experimental Stress Analysis*.

## 16-8. Brittle Lacquer Coatings

Another valuable tool of experimental mechanics is the use of brittle lacquer coatings for stress analysis. Lacquer is sprayed on a surface to be analyzed. The lacquer is designed to crack at a strain level within the elastic range of the material of the structure being analyzed. As a coated structure is being loaded, the first cracks occur at the points at which the "threshold strain" of the coating is reached first. These first cracks indicate the points of maximum tensile strain. By watching the progression of the crack patterns, the engineer can "see" the development and distribution of the strains as loading of the structure is continued. The crack at any point will be perpendicular to the direction of the maximum principal tensile stress at that point. Therefore, the directions of the principal stresses can also be determined.

The lacquer is calibrated by coating both the structure or member to be analyzed and several calibration bars at the same time and in exactly the same manner. After curing and at the time of test, the calibration bars are subjected to known variations of strain and their crack patterns are analyzed so that crack size and appearance may be associated with the corresponding known strain. The threshold strain is also determined by this calibration. The threshold strain or strains at the initiation of cracking is usually in the range from 500 to 1000 microinches per inch. The calibration is usually done with a fixture that deflects the calibration bar, loaded as a cantilever beam, to a known deflection by means of an eccentric roller at the free end.

This method sounds good in principle, but it also has its drawbacks. The lacquers used and their threshold strains are very sensitive to the influences of temperature, humidity, and creep. There is a premium on technique in that it takes an experienced technician working under optimum conditions to apply a uniform and reliable coating. Under the best conditions, fairly good quantitative results are possible. Plus or minus 20 percent for strain magnitudes is about the best accuracy that can usually be expected. The method perhaps has its main value in allowing the engineer to get a rather inexpensive and quick look at the over-all qualitative picture of the strains and their distribution. This analysis can tell him where and in what directions to place strain gages for further and more refined analyses.

Magnaflux Corporation sells the most commonly used lacquers and accessory equipment under the trade name *Stresscoat*. The manual of this corporation is an excellent source of information on the details of the technique. One of the better references describing outstanding work done with this method is *Analysis of Stress and Strain*, by Durelli, Phillips, and Tsao (McGraw-Hill Book Company, Inc., 1958).

## 16-9. Moire Analysis

Moire Analysis is another "field" method which allows an overall view for investigation as does photoelasticity and brittle lacquers. These methods have this advantage over strain gages which give a "point by point" analysis. Moire is usually applied to models (but sometimes to the actual piece) and involves application of a fine pitched grid (or grating) being intimately bonded to the surface to be investigated such that the grid undergoes the same displacements as the surface. When another identical unbonded and undistorted grid is placed over the bonded grid, fringe patterns appear. These moire fringes are loci of points which have the same value of displacements along directions perpendicular to undistorted grid lines. The difference in the displacement components of adjacent fringes is equal to the distance between the lines of the grid, or pitch. Strains can be determined by plotting the displacements and determining slopes. For example

$$\epsilon_x = \frac{\text{grating pitch}}{\text{distance between fringes}}$$

The differentiation of the moire patterns is usually difficult and often lacking in precision and this is a major limitation of the method. This is particularly true when rotations are involved. The method does, however, lend itself well for determination of displacements and its use is relatively simple in this respect. It is important to note that this method gives the surface geometry changes directly without going through any intermediate related property as does photoelasticity and strain gages and, therefore, is not influenced by any changes in these intermediate properties.

The sensitivity of the moire method is dependent on the grating pitch and the present state of the art is such that it is very difficult to make a conveniently usable grating with more than about 1000 lines per inch. Therefore, the sensitivity of the method is limited for most engineering materials. The method is better suited for low modulus materials.

## PROBLEMS

**16-1.** A 120-ohm electrical resistance strain gage with a gage factor of 2.01 undergoes a change of resistance of 0.128 ohm. What is the strain in the direction of the gage axis?

**16-2.** A 120-ohm electrical resistance strain gage is mounted on a solid steel shaft $\frac{1}{2}$ in. in radius with the gage axis 20° from the longitudinal axis of the shaft. What will be the resistance change of the gage, if the shaft is subjected to 3000 in-lb of pure torque? The gage factor is 1.98. Assume elastic behavior.

**16-3.** For the beam shown in the illustration, what is the strain on the surface of the beam where the gages are? The beam is made of a high strength alloy steel and has a rectangular cross section $\frac{1}{8}$ in. high and $\frac{1}{4}$ in. wide. Each gage has a resistance of 120 ohms and a gage factor of 2.03. Gages $C$ and $D$ are mounted on similar unstrained material. What is the voltage drop across the voltmeter $V$, assuming the voltmeter to have a very high resistance?

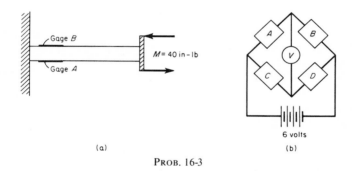

PROB. 16-3

**16-4.** Same as Prob. 16-3 except gage $C$ is mounted beside gage $B$, and gage $D$ is mounted beside gage $A$, all gages being parallel to the longitudinal axis of the beam.

**16-5.** The conditions are the same as in Prob. 16-3, except that gage $C$ is mounted beside gage $A$ with its axis perpendicular to the beam axis and perpendicular to the axis of gage $A$, and gage $D$ is mounted on the top surface beside gage $B$ and its axis is perpendicular to the beam axis and to the axis of gage $B$.

**16-6.** The conditions are the same as in Prob. 16-5, except that gage $C$ is mounted on the top surface and gage $D$ is mounted on the bottom.

**16-7.** Sketch the design and circuitry for a simple strain-gage transducer that will measure the pull on a drawbar between a farm tractor and a plow. Use a readout instrument that will give a permanent record of pull vs. time. Explain carefully how you would calibrate the device.

**16-8.** Sketch the design and circuitry for a strain-gage transducer to measure the impact force of a golf club on a golf ball. A complete record of force vs. time is needed. How would you calibrate the device?

**16-9.** Sketch the design and circuitry for a strain-gage transducer that could be screwed into a spark-plug hole in an automobile engine to measure the pressure in the cylinder. A record of pressure vs. time is desired. How would you calibrate the device?

**16-10.** A simple uniaxial tensile specimen of a photoelastic material is tested in a polariscope. The resulting values recorded for load in pounds vs. fringe order are given below. What is the photoelastic constant for the material? The cross section of the specimen is 0.335 in. wide by 0.275 in. thick.

| Load ($P$)............ | 0 | 8 | 14 | 20 | 28 | 35 | 42 |
|---|---|---|---|---|---|---|---|
| Fringe order ($f$)··· | 0 | 1 | 2 | 3 | 4 | 5 | 6 |

# Appendix

TABLE
   I   Average Mechanical Properties of Selected Engineering Materials
   II   Properties of Plane Areas
  III   Properties of Wide Flange Steel Sections
  IV   Properties of American Standard Steel I Sections
    V   Properties of American Standard Steel Channel Sections
  VI   Properties of Steel Angles with Equal Legs
 VII   Properties of Steel Angles with Unequal Legs
VIII   Pipe Sizes
  IX   American Standard Timber Sizes
    X   Slopes and Deflections of Selected Beams

TABLE I: AVERAGE MECHANICAL PROPERTIES OF SELECTED ENGINEERING MATERIALS

| Material | $E$ $10^6$ psi | $G$ $10^6$ psi | $\mu$ | 0.2 percent Yield Strength (tension) $10^3$ psi | Ultimate Strength $10^3$ psi | Elongation at Rupture in 2 in. Percent | Weight lb/in.$^3$ | Coefficient of Thermal Expansion $10^{-6}$ per °F |
|---|---|---|---|---|---|---|---|---|
| Hot rolled steel (SAE 1020) | 30 | 12 | 0.27 | 36 | 65 (ten) | 30 | 0.283 | 6.5 |
| Structural steel (A-7) | 30 | 12 | 0.27 | 35 | 60–72 (ten) | 30 | 0.283 | 6.5 |
| High-carbon steel (SAE 1090) | 30 | 12 | 0.27 | 67 | 122 (ten) | 10 | 0.283 | 6.5 |
| Alloy steel (SAE 4130) (heat treated) | 30 | 12 | 0.30 | 100 | 125 (ten) | | 0.283 | 6.5 |
| Stainless steel (18-8) | 28 | 9.5 | 0.30 | 80 | 120 (ten) | <1 | 0.284 | 9.6 |
| Gray cast iron (ASTM Class 30) | 14.7 | 5.9 | 0.20 | | 31 (ten) / 124 (comp) | | 0.260 | 6.7 |
| Cast iron (pearlitic malleable) | 26.4 | 10 | | 80 | 100 (ten) / 300 (comp) | 7 | 0.266 | 6.6 |
| Aluminum 1100-0 (annealed) | 10.0 | 3.8 | 0.33 | 3.5 | 11 (ten) | 25 | 0.098 | 13.1 |
| Aluminum alloy 2024-T3 (sheet and plate) | 10.6 | 4.0 | 0.33 | 50 | 70 (ten) | 18 | 0.100 | 12.6 |
| Aluminum alloy 6061-T6 (extruded) | 10.0 | 3.8 | 0.33 | 35 | 38 (ten) | 10 | 0.098 | 13.1 |
| Aluminum alloy 7075-T6 (sheet and plate) | 10.4 | 3.9 | 0.33 | 70 | 80 (ten) | 5 | 0.101 | 12.9 |
| Magnesium alloy (H K31A-H24 sheet) | 6.5 | 2.4 | 0.35 | 23 | 34 (ten) | 4 | 0.0647 | 15 |
| Titanium alloy (6Al-4V sheet) | 15.9 | 6.2 | 0.34 | 120 | 130 (ten) | 10 | 0.160 | 4.6 |
| Brass, hard yellow | 15 | 5.6 | 0.35 | 60 | 74 (ten) | 10 | 0.306 | 10.5 |
| Douglas fir timber (air dry; parallel to grain) | 1.7 | | | | 8.1 (ten) / 7.4 (comp) | | 0.020 | 3.0 |
| Red oak timber (air dry; parallel to grain) | 1.8 | | | | 6.9 (comp) | | 0.025 | 1.9 |
| Lead (rolled) | 2 | 0.7 | 0.43 | 2 | 2.5 (ten) | 50 | 0.410 | 16.4 |
| Tungsten carbide (Carboloy, Grade 999) | 100 | | 0.24 | | 600 (comp) | | | 2.2 |
| Glass (fused silica) | 10.0 | | 0.17 | | 1.3 (ten) / 13 (comp) | Nil | 0.15 | 4.0 |
| Concrete (low strength) | 2 | | 0.15 | | 2 (comp) | | 0.087 | 6.0 |
| Concrete (high strength) | 3 | | 0.15 | | 5 (comp) | | 0.087 | 6.0 |
| Polystyrene (average) | 0.5 | | | | 14 (comp) | 2 | | 70 |
| Polyethylene (average) | 1.8 | | | | 2 (ten) | 350 | 0.033 | 150 |
| Epoxy (cast; average) | 0.65 | | 0.45 | | 7 (ten) / 30 (comp) | 4 | | 33 |
| Rubber (natural; molded) | | | 0.50 | | 3 (ten) | 800 | | 90 |

# TABLE II
### PROPERTIES OF PLANE AREAS

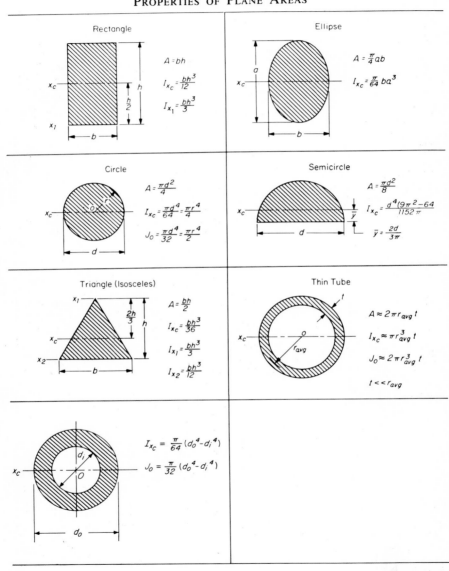

**Rectangle**

$A = bh$

$I_{x_c} = \frac{bh^3}{12}$

$I_{x_1} = \frac{bh^3}{3}$

**Ellipse**

$A = \frac{\pi}{4}ab$

$I_{x_c} = \frac{\pi}{64}ba^3$

**Circle**

$A = \frac{\pi d^2}{4}$

$I_{x_c} = \frac{\pi d^4}{64} = \frac{\pi r^4}{4}$

$J_0 = \frac{\pi d^4}{32} = \frac{\pi r^4}{2}$

**Semicircle**

$A = \frac{\pi d^2}{8}$

$I_{x_c} = \frac{d^4(9\pi^2 - 64)}{1152\,\pi}$

$\bar{y} = \frac{2d}{3\pi}$

**Triangle (Isosceles)**

$A = \frac{bh}{2}$

$I_{x_c} = \frac{bh^3}{36}$

$I_{x_1} = \frac{bh^3}{3}$

$I_{x_2} = \frac{bh^3}{12}$

**Thin Tube**

$A \approx 2\pi r_{avg}\,t$

$I_{x_c} \approx \pi r_{avg}^3\,t$

$J_0 \approx 2\pi r_{avg}^3\,t$

$t << r_{avg}$

$I_{x_c} = \frac{\pi}{64}(d_0^4 - d_i^4)$

$J_0 = \frac{\pi}{32}(d_0^4 - d_i^4)$

## TABLE III
### PROPERTIES OF WIDE FLANGE SECTIONS*

# WF SHAPES

# PROPERTIES FOR DESIGNING

NA

| NOMINAL† SIZE (in.) | WEIGHT PER FOOT (lb) | AREA (in.²) | DEPTH (in.) | FLANGE WIDTH (in.) | FLANGE THICKNESS (in.) | WEB THICKNESS (in.) | AXIS X-X $I$ (in.⁴) | AXIS X-X $S$‡ (in.³) | AXIS X-X $r$ (in.) | AXIS Y-Y $I$ (in.⁴) | AXIS Y-Y $S$‡ (in.³) | AXIS Y-Y $r$ (in.) |
|---|---|---|---|---|---|---|---|---|---|---|---|---|
| 10 × 10 | 112 | 32.92 | 11.38 | 10.415 | 1.248 | .755 | 718.7 | 126.3 | 4.67 | 235.4 | 45.2 | 2.67 |
|  | 100 | 29.43 | 11.12 | 10.345 | 1.118 | .685 | 625.0 | 112.4 | 4.61 | 206.6 | 39.9 | 2.65 |
|  | 89 | 26.19 | 10.88 | 10.275 | .998 | .615 | 542.4 | 99.7 | 4.55 | 180.6 | 35.2 | 2.63 |
|  | 77 | 22.67 | 10.62 | 10.195 | .868 | .535 | 457.2 | 86.1 | 4.49 | 153.4 | 30.1 | 2.60 |
|  | 72 | 21.18 | 10.50 | 10.170 | .808 | .510 | 420.7 | 80.1 | 4.46 | 141.8 | 27.9 | 2.59 |
|  | 66 | 19.41 | 10.38 | 10.117 | .748 | .457 | 382.5 | 73.7 | 4.44 | 129.2 | 25.5 | 2.58 |
|  | 60 | 17.66 | 10.25 | 10.075 | .683 | .415 | 343.7 | 67.1 | 4.41 | 116.5 | 23.1 | 2.57 |
|  | 54 | 15.88 | 10.12 | 10.028 | .618 | .368 | 305.7 | 60.4 | 4.39 | 103.9 | 20.7 | 2.56 |
|  | 49 | 14.40 | 10.00 | 10.000 | .558 | .340 | 272.9 | 54.6 | 4.35 | 93.0 | 18.6 | 2.54 |
| 10 × 8 | 45 | 13.24 | 10.12 | 8.022 | .618 | .350 | 248.6 | 49.1 | 4.33 | 53.2 | 13.3 | 2.00 |
|  | 39 | 11.48 | 9.94 | 7.990 | .528 | .318 | 209.7 | 42.2 | 4.27 | 44.9 | 11.2 | 1.98 |
|  | 33 | 9.71 | 9.75 | 7.433 | .433 | .292 | 170.9 | 35.0 | 4.20 | 36.5 | 9.2 | 1.94 |
| 10 × ¾ | 29 | 8.53 | 10.22 | 5.799 | .500 | .289 | 157.3 | 30.8 | 4.29 | 15.2 | 5.2 | 1.34 |
| 10 × 5¾ | 25 | 7.35 | 10.08 | 5.762 | .430 | .252 | 133.2 | 26.4 | 4.26 | 12.7 | 4.4 | 1.31 |
|  | 21 | 6.19 | 9.90 | 5.750 | .340 | .240 | 106.3 | 21.5 | 4.14 | 9.7 | 3.4 | 1.25 |
| 8 × 8 | 67 | 19.70 | 9.00 | 8.287 | .933 | .575 | 271.8 | 60.4 | 3.71 | 88.6 | 21.4 | 2.12 |
|  | 58 | 17.06 | 8.75 | 8.222 | .808 | .510 | 227.3 | 52.0 | 3.65 | 74.9 | 18.2 | 2.10 |
|  | 48 | 14.11 | 8.50 | 8.117 | .683 | .405 | 183.7 | 43.2 | 3.61 | 60.9 | 15.0 | 2.08 |
|  | 40 | 11.76 | 8.25 | 8.077 | .588 | .365 | 146.3 | 35.5 | 3.53 | 49.0 | 12.1 | 2.04 |
|  | 35 | 10.30 | 8.12 | 8.027 | .493 | .315 | 126.5 | 31.1 | 3.50 | 42.5 | 10.6 | 2.03 |
|  | 31 | 9.12 | 8.00 | 8.000 | .433 | .288 | 109.7 | 27.4 | 3.47 | 37.0 | 9.2 | 2.01 |
| 8 × 6½ | 28 | 8.23 | 8.06 | 6.540 | .463 | .285 | 97.8 | 24.3 | 3.45 | 21.6 | 6.6 | 1.62 |
|  | 24 | 7.06 | 7.93 | 6.500 | .398 | .245 | 82.5 | 20.8 | 3.42 | 18.2 | 5.6 | 1.61 |
| 8 × 5¼ | 20 | 5.88 | 8.14 | 5.268 | .378 | .248 | 69.2 | 17.0 | 3.43 | 8.5 | 3.2 | 1.20 |
|  | 17 | 5.00 | 8.00 | 5.250 | .308 | .230 | 56.4 | 14.1 | 3.36 | 6.7 | 2.6 | 1.16 |

*By permission of the American Institute of Steel Construction.
†Steel I beams are commonly designated by giving their nominal depth, the letter I, then the weight per foot; for example, 18 I 54.7.
‡$S = I/c$ = section modulus.

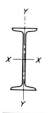

# TABLE IV

## Properties of American Standard
### Steel I Sections*

| Nominal† Size (in.) | Weight per Foot (lb) | Area (in.²) | Depth (in.) | Flange Width (in.) | Flange Thick- ness (in.) | Web Thick- ness (in.) | Axis X-X I (in.⁴) | Axis X-X S‡ (in.³) | Axis X-X r (in.) | Axis Y-Y I (in.⁴) | Axis Y-Y S‡ (in.³) | Axis Y-Y r (in.) |
|---|---|---|---|---|---|---|---|---|---|---|---|---|
| 24 x 7⅞ | 120.0 | 35.13 | 24.00 | 8.048 | 1.102 | .798 | 3010.8 | 250.9 | 9.26 | 84.9 | 21.1 | 1.56 |
|  | 105.9 | 30.98 | 24.00 | 7.875 | 1.102 | .625 | 2811.5 | 234.3 | 9.53 | 78.9 | 20.0 | 1.60 |
| 24 x 7 | 100.0 | 29.25 | 24.00 | 7.247 | .871 | .747 | 2371.8 | 197.6 | 9.05 | 48.4 | 13.4 | 1.29 |
|  | 90.0 | 26.30 | 24.00 | 7.124 | .871 | .624 | 2230.1 | 185.8 | 9.21 | 45.5 | 12.8 | 1.32 |
|  | 79.9 | 23.33 | 24.00 | 7.000 | .871 | .500 | 2087.2 | 173.9 | 9.46 | 42.9 | 12.2 | 1.36 |
| 20 x 7 | 95.0 | 27.74 | 20.00 | 7.200 | .916 | .800 | 1599.7 | 160.0 | 7.59 | 50.5 | 14.0 | 1.35 |
|  | 85.0 | 24.80 | 20.00 | 7.053 | .916 | .653 | 1501.7 | 150.2 | 7.78 | 47.0 | 13.3 | 1.38 |
| 20 x 6¼ | 75.0 | 21.90 | 20.00 | 6.391 | .789 | .641 | 1263.5 | 126.3 | 7.60 | 30.1 | 9.4 | 1.17 |
|  | 65.4 | 19.08 | 20.00 | 6.250 | .789 | .500 | 1169.5 | 116.9 | 7.83 | 27.9 | 8.9 | 1.21 |
| 18 x 6 | 70.0 | 20.46 | 18.00 | 6.251 | .691 | .711 | 917.5 | 101.9 | 6.70 | 24.5 | 7.8 | 1.09 |
|  | 54.7 | 15.94 | 18.00 | 6.000 | .691 | .460 | 795.5 | 88.4 | 7.07 | 21.2 | 7.1 | 1.15 |
| 15 x 5½ | 50.0 | 14.59 | 15.00 | 5.640 | .622 | .550 | 481.1 | 64.2 | 5.74 | 16.0 | 5.7 | 1.05 |
|  | 42.9 | 12.49 | 15.00 | 5.500 | .622 | .410 | 441.8 | 58.9 | 5.95 | 14.6 | 5.3 | 1.08 |
| 12 x 5¼ | 50.0 | 14.57 | 12.00 | 5.477 | .659 | .687 | 301.6 | 50.3 | 4.55 | 16.0 | 5.8 | 1.05 |
|  | 40.8 | 11.84 | 12.00 | 5.250 | .659 | .460 | 268.9 | 44.8 | 4.77 | 13.8 | 5.3 | 1.08 |
| 12 x 5 | 35.0 | 10.20 | 12.00 | 5.078 | .544 | .428 | 227.0 | 37.8 | 4.72 | 10.0 | 3.9 | .99 |
|  | 31.8 | 9.26 | 12.00 | 5.000 | .544 | .350 | 215.8 | 36.0 | 4.83 | 9.5 | 3.8 | 1.01 |
| 10 x 4⅝ | 35.0 | 10.22 | 10.00 | 4.944 | .491 | .594 | 145.8 | 29.2 | 3.78 | 8.5 | 3.4 | .91 |
|  | 25.4 | 7.38 | 10.00 | 4.660 | .491 | .310 | 122.1 | 24.4 | 4.07 | 6.9 | 3.0 | .97 |
| 8 x 4 | 23.0 | 6.71 | 8.00 | 4.171 | .425 | .441 | 64.2 | 16.0 | 3.09 | 4.4 | 2.1 | .81 |
|  | 18.4 | 5.34 | 8.00 | 4.000 | .425 | .270 | 56.9 | 14.2 | 3.26 | 3.8 | 1.9 | .84 |
| 7 x 3⅝ | 20.0 | 5.83 | 7.00 | 3.860 | .392 | .450 | 41.9 | 12.0 | 2.68 | 3.1 | 1.6 | .74 |
|  | 15.3 | 4.43 | 7.00 | 3.660 | .392 | .250 | 36.2 | 10.4 | 2.86 | 2.7 | 1.5 | .78 |
| 6 x 3⅜ | 17.25 | 5.02 | 6.00 | 3.565 | .359 | .465 | 26.0 | 8.7 | 2.28 | 2.3 | 1.3 | .68 |
|  | 12.5 | 3.61 | 6.00 | 3.330 | .359 | .230 | 21.8 | 7.3 | 2.46 | 1.8 | 1.1 | .72 |
| 5 x 3 | 14.75 | 4.29 | 5.00 | 3.284 | .326 | .494 | 15.0 | 6.0 | 1.87 | 1.7 | 1.0 | .63 |
|  | 10.0 | 2.87 | 5.00 | 3.000 | .326 | .210 | 12.1 | 4.8 | 2.05 | 1.2 | .82 | .65 |
| 4 x 2⅝ | 9.5 | 2.76 | 4.00 | 2.796 | .293 | .326 | 6.7 | 3.3 | 1.56 | .91 | .65 | .58 |
|  | 7.7 | 2.21 | 4.00 | 2.660 | .293 | .190 | 6.0 | 3.0 | 1.64 | .77 | .58 | .59 |
| 3 x 2⅜ | 7.5 | 2.17 | 3.00 | 2.509 | .260 | .349 | 2.9 | 1.9 | 1.15 | .59 | .47 | .52 |
|  | 5.7 | 1.64 | 3.00 | 2.330 | .260 | .170 | 2.5 | 1.7 | 1.23 | .46 | .40 | .53 |

* By permission of the American Institute of Steel Construction.
† Steel I beams are commonly designated by giving their nominal depth, the letter I, then the weight per foot; for example, 18 I 54.7.

‡$S = I/c$ = section modulus.

## TABLE V

### Properties of American Standard Steel Channel Sections*

| Nomi-nal† Size (in.) | Weight per Foot (lb) | Area (in.²) | Depth (in.) | Flange Width (in.) | Flange Avg. Thick-ness (in.) | Web Thick-ness (in.) | Axis X-X I (in.⁴) | Axis X-X S‡ (in.³) | Axis X-X r (in.) | Axis Y-Y I (in.⁴) | Axis Y-Y S‡ (in.³) | Axis Y-Y r (in.) | x (in.) |
|---|---|---|---|---|---|---|---|---|---|---|---|---|---|
| 18x4§ | 58.0 | 16.98 | 18.00 | 4.200 | .625 | .700 | 670.7 | 74.5 | 6.29 | 18.5 | 5.6 | 1.04 | .88 |
| | 51.9 | 15.18 | 18.00 | 4.100 | .625 | .600 | 622.1 | 69.1 | 6.40 | 17.1 | 5.3 | 1.06 | .87 |
| | 45.8 | 13.38 | 18.00 | 4.000 | .625 | .500 | 573.5 | 63.7 | 6.55 | 15.8 | 5.1 | 1.09 | .89 |
| | 42.7 | 12.48 | 18.00 | 3.950 | .625 | .450 | 549.2 | 61.0 | 6.64 | 15.0 | 4.9 | 1.10 | .90 |
| 15x3⅜ | 50.0 | 14.64 | 15.00 | 3.716 | .650 | .716 | 401.4 | 53.6 | 5.24 | 11.2 | 3.8 | .87 | .80 |
| | 40.0 | 11.70 | 15.00 | 3.520 | .650 | .520 | 346.3 | 46.2 | 5.44 | 9.3 | 3.4 | .89 | .78 |
| | 33.9 | 9.90 | 15.00 | 3.400 | .650 | .400 | 312.6 | 41.7 | 5.62 | 8.2 | 3.2 | .91 | .79 |
| 12x3 | 30.0 | 8.79 | 12.00 | 3.170 | .501 | .510 | 161.2 | 26.9 | 4.28 | 5.2 | 2.1 | .77 | .68 |
| | 25.0 | 7.32 | 12.00 | 3.047 | .501 | .387 | 143.5 | 23.9 | 4.43 | 4.5 | 1.9 | .79 | .68 |
| | 20.7 | 6.03 | 12.00 | 2.940 | .501 | .280 | 128.1 | 21.4 | 4.61 | 3.9 | 1.7 | .81 | .70 |
| 10x2⅝ | 30.0 | 8.80 | 10.00 | 3.033 | .436 | .673 | 103.0 | 20.6 | 3.42 | 4.0 | 1.7 | .67 | .65 |
| | 25.0 | 7.33 | 10.00 | 2.886 | .436 | .526 | 90.7 | 18.1 | 3.52 | 3.4 | 1.5 | .68 | .62 |
| | 20.0 | 5.86 | 10.00 | 2.739 | .436 | .379 | 78.5 | 15.7 | 3.66 | 2.8 | 1.3 | .70 | .61 |
| | 15.3 | 4.47 | 10.00 | 2.600 | .436 | .240 | 66.9 | 13.4 | 3.87 | 2.3 | 1.2 | .72 | .64 |
| 9x2½ | 20.0 | 5.86 | 9.00 | 2.648 | .413 | .448 | 60.6 | 13.5 | 3.22 | 2.4 | 1.2 | .65 | .59 |
| | 15.0 | 4.39 | 9.00 | 2.485 | .413 | .285 | 50.7 | 11.3 | 3.40 | 1.9 | 1.0 | .67 | .59 |
| | 13.4 | 3.89 | 9.00 | 2.430 | .413 | .230 | 47.3 | 10.5 | 3.49 | 1.8 | .97 | .67 | .61 |
| 8x2¼ | 18.75 | 5.49 | 8.00 | 2.527 | .390 | .487 | 43.7 | 10.9 | 2.82 | 2.0 | 1.0 | .60 | .57 |
| | 13.75 | 4.02 | 8.00 | 2.343 | .390 | .303 | 35.8 | 9.0 | 2.99 | 1.5 | .86 | .62 | .56 |
| | 11.5 | 3.36 | 8.00 | 2.260 | .390 | .220 | 32.3 | 8.1 | 3.10 | 1.3 | .79 | .63 | .58 |
| 7x2⅛ | 14.75 | 4.32 | 7.00 | 2.299 | .366 | .419 | 27.1 | 7.7 | 2.51 | 1.4 | .79 | .57 | .53 |
| | 12.25 | 3.58 | 7.00 | 2.194 | .366 | .314 | 24.1 | 6.9 | 2.59 | 1.2 | .71 | .58 | .53 |
| | 9.8 | 2.85 | 7.00 | 2.090 | .366 | .210 | 21.1 | 6.0 | 2.72 | .98 | .63 | .59 | .55 |
| 6x2 | 13.0 | 3.81 | 6.00 | 2.157 | .343 | .437 | 17.3 | 5.8 | 2.13 | 1.1 | .65 | .53 | .52 |
| | 10.5 | 3.07 | 6.00 | 2.034 | .343 | .314 | 15.1 | 5.0 | 2.22 | .87 | .57 | .53 | .50 |
| | 8.2 | 2.39 | 6.00 | 1.920 | .343 | .200 | 13.0 | 4.3 | 2.34 | .70 | .50 | .54 | .52 |
| 5x1¾ | 9.0 | 2.63 | 5.00 | 1.885 | .320 | .325 | 8.8 | 3.5 | 1.83 | .64 | .45 | .49 | .48 |
| | 6.7 | 1.95 | 5.00 | 1.750 | .320 | .190 | 7.4 | 3.0 | 1.95 | .48 | .38 | .50 | .49 |
| 4x1⅝ | 7.25 | 2.12 | 4.00 | 1.720 | .296 | .320 | 4.5 | 2.3 | 1.47 | .44 | .35 | .46 | .46 |
| | 5.4 | 1.56 | 4.00 | 1.580 | .296 | .180 | 3.8 | 1.9 | 1.56 | .32 | .29 | .45 | .46 |
| 3x1½ | 6.0 | 1.75 | 3.00 | 1.596 | .273 | .356 | 2.1 | 1.4 | 1.08 | .31 | .27 | .42 | .46 |
| | 5.0 | 1.46 | 3.00 | 1.498 | .273 | .258 | 1.8 | 1.2 | 1.12 | .25 | .24 | .41 | .44 |
| | 4.1 | 1.19 | 3.00 | 1.410 | .273 | .170 | 1.6 | 1.1 | 1.17 | .20 | .21 | .41 | .44 |

*By permission of the American Institute of Steel Construction.
†Steel channels are commonly designated by their nominal depth, the symbol ⊔, and the weight per foot; for example, 12 ⊔ 30.0.
‡$S = I/c$ = section modulus.
§Car and shipbuilding channel; not an American Standard.

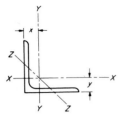

# TABLE VI

### PROPERTIES OF STEEL ANGLES
### WITH EQUAL LEGS*

| SIZE (in.) | THICKNESS (in.) | WEIGHT PER FOOT (lb) | AREA (in.²) | AXIS X-X AND AXIS Y-Y | | | | AXIS Z-Z |
|---|---|---|---|---|---|---|---|---|
| | | | | $I$ (in.⁴) | $S$† (in.³) | $r$ (in.) | $x$ or $y$ (in.) | $r$ (in.) |
| 8 x 8 | 1⅛ | 56.9 | 16.73 | 98.0 | 17.5 | 2.42 | 2.41 | 1.56 |
| | 1 | 51.0 | 15.00 | 89.0 | 15.8 | 2.44 | 2.37 | 1.56 |
| | ⅞ | 45.0 | 13.23 | 79.6 | 14.0 | 2.45 | 2.32 | 1.57 |
| | ¾ | 38.9 | 11.44 | 69.7 | 12.2 | 2.47 | 2.28 | 1.57 |
| | ⅝ | 32.7 | 9.61 | 59.4 | 10.3 | 2.49 | 2.23 | 1.58 |
| | 9⁄16 | 29.6 | 8.68 | 54.1 | 9.3 | 2.50 | 2.21 | 1.58 |
| | ½ | 26.4 | 7.75 | 48.6 | 8.4 | 2.50 | 2.19 | 1.59 |
| 6 x 6 | 1 | 37.4 | 11.00 | 35.5 | 8.6 | 1.80 | 1.86 | 1.17 |
| | ⅞ | 33.1 | 9.73 | 31.9 | 7.6 | 1.81 | 1.82 | 1.17 |
| | ¾ | 28.7 | 8.44 | 28.2 | 6.7 | 1.83 | 1.78 | 1.17 |
| | ⅝ | 24.2 | 7.11 | 24.2 | 5.7 | 1.84 | 1.73 | 1.18 |
| | 9⁄16 | 21.9 | 6.43 | 22.1 | 5.1 | 1.85 | 1.71 | 1.18 |
| | ½ | 19.6 | 5.75 | 19.9 | 4.6 | 1.86 | 1.68 | 1.18 |
| | 7⁄16 | 17.2 | 5.06 | 17.7 | 4.1 | 1.87 | 1.66 | 1.19 |
| | ⅜ | 14.9 | 4.36 | 15.4 | 3.5 | 1.88 | 1.64 | 1.19 |
| | 5⁄16 | 12.5 | 3.66 | 13.0 | 3.0 | 1.89 | 1.61 | 1.19 |
| 5 x 5 | ⅞ | 27.2 | 7.98 | 17.8 | 5.2 | 1.49 | 1.57 | .97 |
| | ¾ | 23.6 | 6.94 | 15.7 | 4.5 | 1.51 | 1.52 | .97 |
| | ⅝ | 20.0 | 5.86 | 13.6 | 3.9 | 1.52 | 1.48 | .98 |
| | ½ | 16.2 | 4.75 | 11.3 | 3.2 | 1.54 | 1.43 | .98 |
| | 7⁄16 | 14.3 | 4.18 | 10.0 | 2.8 | 1.55 | 1.41 | .98 |
| | ⅜ | 12.3 | 3.61 | 8.7 | 2.4 | 1.56 | 1.39 | .99 |
| | 5⁄16 | 10.3 | 3.03 | 7.4 | 2.0 | 1.57 | 1.37 | .99 |
| 4 x 4 | ¾ | 18.5 | 5.44 | 7.7 | 2.8 | 1.19 | 1.27 | .78 |
| | ⅝ | 15.7 | 4.61 | 6.7 | 2.4 | 1.20 | 1.23 | .78 |
| | ½ | 12.8 | 3.75 | 5.6 | 2.0 | 1.22 | 1.18 | .78 |
| | 7⁄16 | 11.3 | 3.31 | 5.0 | 1.8 | 1.23 | 1.16 | .78 |
| | ⅜ | 9.8 | 2.86 | 4.4 | 1.5 | 1.23 | 1.14 | .79 |
| | 5⁄16 | 8.2 | 2.40 | 3.7 | 1.3 | 1.24 | 1.12 | .79 |
| | ¼ | 6.6 | 1.94 | 3.0 | 1.1 | 1.25 | 1.09 | .80 |

*By permission of the American Institute of Steel Construction.
†$S = I/c$ = section modulus.

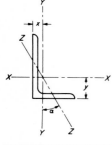

## TABLE VII
### Properties of Steel Angles
### with Unequal Legs*

| Size (in.) | Thick-ness (in.) | Weight per Foot (lb) | Area (in.²) | Axis X-X | | | | Axis Y-Y | | | | Axis Z-Z | |
|---|---|---|---|---|---|---|---|---|---|---|---|---|---|
| | | | | $I$ (in.⁴) | $S$† (in.³) | $r$ (in.) | $y$ (in.) | $I$ (in.⁴) | $S$† (in.³) | $r$ (in.) | $x$ (in.) | $r$ (in.) | Tan $\alpha$ |
| 4 x 3½ | ⅝ | 14.7 | 4.30 | 6.4 | 2.4 | 1.22 | 1.29 | 4.5 | 1.8 | 1.03 | 1.04 | .72 | .745 |
| | ½ | 11.9 | 3.50 | 5.3 | 1.9 | 1.23 | 1.25 | 3.8 | 1.5 | 1.04 | 1.00 | .72 | .750 |
| | ⁷⁄₁₆ | 10.6 | 3.09 | 4.8 | 1.7 | 1.24 | 1.23 | 3.4 | 1.4 | 1.05 | .98 | .72 | .753 |
| | ⅜ | 9.1 | 2.67 | 4.2 | 1.5 | 1.25 | 1.21 | 3.0 | 1.2 | 1.06 | .96 | .73 | .755 |
| | ⁵⁄₁₆ | 7.7 | 2.25 | 3.6 | 1.3 | 1.26 | 1.18 | 2.6 | 1.0 | 1.07 | .93 | .73 | .757 |
| | ¼ | 6.2 | 1.81 | 2.9 | 1.0 | 1.27 | 1.16 | 2.1 | .81 | 1.07 | .91 | .73 | .759 |
| 4 x 3 | ⅝ | 13.6 | 3.98 | 6.0 | 2.3 | 1.23 | 1.37 | 2.9 | 1.4 | .85 | .87 | .64 | .534 |
| | ½ | 11.1 | 3.25 | 5.1 | 1.9 | 1.25 | 1.33 | 2.4 | 1.1 | .86 | .83 | .64 | .543 |
| | ⁷⁄₁₆ | 9.8 | 2.87 | 4.5 | 1.7 | 1.25 | 1.30 | 2.2 | 1.0 | .87 | .80 | .64 | .547 |
| | ⅜ | 8.5 | 2.48 | 4.0 | 1.5 | 1.26 | 1.28 | 1.9 | .87 | .88 | .78 | .64 | .551 |
| | ⁵⁄₁₆ | 7.2 | 2.09 | 3.4 | 1.2 | 1.27 | 1.26 | 1.7 | .73 | .89 | .76 | .65 | .554 |
| | ¼ | 5.8 | 1.69 | 2.8 | 1.0 | 1.28 | 1.24 | 1.4 | .60 | .90 | .74 | .65 | .558 |
| 3½ x 3 | ½ | 10.2 | 3.00 | 3.5 | 1.5 | 1.07 | 1.13 | 2.3 | 1.1 | .88 | .88 | .62 | .714 |
| | ⁷⁄₁₆ | 9.1 | 2.65 | 3.1 | 1.3 | 1.08 | 1.10 | 2.1 | .98 | .89 | .85 | .62 | .718 |
| | ⅜ | 7.9 | 2.30 | 2.7 | 1.1 | 1.09 | 1.08 | 1.9 | .85 | .90 | .83 | .62 | .721 |
| | ⁵⁄₁₆ | 6.6 | 1.93 | 2.3 | .95 | 1.10 | 1.06 | 1.6 | .72 | .90 | .81 | .63 | .724 |
| | ¼ | 5.4 | 1.56 | 1.9 | .78 | 1.11 | 1.04 | 1.3 | .59 | .91 | .79 | .63 | .727 |
| 3½ x 2½ | ½ | 9.4 | 2.75 | 3.2 | 1.4 | 1.09 | 1.20 | 1.4 | .76 | .70 | .70 | .53 | .486 |
| | ⁷⁄₁₆ | 8.3 | 2.43 | 2.9 | 1.3 | 1.09 | 1.18 | 1.2 | .68 | .71 | .68 | .54 | .491 |
| | ⅜ | 7.2 | 2.11 | 2.6 | 1.1 | 1.10 | 1.16 | 1.1 | .59 | .72 | .66 | .54 | .496 |
| | ⁵⁄₁₆ | 6.1 | 1.78 | 2.2 | .93 | 1.11 | 1.14 | .94 | .50 | .73 | .64 | .54 | .501 |
| | ¼ | 4.9 | 1.44 | 1.8 | .75 | 1.12 | 1.11 | .78 | .41 | .74 | .61 | .54 | .506 |

*By permission of the American Institute of Steel Construction.
†$S = I/c$ = section modulus.

## TABLE VIII
### Pipe Sizes*

| | DIMENSIONS | | | | | | COUPLINGS | | | PROPERTIES | | |
|---|---|---|---|---|---|---|---|---|---|---|---|---|
| Nom. Diam. (in.) | Outside Diam. (in.) | Inside Diam. (in.) | Thickness (in.) | Weight per Foot (lb) Plain Ends | Thread & Cplg. | Threads per Inch | Outside Diam. (in.) | Length (in.) | Weight (lb) | $I$ (in.⁴) | $A$ (in.²) | $r$ (in.) |
| Standard | | | | | | | | | | | | |
| 1/8 | .405 | .269 | .068 | .24 | .25 | 27 | .562 | 7/8 | .03 | .001 | .072 | .12 |
| 1/4 | .540 | .364 | .088 | .42 | .43 | 18 | .685 | 1 | .04 | .003 | .125 | .16 |
| 3/8 | .675 | .493 | .091 | .57 | .57 | 18 | .848 | 1 1/8 | 07 | .007 | .167 | .21 |
| 1/2 | .840 | .622 | .109 | .85 | .85 | 14 | 1.024 | 1 3/8 | .12 | .017 | .250 | .26 |
| 3/4 | 1.050 | .824 | .113 | 1.13 | 1.13 | 14 | 1.281 | 1 5/8 | .21 | .037 | .333 | .33 |
| 1 | 1.315 | 1.049 | .133 | 1.68 | 1.68 | 11 1/2 | 1.576 | 1 7/8 | .35 | .087 | .494 | .42 |
| 1 1/4 | 1.660 | 1.380 | .140 | 2.27 | 2.28 | 11 1/2 | 1.950 | 2 1/8 | .55 | .195 | .669 | .54 |
| 1 1/2 | 1.900 | 1.610 | .145 | 2.72 | 2.73 | 11 1/2 | 2.218 | 2 3/8 | .76 | .310 | .799 | .62 |
| 2 | 2.375 | 2.067 | .154 | 3.65 | 3.68 | 11 1/2 | 2.760 | 2 5/8 | 1.23 | .666 | 1.075 | .79 |
| 2 1/2 | 2.875 | 2.469 | .203 | 5.79 | 5.82 | 8 | 3.276 | 2 7/8 | 1.76 | 1.530 | 1.704 | .95 |
| 3 | 3.500 | 3.068 | .216 | 7.58 | 7.62 | 8 | 3.948 | 3 1/4 | 2.55 | 3.017 | 2.228 | 1.16 |
| 3 1/2 | 4.000 | 3.548 | .226 | 9.11 | 9.20 | 8 | 4.591 | 3 5/8 | 4.33 | 4.788 | 2.680 | 1.34 |
| 4 | 4.500 | 4.026 | .237 | 10.79 | 10.89 | 8 | 5.091 | 3 5/8 | 5.41 | 7.233 | 3.174 | 1.51 |
| 5 | 5.563 | 5.047 | .258 | 14.62 | 14.81 | 8 | 6.296 | 4 1/8 | 9.16 | 15.16 | 4.300 | 1.88 |
| 6 | 6.625 | 6.065 | .280 | 18.97 | 19.19 | 8 | 7.358 | 4 1/8 | 10.82 | 28.14 | 5.581 | 2.25 |
| 8 | 8.625 | 8.071 | .277 | 24.70 | 25.00 | 8 | 9.420 | 4 5/8 | 15.84 | 63.35 | 7.265 | 2.95 |
| 8 | 8.625 | 7.981 | .322 | 28.55 | 28.81 | 8 | 9.420 | 4 5/8 | 15.84 | 72.49 | 7.981 | 2.94 |
| 10 | 10.750 | 10.192 | .279 | 31.20 | 32.00 | 8 | 11.721 | 6 1/8 | 33.92 | 125.9 | 9.178 | 3.70 |
| 10 | 10.750 | 10.136 | .307 | 34.24 | 35.00 | 8 | 11.721 | 6 1/8 | 33.92 | 137.4 | 10.07 | 3.69 |
| 10 | 10.750 | 10.020 | .365 | 40.48 | 41.13 | 8 | 11.721 | 6 1/8 | 33.92 | 160.7 | 11.91 | 3.67 |
| 12 | 12.750 | 12.090 | .330 | 43.77 | 45.00 | 8 | 13.958 | 6 1/8 | 48.27 | 248.5 | 12.88 | 4.39 |
| 12 | 12.750 | 12.000 | .375 | 49.56 | 50.71 | 8 | 13.958 | 6 1/8 | 48.27 | 279.3 | 14.58 | 4.38 |
| Extra Strong | | | | | | | | | | | | |
| 1/8 | .405 | .215 | .095 | .31 | .32 | 27 | .582 | 1 1/8 | .05 | .001 | .093 | .11 |
| 1/4 | .540 | .302 | .119 | .54 | .54 | 18 | .724 | 1 3/8 | .07 | .004 | .157 | .15 |
| 3/8 | .675 | .423 | .126 | .74 | .75 | 18 | .898 | 1 5/8 | .13 | .009 | .217 | .20 |
| 1/2 | .840 | .546 | .147 | 1.09 | 1.10 | 14 | 1.085 | 1 7/8 | .21 | .020 | .320 | .25 |
| 3/4 | 1.050 | .742 | .154 | 1.47 | 1.49 | 14 | 1.316 | 2 1/8 | .33 | .045 | .433 | .32 |
| 1 | 1.315 | .957 | .179 | 2.17 | 2.20 | 11 1/2 | 1.575 | 2 3/8 | .47 | .106 | .639 | .41 |
| 1 1/4 | 1.660 | 1.278 | .191 | 3.00 | 3.05 | 11 1/2 | 2.054 | 2 7/8 | 1.04 | .242 | .881 | .52 |
| 1 1/2 | 1.900 | 1.500 | .200 | 3.63 | 3.69 | 11 1/2 | 2.294 | 2 7/8 | 1.17 | .391 | 1.068 | .61 |
| 2 | 2.375 | 1.939 | .218 | 5.02 | 5.13 | 11 1/2 | 2.870 | 3 5/8 | 2.17 | .868 | 1.477 | .77 |
| 2 1/2 | 2.875 | 2.323 | .276 | 7.66 | 7.83 | 8 | 3.389 | 4 1/8 | 3.43 | 1.924 | 2.254 | .92 |
| 3 | 3.500 | 2.900 | .300 | 10.25 | 10.46 | 8 | 4.014 | 4 1/8 | 4.13 | 3.894 | 3.016 | 1.14 |
| 3 1/2 | 4.000 | 3.364 | .318 | 12.51 | 12.82 | 8 | 4.628 | 4 5/8 | 6.29 | 6.280 | 3.678 | 1.31 |
| 4 | 4.500 | 3.826 | .337 | 14.98 | 15.39 | 8 | 5.233 | 4 5/8 | 8.16 | 9.610 | 4.407 | 1.48 |
| 5 | 5.563 | 4.813 | .375 | 20.78 | 21.42 | 8 | 6.420 | 5 1/8 | 12.87 | 20.67 | 6.112 | 1.84 |
| 6 | 6.625 | 5.761 | .432 | 28.57 | 29.33 | 8 | 7.482 | 5 1/8 | 15.18 | 40.49 | 8.405 | 2.20 |
| 8 | 8.625 | 7.625 | .500 | 43.39 | 44.72 | 8 | 9.596 | 6 1/8 | 26.63 | 105.7 | 12.76 | 2.88 |
| 10 | 10.750 | 9.750 | .500 | 54.74 | 56.94 | 8 | 11.958 | 6 5/8 | 44.16 | 211.9 | 16.10 | 3.63 |
| 12 | 12.750 | 11.750 | .500 | 65.42 | 68.02 | 8 | 13.958 | 6 5/8 | 51.99 | 361.5 | 19.24 | 4.34 |

*By permission of the American Institute of Steel Construction.

## TABLE VIII (Continued)

| | DIMENSIONS | | | | | | COUPLINGS | | | PROPERTIES | | |
|---|---|---|---|---|---|---|---|---|---|---|---|---|
| Nom. Diam. (in.) | Outside Diam. (in.) | Inside Diam. (in.) | Thick-ness (in.) | Weight per Foot (lb) Plain Ends | Weight per Foot (lb) Thread & Cplg. | Threads per Inch | Outside Diam. (in.) | Length (in.) | Weight (lb) | $I$ (in.⁴) | $A$ (in.²) | $r$ (in.) |
| | | | | | Double-Extra Strong | | | | | | | |
| ½ | .840 | .252 | .294 | 1.71 | 1.73 | 14 | 1.085 | 1⅞ | .22 | .024 | .504 | .22 |
| ¾ | 1.050 | .434 | .308 | 2.44 | 2.46 | 14 | 1.316 | 2⅛ | .33 | .058 | .718 | .28 |
| 1 | 1.315 | .599 | .358 | 3.66 | 3.68 | 11½ | 1.575 | 2⅜ | .47 | .140 | 1.076 | .36 |
| 1¼ | 1.660 | .896 | .382 | 5.21 | 5.27 | 11½ | 2.054 | 2⅞ | 1.04 | .341 | 1.534 | .47 |
| 1½ | 1.900 | 1.100 | .400 | 6.41 | 6.47 | 11½ | 2.294 | 2⅞ | 1.17 | .568 | 1.885 | .55 |
| 2 | 2.375 | 1.503 | .436 | 9.03 | 9.14 | 11½ | 2.870 | 3⅝ | 2.17 | 1.311 | 2.656 | .70 |
| 2½ | 2.875 | 1.771 | .552 | 13.70 | 13.87 | 8 | 3.389 | 4⅛ | 3.43 | 2.871 | 4.028 | .84 |
| 3 | 3.500 | 2.300 | .600 | 18.58 | 18.79 | 8 | 4.014 | 4⅛ | 4.13 | 5.992 | 5.466 | 1.05 |
| 3½ | 4.000 | 2.728 | .636 | 22.85 | 23.16 | 8 | 4.628 | 4⅝ | 6.29 | 9.848 | 6.721 | 1.21 |
| 4 | 4.500 | 3.152 | .674 | 27.54 | 27.95 | 8 | 5.233 | 4⅝ | 8.16 | 15.28 | 8.101 | 1.37 |
| 5 | 5.563 | 4.063 | .750 | 38.55 | 39.20 | 8 | 6.420 | 5⅛ | 12.87 | 33.64 | 11.34 | 1.72 |
| 6 | 6.625 | 4.897 | .864 | 53.16 | 53.92 | 8 | 7.482 | 5⅛ | 15.18 | 66.33 | 15.64 | 2.06 |
| 8 | 8.625 | 6.875 | .875 | 72.42 | 73.76 | 8 | 9.596 | 6⅛ | 26.63 | 162.0 | 21.30 | 2.76 |
| | | | | | Large O. D. Pipe | | | | | | | |

Pipe 14″ and larger is sold by actual O. S. diameter and thickness. Sizes 14″, 15″, and 16″ are available regularly in thicknesses varying by ¹⁄₁₆″ from ¼″ to 1″, inclusive.

NOTE: All pipe is furnished random length unless otherwise ordered, viz: 12 to 22 ft with privilege of furnishing 5 per cent in 6 to 12 ft lengths. Pipe railing is most economically detailed with slip joints and random lengths between couplings.

# TABLE IX
## AMERICAN STANDARD TIMBER SIZES*

| Nominal Size (in.) | American Standard Dressed Size (in.) | Area of Section (in.²) | Weight per Foot (lb.) | Moment of Inertia (in.⁴) | Section Modulus (in.³) | Nominal Size (in.) | American Standard Dressed Size (in.) | Area of Section (in.²) | Weight per Foot (lb) | Moment of Inertia (in.⁴) | Section Modulus (in.³) |
|---|---|---|---|---|---|---|---|---|---|---|---|
| 2x4 | 1⅝x3⅝ | 5.89 | 1.64 | 6.45 | 3.56 | 10x10 | 9½x9½ | 90.3 | 25.0 | 679 | 143 |
| 6 | 5⅝ | 9.14 | 2.54 | 24.1 | 8.57 | 12 | 11½ | 109 | 30.3 | 1204 | 209 |
| 8 | 7½ | 12.2 | 3.39 | 57.1 | 15.3 | 14 | 13½ | 128 | 35.6 | 1948 | 289 |
| 10 | 9½ | 15.4 | 4.29 | 116 | 24.4 | 16 | 15½ | 147 | 40.9 | 2948 | 380 |
| 12 | 11½ | 18.7 | 5.19 | 206 | 35.8 | 18 | 17½ | 166 | 46.1 | 4243 | 485 |
| 14 | 13½ | 21.9 | 6.09 | 333 | 49.4 | 20 | 19½ | 185 | 51.4 | 5870 | 602 |
| 16 | 15½ | 25.2 | 6.99 | 504 | 65.1 | 22 | 21½ | 204 | 56.7 | 7868 | 732 |
| 18 | 17½ | 28.4 | 7.90 | 726 | 82.9 | 24 | 23½ | 223 | 62.0 | 10274 | 874 |
|  |  |  |  |  |  | 12x12 | 11½x11½ | 132 | 36.7 | 1458 | 253 |
| 3x4 | 2⅝x3⅝ | 9.52 | 2.64 | 10.4 | 5.75 | 14 | 13½ | 155 | 43.1 | 2358 | 349 |
| 6 | 5⅝ | 14.8 | 4.10 | 38.9 | 13.8 | 16 | 15½ | 178 | 49.5 | 3569 | 460 |
| 8 | 7½ | 19.7 | 5.47 | 92.3 | 24.6 | 18 | 17½ | 201 | 55.9 | 5136 | 587 |
| 10 | 9½ | 24.9 | 6.93 | 188 | 39.5 | 20 | 19½ | 224 | 62.3 | 7106 | 729 |
| 12 | 11½ | 30.2 | 8.39 | 333 | 57.9 | 22 | 21½ | 247 | 68.7 | 9524 | 886 |
| 14 | 13½ | 35.4 | 9.84 | 538 | 79.7 | 24 | 23½ | 270 | 75.0 | 12437 | 1058 |
| 16 | 15½ | 40.7 | 11.3 | 815 | 105 |  |  |  |  |  |  |
| 18 | 17½ | 45.9 | 12.8 | 1172 | 134 | 14x14 | 13½x13½ | 182 | 50.6 | 2768 | 410 |
|  |  |  |  |  |  | 16 | 15½ | 209 | 58.1 | 4189 | 541 |
| 4x4 | 3⅝x3⅝ | 13.1 | 3.65 | 14.4 | 7.94 | 18 | 17½ | 236 | 65.6 | 6029 | 689 |
| 6 | 5⅝ | 20.4 | 5.66 | 53.8 | 19.1 | 20 | 19½ | 263 | 73.1 | 8342 | 856 |
| 8 | 7½ | 27.2 | 7.55 | 127 | 34.0 | 22 | 21½ | 290 | 80.6 | 11181 | 1040 |
| 10 | 9½ | 34.4 | 9.57 | 259 | 54.5 | 24 | 23½ | 317 | 88.1 | 14600 | 1243 |
| 12 | 11½ | 41.7 | 11.6 | 459 | 79.9 |  |  |  |  |  |  |
| 14 | 13½ | 48.9 | 13.6 | 743 | 110 | 16x16 | 15½x15½ | 240 | 66.7 | 4810 | 621 |
| 16 | 15½ | 56.2 | 15.6 | 1125 | 145 | 18 | 17½ | 271 | 75.3 | 6923 | 791 |
| 18 | 17½ | 63.4 | 17.6 | 1619 | 185 | 20 | 19½ | 302 | 83.9 | 9578 | 982 |
|  |  |  |  |  |  | 22 | 21½ | 333 | 92.5 | 12837 | 1194 |
| 6x6 | 5½x5½ | 30.3 | 8.40 | 76.3 | 27.7 | 24 | 23½ | 364 | 101 | 16763 | 1427 |
| 8 | 7½ | 41.3 | 11.4 | 193 | 51.6 |  |  |  |  |  |  |
| 10 | 9½ | 52.3 | 14.5 | 393 | 82.7 | 18x18 | 17½x17½ | 306 | 85.0 | 7816 | 893 |
| 12 | 11½ | 63.3 | 17.5 | 697 | 121 | 20 | 19½ | 341 | 94.8 | 10813 | 1109 |
| 14 | 13½ | 74.3 | 20.6 | 1128 | 167 | 22 | 21½ | 376 | 105 | 14493 | 1348 |
| 16 | 15½ | 85.3 | 23.6 | 1707 | 220 | 24 | 23½ | 411 | 114 | 18926 | 1611 |
| 18 | 17½ | 96.3 | 26.7 | 2456 | 281 | 26 | 25½ | 446 | 124 | 24181 | 1897 |
| 20 | 19½ | 107.3 | 29.8 | 3398 | 349 |  |  |  |  |  |  |
|  |  |  |  |  |  | 20x20 | 19½x19½ | 380 | 106 | 12049 | 1236 |
|  |  |  |  |  |  | 22 | 21½ | 419 | 116 | 16150 | 1502 |
| 8x8 | 7½x7½ | 56.3 | 15.6 | 264 | 70.3 | 24 | 23½ | 458 | 127 | 21089 | 1795 |
| 10 | 9½ | 71.3 | 19.8 | 536 | 113 | 26 | 25½ | 497 | 138 | 26945 | 2113 |
| 12 | 11½ | 86.3 | 23.9 | 951 | 165 | 28 | 27½ | 536 | 149 | 33795 | 2458 |
| 14 | 13½ | 101.3 | 28.0 | 1538 | 228 |  |  |  |  |  |  |
| 16 | 15½ | 116.3 | 32.0 | 2327 | 300 | 24x24 | 23½x23½ | 552 | 153 | 25415 | 2163 |
| 18 | 17½ | 131.3 | 36.4 | 3350 | 383 | 26 | 25½ | 599 | 166 | 32472 | 2547 |
| 20 | 19½ | 146.3 | 40.6 | 4634 | 475 | 28 | 27½ | 646 | 180 | 40727 | 2962 |
| 22 | 21½ | 161.3 | 44.8 | 6211 | 578 | 30 | 29½ | 693 | 193 | 50275 | 3408 |

NOTE: All properties and weights given are for dressed size only. The weights given above are based on assumed average weight of 40 lb/cu ft.

*By permission of the American Institute of Steel Construction.

# TABLE X
## Slopes and Deflections of Selected Beams

| Beam | Slope | Deflection (+ upward) |
|------|-------|------------------------|
| **1** | $\theta_1 = -\dfrac{PL^2}{16EI}$ at $x = 0$<br><br>$\theta_2 = +\dfrac{PL^2}{16EI}$ at $x = L$ | $y_{max} = -\dfrac{PL^3}{48EI}$ at $x = L/2$ |
| **2** | $\theta_1 = -\dfrac{wL^3}{24EI}$ at $x = 0$<br><br>$\theta_2 = +\dfrac{wL^3}{24EI}$ at $x = L$ | $y_{max} = -\dfrac{5wL^4}{384EI}$ at $x = L/2$ |
| **3** | $\theta_1 = -\dfrac{Pb(L^2 - b^2)}{6LEI}$ at $x = 0$<br><br>$\theta_2 = +\dfrac{Pa(L^2 - a^2)}{6LEI}$ at $x = L$ | $y_{max} = -\dfrac{Pb(L^2 - b^2)^{3/2}}{9\sqrt{3}LEI}$ at $x = \sqrt{(L^2 - b^2)/3}$<br><br>$y_{center}$ (not max) $= -\dfrac{Pb(3L^2 - 4b^2)}{48EI}$ |
| **4** | $\theta = -\dfrac{PL^2}{2EI}$ at $x = L$ | $y_{max} = -\dfrac{PL^3}{3EI}$ at $x = L$ |
| **5** | $\theta = -\dfrac{wL^3}{6EI}$ at $x = L$ | $y_{max} = -\dfrac{wL^4}{8EI}$ at $x = L$ |
| **6** | Zero at ends<br>Zero at center | $y_{max}$ (at center) $= -\dfrac{wL^4}{384EI}$<br><br>Moment at ends is max.<br><br>$M_{max} = -\dfrac{wL^2}{12}$ |
| **7** | Zero at ends<br>Zero at center | $y_{max}$ (at center) $= -\dfrac{PL^3}{192EI}$<br><br>Moment at ends is max.<br><br>$M_{max} = -\dfrac{PL}{8}$ |

# Answers to Selected Problems

# ANSWERS TO SELECTED PROBLEMS

## Chapter 1

1-1. $\sigma_\theta = 4014$ psi (T), $\tau_\theta = 1220$ psi

1-4. $\sigma_\theta = 3090$ psi (C), $\tau_\theta = 1785$ psi

1-6. $\sigma_\theta = 85,400$ psi (T), $\tau_\theta = 17,140$ psi

1-8.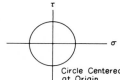

1-10.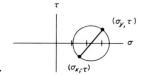

1-12.

1-14.

1-16.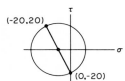

1-18.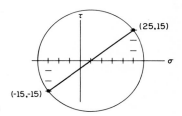

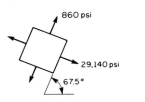

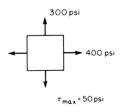

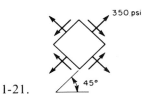

1-19.                                        1-21.

1-24. $\sigma_{\max} = 187{,}200$ psi (T), $\sigma_{\min} = \sigma_z = 0$, $\tau_{\max} = 93{,}600$ psi

1-26. $\sigma_{\max} = \sigma_z = 0$, $\sigma_{\min} = 7000$ psi (C), $\tau_{\max} = 3500$ psi

1-28. $\sigma_{\max} = 360$ (T), $\sigma_{\min} = 360$ (C), $\tau_{\max} = 360$ psi

1-30. $\sigma_{\max} = 5{,}600$ (T), $\sigma_{\min} = 11{,}600$ (C), $\tau_{\min} = 8{,}600$ psi

1-32. $\sigma_{\max} = 26{,}200$ (T), $\sigma_{\min} = 3{,}800$ (T), $\tau_{\max} = 13{,}100$ psi

1-34. $\sigma_{\max} = 565$ (T), $\sigma_{\min} = 405$ (C), $\tau_{\max} = 485$ psi

1-36. $\theta = 6°\;\curvearrowleft$, $\sigma_\theta = 7475$ psi (T) or $\theta = 62.2°\;\curvearrowleft$, $\sigma_\theta = 1475$ psi (C)

1-38. $\theta = 20.7°\;\curvearrowleft$, $\tau_\theta = -7900$ psi or $\theta = 69.8°\;\curvearrowright$, $\tau_\theta = 7900$ psi

1-40. $\sigma_y = 1200$ psi (T)

1-42. $\sigma_\theta = 2800$ psi (C), $\tau_\theta = -9600$ psi

1-43. $\sigma_\theta = 6000$ psi (T), $\tau_\theta = 0$

1-45. $\sigma_\theta = 976$ psi (T), $\tau_\theta = 91$ psi

1-47. $\sigma_{\max} = 25{,}000$ psi (T), $\sigma_{\min} = 15{,}000$ psi (C), $\tau_{\max} = 20{,}000$ psi

## Chapter 2

2-2. $\epsilon = 0.022$ in./in.

2-4. *a)* $\epsilon = 0.00436$ in./in., *b)* $\epsilon = 0.00397$ in./in.

2-6. $e_B = 1'' + e_A$

$e_c = \dfrac{9}{6}(1) + e_A$

2-8. $\Delta\epsilon_{AC} = 0.112$ in./in.

$\Delta\epsilon_{BD} = -0.128$ in./in.

2-10. $\Delta\epsilon = 0.005$ in./in.

2-12. $\epsilon_{AP} = \dfrac{[(L - X)^2 + (\sqrt{3}L + Y)^2]^{1/2} - 2L}{2L}$ in./in.

For small $X$, $Y \ll L$

$\epsilon_{AP} \approx \dfrac{\sqrt{3}\,Y - X}{4L}$ in./in.

2-14. *a)* $\gamma_{xy} = \dfrac{e_0}{h}$ at $(0,0)$ and $(L/3, h/2)$

    *b)* $\gamma_{xy} = 0$ at $(0,0)$

        $\bar{\gamma}_{xy} = \dfrac{e_0}{h}$ at $(L/3, h/2)$

2-15. $\gamma_{r\theta} = \dfrac{a}{r}\left(\dfrac{b\phi}{b-a}\right)$

2-18. $\epsilon_{max} = 2400\,\mu$ in./in., $\epsilon_{min} = 0$, $\gamma_{max} = 2400\,\mu$ in./in.

2-20. $\epsilon_{max} = 2100\,\mu$ in./in., $\epsilon_{min} = -2900\,\mu$ in./in., $\gamma_{max} = 5000\,\mu$ in./in.

2-22. $\epsilon_{max} = 2400\,\mu$ in./in., $\epsilon_{min} = 0$, $\gamma_{max} = 2400\,\mu$ in./in.

2-24. $\epsilon_{max} = 1487\,\mu$ in./in., $\epsilon_{min} = -736\,\mu$ in./in., $\gamma_{max} = 2223\,\mu$ in./in.

2-26. $\epsilon_{max} = 4290$, $\epsilon_{min} = -3740$

2-28. $\epsilon_{max} = 5690$, $\epsilon_{min} = -2930$

2-30. $\epsilon_{max} = 1990$, $\epsilon_{min} = -790$

2-32. $\epsilon_{max} = 180$, $\epsilon_{min} = -680$

## Chapter 3

3-6. $\sigma_{PL} = 8100$ psi, $E = 1.4 \times 10^6$ psi

3-7. $E_T = 2.08 \times 10^5$ psi, $E_s = 10.62 \times 10^5$ psi

3-10. $E_T = 3.26 \times 10^6$ psi, $E_s = 5.47 \times 10^6$ psi

3-12. $G = 2.22 \times 10^6$ psi

3-16. $E_s = 15.6 \times 10^6$ psi, $E_T = 10.84 \times 10^6$ psi

3-18. $\mu = 0.3$

3-21. $\sigma_{max} = 64,800$ psi, $\sigma_{min} = 21,600$ psi, $\tau_{max} = 21,600$ psi

       min    24.2°

3-22. $\sigma_{max} = 14,400$ psi, $\sigma_{min} = -14,400$ psi, $\tau_{max} = 14,400$ psi

       16.8°  min

3-24. $\sigma_{max} = 20,200$ psi, $\sigma_{min} = -33,600$ psi, $\tau_{max} = 26,880$ psi

       max    31.8°

3-26. $\sigma_{max} = -6,940$ psi, $\sigma_{min} = -53,100$ psi, $\tau_{max} = 23,080$

       18.4°  min

3-32. Assuming $E$ positive, for $G$ to be positive

$$G = \frac{E}{2(1+\mu)} \implies \mu > -1$$

Similarly, for the bulk modulus $K$ to be positive

$$K = \frac{E}{3(1 - 2\mu)} \implies \mu < \frac{1}{2}$$

## Chapter 4

4-1. $\sigma_{ys} = 52,800$ psi

4-4. $R = 105.8$ in-lb/in.$^3$

4-6. $R = 20.2$ in-lb/in.$^3$

     $T = 123.4$ in-lb/in.$^3$

4-9. b) $U = 358$ in-lb/in.$^3$ (tension)

       $U = 310$ in-lb/in.$^3$ (comp)

    d) $U = 630$ in-lb/in.$^3$

4-10. $U = 181$ in-lb/in.$^3$

## Chapter 6

6-1. a) 200 lb compression

    b) 300 lb compression

    c) 40  lb tension

6-3. *a)* $F_n = 145.5$ lb compression, $F_s = 122.1$ lb $\searrow$

*b)*

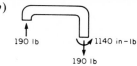

190 lb     1140 in-lb

190 lb

*c)* 570 in-lb

190 lb

190 lb

6-5. *a)* $F_n = 533$ lb (T), $V = 500$ lb, $M = 1500$ ft-lb
    *b)* $F_n = 800$ lb (C), $V = 133$ lb, $M = 400$ ft-lb
    *c)* $F_n = 300$ lb (C), $V = 400$ lb, $M = 400$ ft-lb

6-7. *a)*   2000   2000 lb

1000 in-lb

    *b)*   2000   2000 lb

    *c)* Same as (*b*)

6-9. *a)* $V = 250$ lb, $M_x = 500$ in-lb
    *b)* $V = 250$ lb, $M_x = 1500$ in-lb, $M_y = 2000$ in-lb
    *c)* $V = 250$ lb, $M_x = 1500$ in-lb, $M_y = 3000$ in-lb

6-11. *a)* $F_n = 300$ lb (T), $V_y = 100$ lb, $V_z = 200$ lb, $M_y = 800$ in-lb,
           $m_z = 800$ in-lb, $M_x = 800$ in-lb
    *b)* $F_n = 100$ lb (T), $V_x = 300$ lb, $V_z = 200$ lb, $M_x = 1200$ in-lb,
       $M_z = 1800$ in-lb, $M_y = 0$
    *c)* $F_n = 100$ lb (T), $V_x = 300$ lb, $V_z = 200$ lb, $M_x = 2400$ in-lb,
       $M_z = 3600$ in-lb, $M_y = 0$

6-13. *a)* $V = 5.7$ K, $M = 84.7$ in-K, $T = 6$ in-K
    *b)* $V = 5.3$ K, $M = 103.6$ in-K, $T = 3$ in-K
    *c)* $V = 13.3$ K, $M = 26.6$ in-K, $T = 0$

6-15. *a)* $F_n = \dfrac{3850}{2} \sqrt{2}$ lb (C), $V = \dfrac{15000}{2} \sqrt{2}$ lb, $M = \dfrac{45000}{2} \sqrt{2}$ ft-lb
    *b)* $F_n = 3621$ lb (C), $V = 2121$ lb, $M = 4320$ ft-lb
    *c)* $F_n = 7925$ lb (T), $V = 8575$ lb, $M = 34,300$ ft-lb

6-17. $M_{max} = \dfrac{WL^2}{8}$

6-19. $V_{max} = \dfrac{3}{8} WL$, $M_{max} = \dfrac{9}{128} WL^2$

6-21. $V_{max} = \dfrac{1}{3} WL$, $M_{max} = \dfrac{\sqrt{3}}{27} WL^2$

6-23. $T_{max} = 7T$

6-25. $M_{max} = 3PR$

6-27. $V_{max} = \dfrac{8}{7} P$, $M_{max} = \dfrac{32}{7} Pa$

6-29. $V_{\max} = 2P, M_{x\max} = \dfrac{PL}{8}, M_{z\max} = \dfrac{4}{3} PL$

6-31. Maximum sheer $= 4P$
Maximum compression $= 8P$
Maximum bending moment $= \dfrac{3}{4} Ph$

6-33. $V_{\max} = \dfrac{2M_{\mathrm{o}}}{L}, M_{\max} = M_{\mathrm{o}}$

6-37.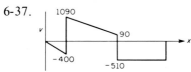

6-39.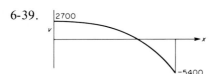

6-41.

6-43.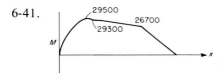

6-45.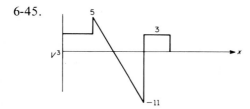

6-49.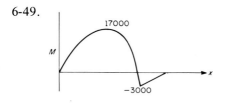

6-51.

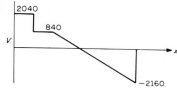

6-53.

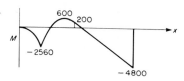

6-55.

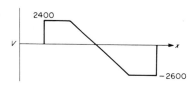

6-57.

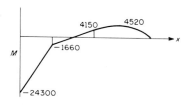

6-60.

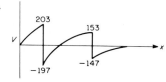

## Chapter 8

8-1. $P = 576{,}000$ lb, $y_o = 5.33$ in.

8-3. $P = 23{,}040$ lb, $y_o = 3.33$ in.

8-5. $P = 17{,}440$ lb (T), $y_o = 1.4$ in.

8-7. $P = 135 \times 10^4$ lbs, $y_o = 2.78$ in.

8-9. $P = 256 \sqrt{2} \times 10^3$ lb, $y_o = 4.8$ in.

8-11. $\sigma = \dfrac{pD}{4t}$

8-14. $F_n = 1785$ lb/in.

8-16. a) $\sigma_1 = 9190$ psi, $\sigma_2 = 11,060$ psi
b) $\sigma_L = 6,750$ psi, $\sigma_c = 13,500$ psi
c) $p = 140.6$ psi

8-17. a) $\sigma_H = 15.03 \times 10^3$ psi
b) $\sigma_B = 376$ psi

8-18. $e = 0.0054$ in.

8-20. $e = \dfrac{WL}{2AE}$

8-22. $\epsilon_L = 0.2$ in.
$\Delta V = 0.2$ in.$^3$

8-23. a) $L_s = 19.6$ in., $L_B = 19.6$ in., $L_A = 8.8$ in.
b) $P = 78.8 \times 10^3$ lbs

8-26. $e = 0.0106$ in.

8-27. $e_{total} = 0.00332$ in.

8-29. $e = 0.000216$ in.

8-31. a) $e = 0.05$ in.,  b) $e = 0.25$ in.

8-33. $b = 1.77$ in.

8-35. area $= \frac{1}{3}$ sq. in. based on $\sigma_{ys}$

8-37. approx. $4\frac{1}{2}$ hrs

8-38. $\sigma_s = 31,000$ psi, $\sigma_{al} = 13,000$ psi

8-40. a) $T = 7.8°F$,  b) $\sigma = 10,280$ psi

8-42. $P = 45,000$ lbs

8-44. $D = 7.350$ in.

8-46. a) $\sigma_A = 5450$ psi, b) $\sigma_B = 80,000$ psi

8-48. $D = 1$ in. based on shear stress

8-50. $N = 4$ rivets due to shear

8-52. a) $\epsilon_B = 768\,\mu$ in./in.,  b) $e = 0.0243$ in.

8-54. a) $\epsilon_B = 1037$,  b) $e = 0.132$

8-56. $\epsilon_A = 223\,\mu$ in./in., $\epsilon_B = 1088\,\mu$ in./in.

8-57. a) $\Delta R = \dfrac{pR^2}{2tE}(1 - \mu)$,  b) $\Delta V = \dfrac{2p\pi R^4}{tE}(1 - \mu)$

8-58. a) $p = 200$ psi,  b) $p = 200$ psi

8-61. $e = \dfrac{wb}{2E}\left[\dfrac{a(a + 2b)}{2b} - b\ln\left(\dfrac{a + b}{b}\right)\right]$

8-64. a) $e = \dfrac{9}{80\pi}\ln 5$,  b) $e = \dfrac{8}{5}\dfrac{(5.06)}{\pi^2}$

8-65. a) $\sigma = \dfrac{\rho\omega^2}{2}[L^2 - l^2]$,  b) $\Delta L = \dfrac{\rho\omega^2 L^3}{3E}$

8-68. reading $= 0.1675$ in.

## Chapter 9

9-3. $T = .626\pi r^3 \tau_{max}$

9-4. $T = \dfrac{2}{3}\pi\tau_{max} r^3(1 - c^3)$

9-6. $T = \frac{4}{7} \pi r^3 \tau_{max}$

9-7. $T = \frac{\tau_r J}{r}, T = \frac{\gamma_r GJ}{r}$

9-9. $T = \frac{2\pi n r^3}{1 + 3n} \tau_r, T = \frac{2\pi n r^3}{1 + 3n} (G\gamma_r)^{1/n}$

9-11. $\tau_{1/2} = 17,400$ psi

9-12. $T = 39.3$ in-lb

9-14. $\frac{\Delta T}{T} = 25\%$

9-15. $T = 293,000$ in-lb

9-17. $\sigma_{max} = \tau_{max} = 12,000$ psi

9-19. $\theta = 0.053$ rad, $T = 610,000$ in-lb

9-21. $T = 48,900$ in-lb, $L_{st} = 6.74$ in., $L_{al} = 11.38''$, $L_{Br} = 53.90$ in

9-23. a) hp $= 935$, b) based on fully plastic condition hp $= 291$

9-25. $\Delta\theta = 0.0246$ radian

9-27. $\Delta\theta = 0.0173$ radian

9-29. $L = 300\pi$ in

9-30. $L_{Br} = 92.8$ in., $L_{st} = 3.2$ in.

9-31. hp $= 12.46$

9-32. hp $= 33.3$

9-34. a) hp $= 3803$, b) hp $= 2717$

9-36. a) $\tau_{L/4} = \frac{3}{2} \frac{TL}{\pi R^3}$, b) $\theta = \frac{TL^2}{\pi R^4 G}$

9-38. $\tau_{st} = 4160$ psi

9-40. $T = 78,400$ in-lb

9-42. $\tau_{max} = 10,390$ psi, $\sigma_{max} = 15,375$ psi

9-44. $\tau_{max} = 3350$ psi, $\sigma_{max} = 5635$ psi, $\sigma_{min} = 1065$ psi (C)

9-46. $\epsilon = 821 \mu$ in./in.

9-48. $\tau = 18,300$ psi

9-50. $T = 576$ in-lb

9-52. $\tau = \frac{11}{6} \frac{\rho^{2/3} T}{R^{11/3} \pi}$

9-53. $J$ is approximately 12.7% larger

9-54. $\theta = \frac{TL^2}{6 J G}$

9-57. $r_0 = 2.20$ in., $r_i = 1.34$ in.

9-60. $\Delta\theta = \frac{PR}{6\pi L} \left[ \frac{1}{R_1^2} - \frac{1}{R_2^2} \right]$

## Chapter 10

10-1. $I_a = .0438 D^4, I_b = .0683 D^4$; therefore (b) is better.

10-2. $M = 227,000$ in-lb

10-4. $\sigma_{max} = 1100$ psi

10-6. $\sigma_{max} = 11{,}800$ psi

10-8. $p = 5000$ lbs

10-10. $M_{fp}/M_{el} = 1.4$

10-12. $M_{fp}/M_{el} = \dfrac{16}{3\pi}$

10-14. $M_{fp} = 1.44 \times 10^6$ in-lb

10-15. $K = 3/4$

10-16. $M = 8.8 \times 10^5$ in-lb

10-18. $\tau_{②} = 12$ ksi, $\tau_{③} = 12.5$ ksi, $\tau_{④} = 8$ ksi

10-22. spacing $= 1.137$ in.

10-23. $\tau_{mix} = 375$ psi

10-24. $\sigma_{mix} = 884$ psi, $\tau_{max} = 245$

10-26. $\sigma_t = 1730$ psi at wall, $\sigma_c = 1380$ psi 15′ from wall, $\tau = 152$ psi at wall.

10-27.

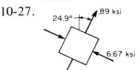

10-29. $\sigma_t = 6430$ psi at right end, $\sigma_c = 3460$ psi at right end, $\tau = 98.8$ psi

10-31. $2'' \times 10''$ nominal size

10-32. a) $p = 89{,}800$ lb failure due to compressive stress

     b) $p = 241{,}000$ lb

     c) $p = 65{,}600$ lb

10-33. $W = 8160$ lbs failure due to formation of plastic hinge

10-35. $\sigma = \dfrac{4p}{d^2}$ comp.

10-37. $p_{max} = 7750$ due to compressive stress

10-38. $\sigma_{mix} = 5250$ psi circumferential stress due to pressure

10-40. $\sigma_c = 4460$ psi, $\sigma_t = 7830$ psi, $\tau_m = 4500$ psi

10-42. $p = \dfrac{6}{17}\,\sigma_m\,\dfrac{\pi r^3}{L}$

10-43. $\sigma_{max} = 26.1$ ksi comp., $\tau_{mix} = 13.7$ ksi

10-45. $\delta = -\dfrac{wL^4}{8EI}$, w. lb/in., $E$ psi, $L$ in., $I$ in.$^4$

10-47. $W = 144$ lbs

10-48. $\dfrac{b}{a} = 1.96$, $\dfrac{c}{a} = 1.74$, $\dfrac{b}{c} = 1.125$

10-50. $y_A = -0.238$ in.

10-51. $y_A = -0.440$ in.

10-53. $\delta_{VA} = -\dfrac{5PL^3}{48EI} \downarrow$, $\delta_{HA} = \dfrac{PL^3}{16EI} \rightarrow$

10-54. $\delta = 0, \theta_m = -0.002$ radian

10-56. $\delta = \dfrac{3\,wL^4}{8\,(3\,EI + k\,L^3)}$

10-57. $M_0 = 81,000$ in-lb, $Y_{max} = 5.40$ in.

10-58. $y_A = 1.37$ in.

10-59. $y = .3935$ in.

10-62. $\delta = 2.62$ in.

10-64. $\delta_t = \dfrac{5\,w\,L^4}{32\,E\,b^4}$ always vertical

10-65. $R = P/2$, $M_0 = \dfrac{PL}{8}$, $\delta = -\dfrac{PL^3}{192\,EI}$

10-67.

10-70. $R_1 = {}^3/_{16}\ wL$, $R_2 = {}^5/_8\ wL$, $R_3 = {}^3/_{16}\ wL$

10-71. $P = 1660$ lbs

10-73. $M_{fp}/M_{el} = 16/3\,\pi$

10-74. $M = 2.258 \times 10^6$ in-lb

10-75. $L = 15.78$ ft

10-77. $\sigma_m = 528$ psi, $\tau_H = 54.2$ psi, $\tau_{mix} = {}^{528}/_2$ psi

10-78. $r = 1.18$ in.

10-80. $\delta = \dfrac{7}{128}\dfrac{wL^4}{(3\,EI + kL^3)}$

10-82. $\delta = \dfrac{PL^3}{32\,EI}$

10-83. $\delta_p = \dfrac{1}{3}\dfrac{WL^3}{EI}$

10-84. capacity $\leq 20,000$ lbs

10-85. $M_0 = 6000$ ft-lbs

10-87. $\delta_x = .146$, $\delta_y = .182$

10-89. $R_1 = \dfrac{4}{27}\,P$, $R_2 = \dfrac{46}{27}\,P$

# Chapter 11

11-1. For small deflections, $P = \dfrac{4\beta}{L}$ lb

11-3. $P = \dfrac{kL}{16}$ lb

11-5. $L = 75''$

11-6. $L/r = 103$

11-8. $L/r = 108$

11-9. $D = 10^{11}$

11-12. $P = 15$ lb

11-14. $P = 9500$ lb

11-16. $d = 7.79''$

11-18. $a)$ 49.7, $b)$ 44.5, $c)$ 49.7

11-20. $W = 400$ lb

11-22. $\Delta t = 3.10°$

11-23. $a = 7.2$ in.$^2$

11-24. 4 turns

11-27. $E = 20 \times 10^6$ psi, $t = 81.4°$

11-29. $b = 1.25''$, $h = .622$ in.

11-30. $FS = 1.33$

11-32. $a = 46.9$ in.$^2$

11-33. $a)$ $l/r = 87$, $b)$ no, since min $l/r = 111$

11-35. $P = 823$ lbs

11-37. $l = 14.52$ in.

11-39. $P \approx 28,800$ lbs

11-43. $D_0 = 4.1$ in.

11-45. $P = 31,000$ lb

11-46. $P = 7500$ lb

11-47. $P = 240,000$ lb

11-48. $P = 636$ lb

11-49. $P = 33,100$ lb

11-52. $P = 233,000$ lb

# Chapter 12

12-2. Shear center at juncture of tee

12-3. Shear center at centroid

12-7. Shear center on horizontal web at $\dfrac{t_2 b_2{}^3}{12 I} a$ from left flange

12-8. $S = 1.01$ in. to left of web

12-9. $s = 0.412$ in. to right of web, $\sigma_{max} = 13,150$ psi

12-11. $\sigma_t = 989$ psi, $\sigma_c = -503$, error $= 179\%$

12-12. $\sigma_t = 1543$, $\sigma_c = -650$, error $= 40.3\%$

12-13. $b = 1.24$ in.

12-15. $R = 9.67$ in.

12-17. $\sigma_{max} = -20,300$ psi at inside

12-19. $\sigma_c = 5333$ psi, $\sigma_t = 727$ psi

12-20. $M = 362,500$ in-lb

12-22. $M = 53,100$ in-lb

12-24. $a)$ $t = 0.31$ in., $b)$ $w = 4.97$ in.

12-25. $4'' \times 16''$ nominal size, $\delta_w/\delta_c = .387$

12-27. $\sigma_c = 1085$ psi, $\sigma_s = 16{,}270$ psi

12-29. $M = 663{,}000$ in-lb

## Chapter 13

13-1. $U_T = \dfrac{\sigma^2}{2E}$, $\quad U_D = \dfrac{\sigma^2}{3E}(1 + \mu)$, $U_V = \dfrac{\sigma^2}{6E}(1\text{-}2\,\mu)$

13-2. $U_T = \dfrac{1}{2E}(\sigma_x^2 + \sigma_y^2) - \dfrac{\mu}{E}(\sigma_x\,\sigma_y) + \dfrac{1}{2G}(\tau_{xy}^2)$

13-3. $U_D = \dfrac{1}{3E}(1 + \mu)[\sigma_x^2 + \sigma_y^2 + \sigma_z^2 - \sigma_x\sigma_y - \sigma_y\sigma_z - \sigma_x\sigma_z] + \dfrac{1}{2G}\tau_{xy}^2$ where $\sigma_z = \mu(\sigma_x + \sigma_y)$

13-6. Max $\sigma = \tau_{PL}$, $\quad U_T = \dfrac{\tau_{PL}^2}{2G}$, $\tau_{\max} = \tau_{PL}$

Max $\epsilon = \dfrac{\tau_{PL}}{2G}$, $\sigma_e = \sqrt{6}\,\tau_{PL}$

13-7. Axial strain theory is least conservative based on uniaxial data. Total elastic energy theory is most conservative.

13-9. Failure should be based on maximum shear stress theory or energy of distortion theory with limiting value taken from a pure torsion test. (Refer to Prob. 13-6)

$$\tau_{yP} = \sqrt{L^2 + b^2}\left(\frac{2P}{\pi r^3}\right) \text{ max shear stress}$$

$$\tau_{yP} = \frac{\sqrt{4L^2 + 3b^2}}{\sqrt{3}}\left(\frac{2P}{\pi r^3}\right) \text{ energy of distortion theory}$$

13-10. Failure based on fracture with limiting value taken from uniaxial data.

$$\sigma_{ult} = \frac{P}{\pi r^2}\left[\frac{1}{2} + \sqrt{\left(\frac{1}{2}\right)^2 + \left(\frac{2}{r}\frac{T}{P}\right)^2}\right]$$

13-11. $P = 1490$ lb based on maximum normal stress theory

$P = 1460$ lb based on maximum axial strain theory

13-12. Maximum sheer stress theory $r \approx 1.32''$

Energy of distortion theory $r \approx 1.265''$

## Chapter 14

14-3. $\dfrac{\sigma_3}{\sigma_1} = 2\sqrt{\dfrac{2}{5}}$, $\dfrac{\sigma_2}{\sigma_1} = \dfrac{1}{2}\sqrt{\dfrac{2}{5}}$

14-4. $\dfrac{\tau_2}{\tau_1} = \dfrac{1}{\sqrt{34}}$, $\dfrac{\tau_3}{\tau_1} = \dfrac{8}{\sqrt{34}}$

14-5. $W = \dfrac{w^2 L}{6\,EA}$

14-6. $e = \dfrac{wL}{2\,EA}$

14-8. $W = \dfrac{tL^3}{6JG}$

14-9. $\theta = \dfrac{tL^2}{2\,JG}$

14-11. $R_1 = R_2 = P/2, \quad M_0 = \dfrac{PL}{8}$

14-13. $y_A = 1.31$ in.

14-18. $R_1 = \dfrac{3}{8}\, wL,\, R_2 = \dfrac{5}{8}\, wL,\, M_0 = -\dfrac{wL^2}{8},$

14-20. $R_1 = -\dfrac{3\,wb^2}{2L}\left[\dfrac{b^2}{12L^2} - \dfrac{1}{2}\right]$

$R_2 = \dfrac{3}{2}\dfrac{wb^2}{L}\left[\dfrac{b^2}{12\,L^2} - \dfrac{1}{2} + \dfrac{2\,L}{3\,b}\right]$

$M_0 = \dfrac{wb^2}{4}\left[\dfrac{b^2}{2\,L^2} - 1\right]$

14-21. $R_1 = \dfrac{13}{32}\, wL,\, R_2 = \dfrac{3}{32}\, wL,\, M_1 = \dfrac{-11}{192}\, wL^2,\, M_2 = \dfrac{-5}{192}\, wL^2$

14-24. $\delta_v = 0.0139$ in., $\delta_H = 0$

14-26. $\delta_H = 0.0416$ in., $\delta_v = 0.191$

## Chapter 15

15-1. $\sigma = 19{,}000$ psi

15-3. $a)\ \sigma = 35{,}000$ psi, $\quad b)\ \sigma = 38{,}400$ psi

15-5. $\sigma_{max} = 7080$ psi at groove

15-7. $P = 1040$ lb

15-9. time $= 6.93$ hrs

15-11. $T = 1770$ in-lb

15-13. $P = 40{,}200$ lb

15-14. F.S. $= 1.58$

15-15. $b = 1.78$ in.

15-17. $L = 11.6$ in.

15-19. time $= 40$ hrs

15-20. $h = 1.925$ in.

15-22. $\Delta = 1.282$ in.

15-24. $\delta_D = -39.8$ in., $\sigma = 17{,}280$ psi, $\Delta_S = -4.14$ in.

15-25. Yes

15-26. $h = 13.3$ in.
15-27. $\tau = 2,135$ psi
15-29. $h = 0.058''$

## Chapter 16

16-1. $\epsilon = 532\,\mu$ in./in.
16-2. $\Delta R = 0.097$ ohms
16-3. $\Delta V = 0.0125$ volts
16.4. $\Delta V = 0.025$ volts
16-5. $\Delta V = 0.0163$ volts
16-6. $\Delta V = 0.0088$ volts
16-10. $C = 20.9$ lb/in-fringe

# Index

# INDEX

Accelerometer, 395
Alternating stress, 372
Angle of twist, 184
Angular deformation, 184
Axial deformation, 147
Axially loaded members, 134
   statically indeterminate, 151

Beams
   curved, 314
   equivalent, 319
   nonhomogeneous, 319
   statically indeterminate, 246
Bearing stresses, 154
Bending
   elastic, 212
   modulus, 232
   plane of, 206
   pure, 205, 206
Bifurcation point, 277
Birefringent, 400
   coatings, 402
Body-contact, 89
Bonded gages, 387
Brittle lacquer coatings, 403
Brittleness, 67
Bulk modulus, 60, 336
Buckling, 130, 273
   elastic, 277
   inelastic, 286
   load, 275

Castigliano's Theorem, 350
Centroidal loading, 138
Charpy test, 378
Circumferential strains, 143
Circumferential stress, 144
Coatings; see Brittle lacquer coatings
Columns, 273
Complementary work, 350
Compression test, 67
Concentrated load, 89
Connectors, 154
Constitutive equation, 52
Creep, 81, 374
   limit, 375
   rate, 375
   strength, 375
   tertiary, 375
   test, 375
Cubical dilatation, 60, 335
Curvature, 231

Curved beams, 314
Cylinders, thick-walled, 144

Dead load, 91
Deflection
   curve, 230
   by direct integration, 233
   geometric relationships for, 230
   moment-area method for, 240
   use of singularity functions, 237
   by superposition, 244
Deformation, 26
   axial, 28
   shear, 29
Delta function, 107
Design procedure, 127
Dilatation, 60
Dirac delta function, 107
Displacement, 26
   transducer, 395
Distortion, 26, 27
   energy of, 337
Distortional elastic energy, 335
Distributed load, 89
Double modulus theory, 286
Doublet function, 110
Ductility, 67
Dummy gage, 391
Dynamic loads, 90

Effective stress concentration factor, 364
Elastic
   action, 65, 76
   bending, 212
   buckling, 277
   column, 277
   constants, 58
   deformation, 65
   energy
      distortional, 335
      volumetric, 335
   flexure formula, 215
   limit, 65, 67
   range, 65
   strain energy, 349
   torsion formula, 182
Elasticity, theory of, 20
Electrical resistance strain gages, 387
Elevated temperatures, 374
Empirical column formulas, 292
End conditions, columns, 283
Endurance limit, 370

Energy
  of distortion, 337
  loads, 70, 91
  strain, 70
Engesser equation, 288, 289
Equivalent beam, 319
Equivalent static load, 379
Euler elastic buckling load, 280, 281
Experimental mechanics, 386
Experimental stress analysis, 386

Factor of safety, 128
Factors, stress concentration, 362
Failure, 128
  definition of, 128
  load, 128
  modes of, 129
  theories of, 329, 338
Fatigue, 367
  life, 369
  limit, 370
  loading, 81
  strength, 370
  tests, 370
Foil gage, 387
Fracture, 129
Friction load, 89
Fringe, 401
  order, 401
Flexural rigidity, 232
Flexure formula, elastic, 215
Flexure stresses, 211
Fluctuating loads, 372
Fluctuating stress, 372

Gage
  factor, 389
  high-temperature, 389
General yielding, 130
Goodman diagram, 372
Gravity load, 89

High-temperature gage, 389
High-velocity impact loads, 378
Hollow thin-walled torsion members, 191
Hooke's law, 50
  generalized, 55
Hoop stress, 145
Huber-Hencky-von Mises theory, 340
Hydrostatic
  compression, 341
  stress, 339
  tension, 341

Ideally plastic behavior, 215
Impact
  loads, 91, 377
  test, 378
Inelastic
  buckling, 286
    load, 288
  deformation, 130
Inertial forces, 89
Instability, 275
Integration, deflection by, 233
Isochromatics, 401
Isoclinics, 402
Izod test, 378

Kelvin, Lord, 388

Lacquer coating; *see* Brittle lacquer coatings
Life, fatigue, 369
Limit
  creep, 375
  endurance, 370
  fatigue, 370
Line load, 89
Linear stress distribution, 213
Live load, 91
Load
  buckling, 275
  energy, 70
  failure, 128
  ficticious, 353
  fluctuating, 372
  impact, 377
  nonstatic, 361
  repeated, 129
  transducer, 395
  working, 128
Loading
  centroidal, 138
  repeated, 361
Longitudinal shear stress, 222
Low-velocity impact load, 379

Magnetic loads, 90
Mathematical model, 51
Membrane analogy, 191
Mirror-galvanometer oscillograph, 395
Model, mathematical, 51
Modes of failure, 129
Modulus
  bulk, 60
  of elasticity, 50
  of resilience, 72
  of rigidity, 51
Mohr's circle, 12
  for plane strain, 36

Moire analysis, 404
Moment diagrams, 100, 104
Moment of inertia, 214
Moment-area method for deflection, 240
Monochromatic light, 401

Necking, 68
Neutral
    axis, 208
    equilibrium, 376
    surface, 208
Noncylindrical torsion members, 190
Nonhomogeneous beams, 319
Nonstatic loads, 361
Null-balance indicator, 392

Octahedral shear stress theory, 340

Photoelastic coatings, 402
Photoelasticity, 399
Plane of bending, 206
Plastic
    action, 67, 77
    buckling, 273
    deformation, 67
Point load, 89
Poisson effect, 210
Poisson's ratio, 54
Polariscope, 399
Pressure
    load, 89
    transducer, 395
    vessels, 143
Principal planes, 15
Principal stress, 15
Proportional limit, 50
Pure bending, 205, 206, 227

Quasi-static deformation, 48

Radial stress, 146
Ramberg and Osgood, 54
Rate, creep, 375
Relative retardation, 400
Relaxation, 81, 374, 376
    test, 377
Resilience, modulus of, 72
Resistance gage, 388
Resisting torque, 176
Rivets, 154
Rosette
    gage, 399
    strain, 39
        gage, 40

Rotating-beam test, 369
Rupture
    strength, 375
    stress, 68

Safety, definition of, 128
Secant modulus, 53
Shanley, F. R., 288
Shear
    center, 310, 311
    diagram, 100, 104
    flow, 192, 221, 310
    modulus, 51
Shock loads, 377
Singularity functions, 107
Slenderness, ratio, 281
*S-N* curve, 369
Soderberg diagram, 372
SR-4 strain gages, 390
Stable equilibrium, 275
Statically indeterminate axially-loaded
    members, 151
Statically indeterminate beams, 246
Step function, 109
Stiffness, 67
Strain
    axial, 27
    compressive, 28
    concentrations, 142, 361
    definition of, 27
    energy, 70, 329
        density, 71, 330
    gages, 387
        electrical resistance, 387
        rosette, 40
    hardening, 80
    measurement, 39, 387, 392
    plane, 31
    principal, 37
    rosette, 39
    shear, 28
    tensile, 28
    waves, 377
Strain-gage transducers, 395
Strength
    creep, 375
    fatigue, 370
Stress, 3
    alternating, 372
    analysis, experimental, 386
    biaxial, 7
    circumferential,
    concentration, 142, 361
    concentration factor, 362
        effective, 364
        theoretical, 363
    flexural, 211
    fluctuating, 372
    history, 370

Stress (*continued*)
  hoop, 145
  normal, 5
  plane, 8
  principal, 15
  radial, 146
  raiser, 368
  shear, 5, 223
  triaxial, 7
  true, 73
Stresscoat, 404
Stress-strain curve, 48
Superposition, deflection, 244

Tangent modulus, 53
  equation, 288
Tangential deviation, 241
Temperature, elevated, 374
Temperature-compensating gage, 391
Tertiary creep, 375
Tests
  creep, 375
  fatigue, 370
  impact, 378
  relaxation, 377
Theoretical stress-concentration factor,
    363
Theories of failure, 329, 338
Theory of elasticity, 20
Thermal loads, 90
Thermal shock, 90
Thick-walled cylinders, 144
Thin-walled pressure vessels, 143
Time-dependent properties, 361, 374, 377
Torque
  resisting, 176
  transducer, 395

Torsion, 173
  angle of twist, 184
  elastic, 181
    formula, 182
  inelastic, 182
  member
    hollow thin-walled, 186, 191
    noncylindrical, 190
  plastic, 182
Toughness, 72
Transducer, strain-gage, 395
Transverse loads, 205
Transverse shear stresses, 222
True stress, 73

Ultimate strength, 67
Uniform stress distribution, 139
Unstable equilibrium, 275
Unsymmetrical sections, 224
  bending, 224
Unsymmetrical shear loading, 309

Volumetric elastic energy, 335

Welds, 154
Wheatstone-bridge circuit, 390
Wire gage, 387
Work, 346, 379
Work hardening, 80
Working load, 128

Yield point, 68
Yield strength, 68
Yielding, 77
  general, 130